MACMILLAN/McGRAW-HILL

MATHEMATICS IN ACTION

Alan R. Hoffer   Steven J. Leinwand   Gary L. Musser
Martin L. Johnson Richard D. Lodholz   Tina Thoburn

MACMILLAN/McGRAW-HILL SCHOOL PUBLISHING COMPANY
New York / Chicago / Columbus

CONSULTANTS

Zelda Gold, Mathematics Advisor, Los Angeles Unified School District, Panorama City, California • Audrey Friar Jackson, Math Specialist K-6, Parkway School District, Chesterfield, Missouri • Susan Lair, Department Chairperson, Wedgwood Middle School, Fort Worth Independent School District, Texas • Gail Lowe, Principal, Conejo Valley United School District, Thousand Oaks, California

ACKNOWLEDGMENTS

The publisher gratefully acknowledges permission to reprint the following copyrighted material:

Chart on "The Planets" from USBORNE BOOK OF FACTS AND LISTS, Omnibus Edition, EDC Publishing, Tulsa, Oklahoma. Used by permission.

"Lowest Annual Rainfall in the United States"; "Milk Cows on Farms in the United States"; "Money Spent by Americans on Tickets to Sports Events"; "Number of People per Square Mile in Texas"; "Seating Capacities of Baseball Stadiums"; "United States Postal Rates"; "Weather in Southern Cities in the United States"; and "Workers in Manufacturing in the United States" from THE WORLD ALMANAC, 1989. Copyright © Newspaper Enterprise Association. New York, N.Y. 10166. Used by

"Sizes of Modern and Prehistoric Mammals," adapted from COMPARISONS, The Diagram Group. Copyright © 1980. Published by St. Martin's Press and used with their permission.

COVER DESIGN B B & K Design Inc.   **COVER PHOTOGRAPHY** Scott Morgan

ILLUSTRATION   Clara Aich; 106, 107 • Armando Baez; 16, 17, 52, 53, 62, 63, 192, 201, 376, 377 • George Baquero; 38, 505 • Shirley Beckes; 492, 493 • Lloyd Birmingham; 310 • Rick Brown; 152, 153 • Jack Brusca; 542, 548, 549, 550, 568 • Pat Chapin; 186, 264, 372 • Circa 86, Inc.; 91, 506, 511, 514, 516 • Deborah Danilla; 361 • Bruce Day; 2, 3 • Nancy Didion; 316 • Kathy Diefendorf; 326, 454 • Eldon Doty; 437, 458 • Mary Young Duarte; 266, 267 • Pat Duffy; 270, 414 • Ted Enik; 272 • Gregg Fitzhugh; 418 • Barbara Friedman; 104, 148, 149, 240, 241, 452 • Rick Geary; 194, 195 • Gershom Griffith; 138, 188 • Meryl Henderson; 108 • Steve Henry; 322, 323 • Dave Joly; 6, 7 • Ron Jones; 72, 354 • Mark Kaplan; 540, 545, 546, 562, 563, 574 • Knight / Train Design; 468 • Terry Kovalcik; 232, 462, 465 • Lingta Kung; 398, 400 • Hal Lōse; 278, 279 • Benton Mahan; 135 • Claude Martinot; 4, 5 • Fred Marvin; 482, 483 • Frank Mayo; 230 • Pat Merrell; 28, 29, 228, 229, 356 • MKR Design; Handmade props • Leo Monahan; 1 • Ann Neumann; 95, 208 center, 234, 335, 360, 379, 492 • Cheryl Kirk Noll; 313, 315, 424 • Michael O'Reilly; 144, 145, 247, 264, • Hima Pamoedjo; 56, 57, 246, 547 • Bob Pasternak; 25, 118, 119, 282, 283, 352, 412, 413 • Scott Pollack; 450, 451 • Lisa Pomerantz; 114, 358 • Real Comics; 8, 9, credits on page • Ed Sauk; 290, 291 • Roz Schanzer; 332, 333 • Phil Scheuer; 286 • S. D. Schindler; 219, 307 • David Shelton; 368, 369 • Barton Stabler; 22, 23 • Susan Swan; 10 • Marc Taffet; 426 • George Ulrich; 242, 243, 343, 345, 403 • Joe Veno; 484 • Rhonda Voo; 169 • Anna Walker; 68, 69, 146, 147 • Marc Yankus; 498, 499 • Tim Young; 74, 208 top and bottom • Rusty Zabransky; 36, 37, 64, 78, 122, 123, 166, 167, 193, 200, 202, 206, 248, 294, 336, 337, 364, 370, 380, 381, 424, 425, 466, 467, 486, 487, 496, 502 • Ron Zalme; 96 • Jerry Zimmerman; 102, 143, 196, 382, 406, 407, 420, 442, 443, 460, 538, 543, 572, 580

PHOTOGRAPHY   Allsport / Tony Duffy, 100 • Peter Arnold / H.R. Bramaz, 251 • Ken Cavanagh for Macmillan / McGraw-Hill School Division, 39, 81, 125, 244, 383, 469 • Bruce Coleman Inc. / Keith Gunnar, 64R; David Madison, 98T • Culver Pictures Inc., 120BR • FPG / A. Stone, 151 • Focus On Sports, 120TR • Grant Heilman Photography Inc. / Lefever / Grushow, 71 • Richard Haynes, Jr., 33, 59 • Nancy Hays, 66 • Michal Heron, 396, 397, 408, 456 • The Image Bank / Walter Bibikow, 168T; Gary Faber, 269; Robin Forbes, 80B; Gary Gay, 284B; Janeart Ltd., 140; Steve Krongard, 536; Peter Miller, 284T; Toby Rankin, 551; Michael Salas, 124L; Al Satterwhite, 33; • Jurgen Vogt, 312 • The Image Works / Mark Antman, 504 • International Stock Photography Ltd., 466 • LGI Photo Agency / Duncan Rabin, 80 • Monkmeyer Press / Rameshwar Das, 67 • Warren Ogden, 564 • Stephen Ogilvy, 58, 78, 79, 122, 183, 223, 226, 244, 245, 336, 370, 378, 408, 409, 410, 422, 424, 496 • Omni-Photo Communications, Inc. / Ken Karp, 19, 60, 70, 164, 182, 202, 276, 277, 292, 363, 404, 405, 411, 440, 441, 443, 446, 447, 492, 493, 494, 495, 498, 502, 503, 552; John Lei, 12, 13, 14, 18, 30, 70, 122, 184, 207, 226, 245, 280, 281, 328, 330, 354, 364, 365, 396, 445, 502 • Photo Researchers / Jules Bucher, 204; Farrel Grehan, 190; George Jones, 156; Russ Kinne, 338; Paolo Koch, 168B; Susan McCartney, 27; Lawrence Migdale, 224; David Parker / Science Photo Library, 154; Nicholas de Sciose, 222L; Lee Snyder, 504T; James Steinberg, 160T • The Picture Cube / Franz Kraus, 222R. Research Plus / Kurt Anderson, 23, 221, 249, 467; Gene Anthony, 336; Laurence Bartone, 51, 79, 181, 207, 262, 295, 309, 337, 351, 381, 395, 425; Lawrence Migdale, 93, 439; Karen Rantzman, 11, 37, 137, 167, 481, 503 • Joseph Sachs, 206, 207, 294, 295, 297, 362 • Sipa Press / Lee Celano, 124 • Sonlight Images for M / M, 561, 565, 570, 576 • Stock Boston / Jacques Charles, 150C • The Stock Market / Berenholtz, 324; James Blank, 30; Byron, 467; Rich Chisholm, Assoc., 339; Michael Kevin Daly, 268; Gabe Palmer, 160B; Ken Straiton, 296 • Texastock / Ralph Barrera Photo, 54L • Viesti Associates, Inc. / Dan Barba, 35A, 198B, 199, 205, 236, 237, 274, 367, 402-403B, 416, 449; Roger Holden, 150L; Skye Mason, 110-111; Ginny Ganong Nichols, 21, 34, 76, 158, 159, 288, 289, 320R, 320B, 321, 366, 374, 402T, 448, 488; Bill Reaves, 489; Joe Viesti, 77, 117, 320L, 375, 417 • Bill Waltzer, 55, 65, 157 • Dan de Wilde, 204-205 • Woodfin Camp / Oliver Rebbot, 242L

Copyright © 1992 Macmillan/McGraw-Hill School Publishing Company

All rights reserved. No part of this book may be reproduced or transmitted in any form or by any means, electronic or mechanical, including photocopying, recording, or by any information storage and retrieval system, without permission in writing from the Publisher.

Macmillan/McGraw-Hill School Division
866 Third Avenue
New York, New York 10022

Printed in the United States of America
ISBN 0-02-109005-X
9 8 7

Contents

Thinking Mathematically
LETTER TO STUDENTS	1
APPLYING MATHEMATICS	2
USING NUMBER CONCEPTS	4
VISUAL REASONING	6
MEASURING	7
COLLECTING AND INTERPRETING DATA	8
LOGICAL REASONING	10

CHAPTER 1 Understanding Numbers 11
MATH CONNECTIONS ALGEBRA / GRAPHING / STATISTICS / PROBLEM SOLVING

Using Tables	12
Frequency Tables	14
Numbers to Hundred Thousands	16
Millions and Billions	18
PROBLEM SOLVING USING THE FIVE-STEP PROCESS	20
THINKING MATHEMATICALLY INFORMAL ALGEBRA: INVESTIGATING PATTERNS	22
Comparing and Ordering Numbers	24
Rounding Numbers	26
Coordinate Graphing	28
Interpreting Line Graphs	30
Making Line Graphs	32
PROBLEM SOLVING STRATEGY: USING NUMBER SENSE	34
DECISION MAKING JOINING AN AFTER-SCHOOL CLUB	36
CURRICULUM CONNECTION SOCIAL STUDIES	38
TECHNOLOGY COMPUTER EXPLORATION: POWERS AND EXPONENTS	39
EXTRA PRACTICE	40
PRACTICE ***PLUS***	44
Chapter Review	46
Chapter Test	48
ENRICHMENT FOR ALL BASE FIVE	49
Cumulative Review	50

DECISION MAKING
Joining an After-School Club,
pages 11 and 36

DECISION MAKING
Making a Class Video,
pages 51 and 78

CHAPTER 2 / Using Addition and Subtraction 51
MATH CONNECTIONS ALGEBRA / MEASURING LENGTH / TIME / PROBLEM SOLVING

Informal Algebra: Meaning of Addition and Subtraction	52
Estimating Sums, Differences by Rounding	54
Front-End Estimation: Sums and Differences	56
Mental Math: Adding Whole Numbers	58
Adding Whole Numbers	60
Mental Math: Subtracting Whole Numbers	62
Subtracting Whole Numbers	64
PROBLEM SOLVING STRATEGY: CHOOSING THE OPERATION	66
THINKING MATHEMATICALLY INFORMAL ALGEBRA: USING NUMBER CONCEPTS	68
Measuring and Estimating Length	70
Measuring Length	72
Time and Elapsed Time	74
PROBLEM SOLVING IDENTIFYING EXTRA INFORMATION	76
DECISION MAKING MAKING A CLASS VIDEO	78
CURRICULUM CONNECTION ART	80
TECHNOLOGY COMPUTER SPREADSHEET: TRAVEL COSTS	81
EXTRA PRACTICE	82
PRACTICE PLUS	86
Chapter Review	88
Chapter Test	90
ENRICHMENT FOR ALL OTHER NUMBER SYSTEMS	91
Cumulative Review	92

CHAPTER 3 / Adding and Subtracting Decimals ... 93
MATH CONNECTIONS GRAPHING / STATISTICS / PROBLEM SOLVING

Tenths and Hundredths	94
Thousandths	96
Comparing and Ordering Decimals	98
Rounding Decimals	100
Estimating Decimal Sums and Differences	102
Adding Tenths and Hundredths	104
THINKING MATHEMATICALLY EXPERIMENTING AND PREDICTING	106
Subtracting Tenths and Hundredths	108
PROBLEM SOLVING STRATEGY: CHOOSING THE OPERATION	110
More Estimating: Sums and Differences	112
Adding and Subtracting Thousandths	114
PROBLEM SOLVING STRATEGY: USING ESTIMATION	116
Interpreting Bar Graphs	118
Making Bar Graphs	120

DECISION MAKING
Buying a New or
Used Item,
pages 93 and 122

DECISION MAKING BUYING A NEW OR USED ITEM	122
CURRICULUM CONNECTION SCIENCE	124
TECHNOLOGY COMPUTER GRAPHING: REACTION TIME	125
EXTRA PRACTICE	126
PRACTICE PLUS	130
Chapter Review	132
Chapter Test	134
ENRICHMENT FOR ALL NUMBER SEQUENCES	135
Cumulative Review	136

CHAPTER 4 / Multiplying Whole Numbers 137
MATH CONNECTIONS ALGEBRA / PERIMETER / AREA / CIRCUMFERENCE / PROBLEM SOLVING

Informal Algebra: Meaning of Multiplication	138
Basic Facts and Strategies	140
Mental Math: Using Multiplication Patterns	142
Estimating Products	144
Multiplying by a 1-Digit Number	146
Multiplying by a 2-Digit Number	148
PROBLEM SOLVING FINDING NEEDED INFORMATION	150
THINKING MATHEMATICALLY INVESTIGATING PATTERNS	152
More Multiplying by a 2-Digit Number	154
Multiplying Large Numbers	156
PROBLEM SOLVING STRATEGY: SOLVING A MULTISTEP PROBLEM	158
Perimeter	160
Area	162
Circumference	164
DECISION MAKING PLANNING A DAY TRIP	166
CURRICULUM CONNECTION SCIENCE	168
TECHNOLOGY CALCULATOR: FINDING FACTORS	169
EXTRA PRACTICE	170
PRACTICE PLUS	174
Chapter Review	176
Chapter Test	178
ENRICHMENT FOR ALL ENLARGING FIGURES ON A COORDINATE GRAPH	179
Cumulative Review	180

DECISION MAKING
Planning a Day Trip,
pages 137 and 166

CHAPTER 5 / Dividing Whole Numbers—1-Digit Divisors 181
MATH CONNECTIONS ALGEBRA / GRAPHING / STATISTICS / PROBLEM SOLVING

Meaning of Division	182
Informal Algebra: Basic Division Facts	184

Mental Math: Using Division Patterns 186
Estimating Quotients 188
Dividing 2- and 3-Digit Numbers 190
Dividing 4- and 5-Digit Numbers 192
THINKING MATHEMATICALLY INVESTIGATING PATTERNS 194
Zeros in the Quotient 196
PROBLEM SOLVING STRATEGY: USING ESTIMATION 198
Using Pictographs 200
Mean, Median, Mode, and Range 202
PROBLEM SOLVING STRATEGY: GUESS, TEST, AND REVISE 204
DECISION MAKING DETERMINING PRICES 206
CURRICULUM CONNECTION LITERATURE 208
TECHNOLOGY CALCULATOR: ORDER OF OPERATIONS 209
***E**XTRA* **PRACTICE** 210
PRACTICE *PLUS* 214
Chapter Review 216
Chapter Test 218
ENRICHMENT FOR ALL DIVISIBILITY 219
Cumulative Review 220

DECISION MAKING
Determining Prices,
pages 181 and 206

CHAPTER 6
Dividing Whole Numbers— 2-Digit Divisors 221
MATH CONNECTIONS TIME / VOLUME / ALGEBRA / PROBLEM SOLVING

Mental Math: Using Division Patterns 222
Estimating Quotients 224
Division with 2-Digit Divisors 226
Dividing with 2-Digit Divisors 228
Changing Estimates 230
Zeros in the Quotient 232
Dividing Larger Numbers 234
PROBLEM SOLVING STRATEGY: CHOOSING THE OPERATION 236
THINKING MATHEMATICALLY USING NUMBER CONCEPTS 238
Changing Measures of Time 240
PROBLEM SOLVING STRATEGY: FINDING A PATTERN 242
Volume 244
Volume of a Rectangular Prism 246
DECISION MAKING BUYING MAGAZINES 248
CURRICULUM CONNECTION MUSIC 250
TECHNOLOGY COMPUTER SPREADSHEET INFORMAL ALGEBRA:
   OPERATION RULES 251
***E**XTRA* **PRACTICE** 252
PRACTICE *PLUS* 256
Chapter Review 258

DECISION MAKING
Buying Magazines,
pages 221 and 248

Chapter Test.. 260
ENRICHMENT FOR ALL MAGIC SQUARES....................... **261**
Cumulative Review.. 262

CHAPTER 7 / Multiplying and Dividing Decimals.. 263
MATH CONNECTIONS MEASUREMENT / MASS AND CAPACITY / PROBLEM SOLVING

Mental Math: Multiply and Divide Decimals................... 264
Estimating Products.. 266
Multiplying Whole Numbers and Decimals...................... 268
Multiplying Decimals by Whole Numbers....................... 270
More Multiplying... 272
PROBLEM SOLVING STRATEGY: WORKING BACKWARD.............. **274**
The Products of Two Decimal Factors.......................... 276
THINKING MATHEMATICALLY APPLYING MATHEMATICS............ **278**
Division of a Decimal by a Whole Number..................... 280
Dividing Decimals by Whole Numbers.......................... 282
Zeros in Division.. 284
More Dividing Decimals....................................... 286
PROBLEM SOLVING STRATEGY: MAKING A TABLE................ **288**
Measuring Metric Capacity and Mass........................... 290
Renaming Metric Measures..................................... 292
DECISION MAKING BUYING SHOES............................. **294**
CURRICULUM CONNECTION SCIENCE............................ **296**
TECHNOLOGY CALCULATOR: COOKING WITH A CALCULATOR........ **297**
EXTRA **PRACTICE**... **298**
PRACTICE *PLUS*... **302**
Chapter Review... 304
Chapter Test... 306
ENRICHMENT FOR ALL ORDER OF OPERATIONS................... **307**
Cumulative Review.. 308

CHAPTER 8 / Geometry 309
MATH CONNECTIONS ALGEBRA / PROBLEM SOLVING

Geometry Around Us... 310
Measuring Angles... 312
Angles... 314
Plane Figures.. 316
Triangles and Quadrilaterals................................. 318
PROBLEM SOLVING STRATEGY: DRAWING A DIAGRAM............. **320**

DECISION MAKING
Buying Shoes,
pages 263 and 294

DECISION MAKING
Paying for a School Newspaper,
pages 309 and 336

*T*HINKING MATHEMATICALLY VISUAL REASONING 322
Symmetry 324
Slides, Flips, Turns 326
Congruent Figures 328
Similar Figures 330
*P*ROBLEM SOLVING STRATEGIES REVIEW 332
Space Figures 334
*D*ECISION MAKING PAYING FOR A SCHOOL NEWSPAPER 336
CURRICULUM CONNECTION ART 338
TECHNOLOGY COMPUTER: EXPLORING ANGLES 339
EXTRA **PRACTICE** 340
PRACTICE *PLUS* 344
Chapter Review 346
Chapter Test 348
ENRICHMENT FOR ALL INFORMAL ALGEBRA: VENN DIAGRAMS 349
Cumulative Review 350

CHAPTER 9 Understanding Fractions and Mixed Numbers 351
MATH CONNECTIONS MEASUREMENT / LENGTH / PROBLEM SOLVING

Fractions 352
Equivalent Fractions 354
Greatest Common Factor 356
Simplifying Fractions 358
Mixed Numbers 360
Renaming Mixed Numbers 362
Rounding Mixed Numbers 364
*P*ROBLEM SOLVING STRATEGY: MAKING AN ORGANIZED LIST 366
*T*HINKING MATHEMATICALLY LOGICAL REASONING 368
Least Common Multiple 370
Comparing Fractions and Mixed Numbers 372
*P*ROBLEM SOLVING INTERPRETING THE QUOTIENT AND REMAINDER 374
Estimating Length 376
Measuring Length 378
*D*ECISION MAKING BUILDING A MODEL AIRPLANE 380
CURRICULUM CONNECTION SOCIAL STUDIES 382
TECHNOLOGY COMPUTER: EXPLORING QUADRILATERALS 383
EXTRA **PRACTICE** 384
PRACTICE *PLUS* 388
Chapter Review 390

DECISION MAKING
Building a Model Airplane,
pages 351 and 380

Chapter Test..392
ENRICHMENT FOR ALL PRIME FACTORIZATION.................**393**
Cumulative Review....................................394

CHAPTER 10 / Adding and Subtracting Fractions .. 395
MATH CONNECTION PROBLEM SOLVING

Sums and Differences of Fractions.........................396
Adding Fractions: Like Denominators.......................398
Adding Fractions: Unlike Denominators.....................400
PROBLEM SOLVING STRATEGY: USING NUMBER SENSE **402**
Subtracting Fractions....................................404
THINKING MATHEMATICALLY VISUAL REASONING **406**
Estimating Sums and Differences..........................408
Adding, Subtracting Mixed Numbers........................410
Adding Mixed Numbers....................................412
Adding Mixed Numbers with Renaming.......................414
PROBLEM SOLVING STRATEGY: SOLVING A SIMPLER PROBLEM **416**
Subtracting Mixed Numbers................................418
Renaming to Subtract Mixed Numbers.......................420
Subtracting Mixed Numbers with Renaming..................422
DECISION MAKING REDECORATING A ROOM **424**
CURRICULUM CONNECTION MUSIC.........................**426**
TECHNOLOGY CALCULATOR: FIND THE PATH **427**
***EXTRA* PRACTICE** **428**
PRACTICE *PLUS* **432**
Chapter Review..434
Chapter Test...436
ENRICHMENT FOR ALL PREDICTING FROM A SAMPLE **437**
Cumulative Review......................................438

DECISION MAKING
Redecorating a Room,
pages 395 and 424

CHAPTER 11 / Multiplying and Dividing Fractions .. 439
MATH CONNECTIONS MEASURING CAPACITY AND WEIGHT /
AREA / PROBLEM SOLVING

Multiplying Fractions....................................440
More Multiplying Fractions...............................442
Multiplying Fractions and Whole Numbers..................444
Division: Whole Numbers by Fractions.....................446
PROBLEM SOLVING STRATEGY: USING DIFFERENT STRATEGIES **448**
THINKING MATHEMATICALLY USING NUMBER CONCEPTS **450**

Interpreting Circle Graphs	452
Area of a Right Triangle	454
Area of a Parallelogram	456
PROBLEM SOLVING STRATEGY: SOLVING A SIMPLER PROBLEM	**458**
Measuring Customary Capacity and Weight	460
Renaming Customary Measures	462
Temperature	464
DECISION MAKING PLANNING A BARBECUE	**466**
CURRICULUM CONNECTION THE CONSUMER	468
TECHNOLOGY COMPUTER GRAPHING: CIRCLE GRAPH	469
EXTRA PRACTICE	**470**
PRACTICE *PLUS*	**474**
Chapter Review	476
Chapter Test	478
ENRICHMENT FOR ALL AREA OF IRREGULAR-SHAPED REGIONS	**479**
Cumulative Review	480

DECISION MAKING
Planning a Barbecue,
pages 439 and 466

CHAPTER 12 / Ratio and Probability 481
MATH CONNECTION PROBLEM SOLVING

Ratios	482
Equal Ratios	484
Scale Drawings	486
PROBLEM SOLVING STRATEGIES REVIEW	**488**
PROBLEM SOLVING STRATEGY: CONDUCTING AN EXPERIMENT	**490**
Experiments in Probability	492
Probability	494
Independent Events	496
THINKING MATHEMATICALLY EXPERIMENTING AND PREDICTING	**498**
Predicting Outcomes	500
DECISION MAKING CHOOSING A SUMMER CAMP	**502**
CURRICULUM CONNECTION HEALTH	504
TECHNOLOGY COMPUTER SIMULATION: PROBABILITY EXPERIMENTS	505
EXTRA PRACTICE	**506**
PRACTICE *PLUS*	**510**
Chapter Review	512
Chapter Test	514
ENRICHMENT FOR ALL PERCENT	**515**
Cumulative Review	516
Data Bank	517
Glossary	523
Table of Measures	528
Index	529

DECISION MAKING
Choosing a Summer Camp,
pages 481 and 502

Thinking MATHEMATICALLY

What do you think mathematics is? Do you think it is adding, subtracting, multiplying, and dividing? If you said yes, you are right. But mathematics is more than that. It is looking for patterns, collecting and organizing data, and using logic. Perhaps most important of all, mathematics is also problem solving. It is something that you do every day.

In this short chapter you will be doing the things mentioned above and more. There are some interesting puzzles and activities for you to try, and each one will show you another way of thinking mathematically. So turn the page and have fun.

Alan Hoffer
Martin Johnson
Richard Lodholz
Steve Leinwand
Gary L. Musser
Tina Thoburn

Take Me Out to the BALL GAME

Applying Mathematics

A. Suppose you are working at the refreshment stand at a baseball game. Some customers have placed the orders below. How much does each order cost?

REFRESHMENTS
HAMBURGER..... 70¢
JUICE .... 50¢
POPCORN .... 30¢ a box
PEANUTS .... 25¢ a bag
LIMIT: NO MORE THAN 3 OF EACH ITEM

1. 1 hamburger
   1 juice
   1 bag of peanuts

2. 2 hamburgers
   2 juices
   1 box of popcorn

3. 3 juices
   2 boxes of popcorn

4. 1 hamburger
   2 juices
   3 boxes of popcorn
   2 bags of peanuts

5. Make up your own order. How much does it cost?

6. Melinda spent exactly $2.00. What could she have bought?

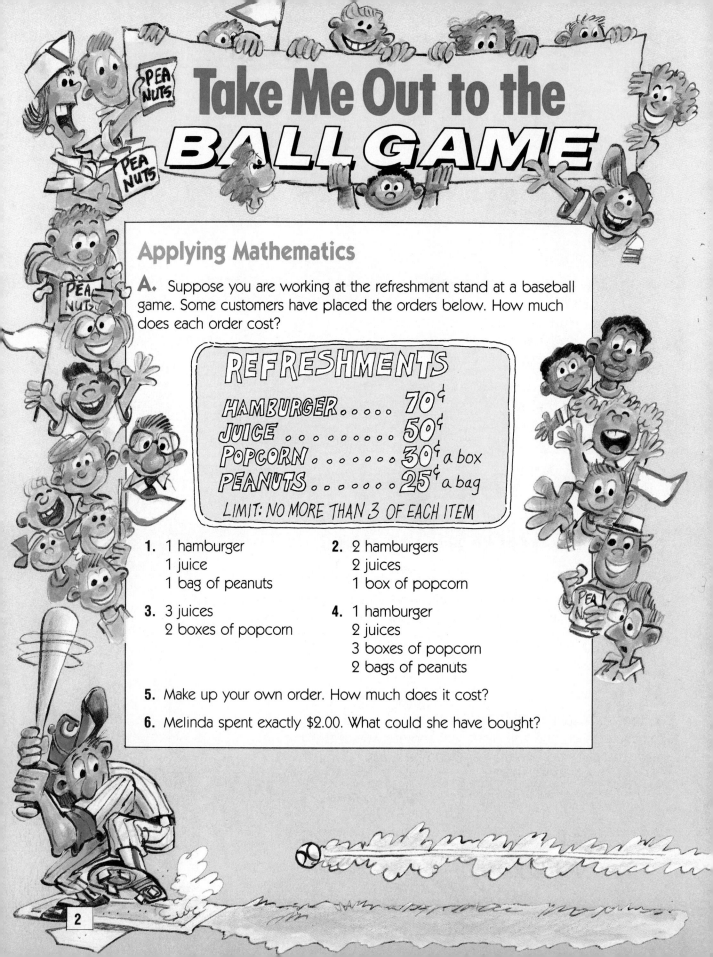

B. Suppose you are selling food at the refreshment stand.

7. Sean pays for 1 hamburger, 3 juices, and 1 bag of peanuts with 3 one-dollar bills.
   a. How much did Sean spend?
   b. How much change should Sean get?
   c. What coins would you give him to make the change?

8. Lisa pays for one of each item with a five-dollar bill.
   a. How much did she spend?
   b. How much change should she get?
   c. What bills and coins would you give her to make the change?

9. Ronnie wants to buy 2 juices and 2 boxes of popcorn. He has one dollar.
   a. What is the problem?
   b. What can Ronnie do to solve his problem?

C. Now suppose you are a customer at the refreshment stand.

10. What is the most you can spend at one time? (*Hint:* Remember the limit.)

11. List all the different orders you can make that have a total of $1.00 or less.

12. Suppose you have exactly $1.70.
    a. Can you spend the $1.70 exactly at the refreshment stand? How?
    b. Is there more than one way to spend exactly $1.70? List the ways.

NUMBER

Using Number Concepts

Have some fun with numbers. First make a set of number cards and an addition board.

Cards:
Cut nine rectangles from a sheet of paper. Make each rectangle about the size of a stamp. Write the numbers 1 to 9 on the rectangles.

Addition board:
On a sheet of paper, draw a board like the one at the right. The spaces should be about the same size as your cards.

GAMES

1. Pick any five cards. Which cards did you pick? Arrange the cards on your addition board to get the greatest sum. How did you arrange the cards?

2. Pick the five cards that would give you the greatest possible sum. How would you arrange these cards? What is the greatest possible sum?

3. Pick the five cards that would give you the least possible sum. How would you arrange these cards? What is the least possible sum?

4. Make sums of exactly 500 using any five cards. How many different ways did you get 500? List them.

   Now make a multiplication board like the one at the right.

5. Pick any four cards. Which cards did you pick? Can you arrange them on the board to get the product shown? How did you arrange the cards?

6. Try again with four other cards. How many different answers did you find? List them.

7. Think of a puzzle involving a subtraction board. Make up your own board and rules. Try out your puzzle with a partner.

Thinking Mathematically

TAKING SHAPE

Visual Reasoning

Cut out 6 equal-sized squares.

Pretend that each square is a lunch table and that only one person can sit at each side.

1. What is the greatest number of people that can sit at the 6 tables? Explain how you figured this out.

Suppose the tables are pushed together to form larger tables.

A.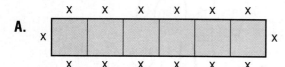

B.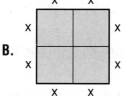

2. How many people can sit at arrangement A?

3. How many people can sit at arrangement B?

Make six other arrangements (C–H) using all 6 tables.

4. Draw the arrangements on a chart like this.

Arrangement	Number of People
C.	
D.	

5. How many people can sit at each arrangement? Write the answers on the chart.

6. At which arrangement can the least number of people sit?

7. At which arrangement can the greatest number of people sit?

8. Did you make an arrangement where exactly 12 people can sit? If not, try to make one.

9. Did you make an arrangement where exactly 18 people can sit? If not, try to make one.

10. Are there any arrangements that are not practical? Explain.

A HEARTBEAT IN TIME

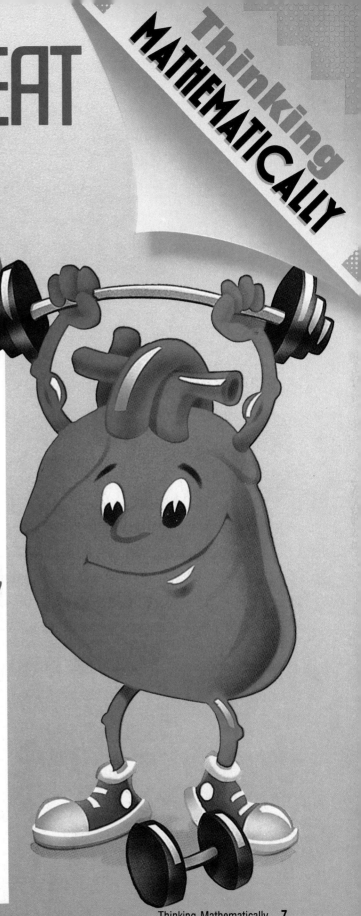

Measuring

Your heart is like a busy machine. Have you ever thought about how hard it works?

Try this activity and find out. Feel your pulse on your wrist or on the side of your neck. Your pulse matches the beating of your heart. It tells how many times your heart is beating.

1. Use a clock or watch with a second hand to find how many times your heart beats in one minute.

2. Use your calculator to find about how many times your heart:
   a. beats in one hour.
   b. beats in one day.
   c. beats in one week.
   d. beats in one year.
   e. had beaten by your fifth birthday.
   f. has beaten since you were born.

3. About how old were you when your heart had beaten for the one millionth time?

4. Use your calculator to find about how long it takes you to count to one million. (*Hint:* How high can you count in one minute?)

Family Ties

Collecting and Interpreting Data

Every ten years our government takes a survey, called a census, of the country's population. One of the things it finds out is:

HOW LARGE IS THE TYPICAL AMERICAN FAMILY?

Suppose your class represents the general population. How would you go about answering the question? Here is one way.

GETTING STARTED

1. Decide who should be included as a family member.

2. Guess how many people are in the typical American family.

3. List all the guesses made. Discuss whether they are reasonable.

COLLECTING DATA

4. Take a class census to find out how many people are in each student's family.

5. Write the answers in a table like this:

Family Size	Tally	Number of Families
2		
3		
4		
5		

Thinking MATHEMATICALLY

DISPLAYING DATA

6. Use the data in your table to make a bar graph on a grid like this.

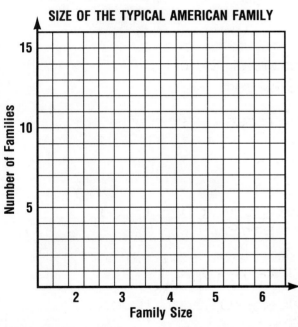

SIZE OF THE TYPICAL AMERICAN FAMILY

Number of Families vs. Family Size

ANALYZING DATA

7. What does the bar graph show you?

8. Write three conclusions you can draw from the bar graph.

9. What do you now think is a reasonable estimate for the number of people in the typical American family? Explain.

10. **What if** you saw the headline "Family Size 3.8"? What do you think the data means?

WHO'S WHO?

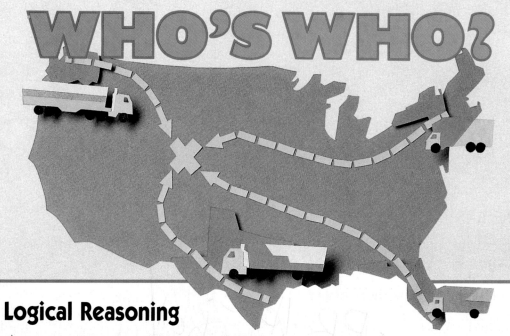

Logical Reasoning

A. Ben, Carlos, Darlene, and Erica are new children in your neighborhood. They have moved from New York, Texas, Washington, and Florida. You find out that:

- Ben moved from Washington.
- Erica moved from New York.
- Carlos did not move from Texas.

From which state did Darlene move?

Use a chart to organize your information. Write ✓ for *yes* and X for *no*. The first clue is already on the chart.

	New York	Texas	Washington	Florida
Ben			✓	
Carlos				
Darlene				
Erica				

1. From which state did each child move?
2. Explain your reasoning.

B. The ages of Ben, Carlos, and Erica add to 26. Ben is not the oldest. Carlos is 11 years older than the youngest child. Erica is 3 years old. How old is Ben?

Use the chart to organize your information.

	Name	Age
Oldest		
Middle		
Youngest		

3. Give the name and age of the oldest, middle, and youngest child.
4. Explain your reasoning.

Understanding Numbers

CHAPTER 1

MATH CONNECTIONS: ALGEBRA • GRAPHING • STATISTICS • PROBLEM SOLVING

HOME	3	4
VISITORS	3	2

1. What information do you see in this picture?
2. How can you use the information?
3. Why are there numbers on the uniforms?
4. Write a problem using the information.

DEVELOPING A CONCEPT
Using Tables

At the beginning of the year, Laura's class members decided to write a class biography. They began their project by gathering information about students' birthdays—the month, day, year, and place they were born.

WORKING TOGETHER

1. How could you gather this information from your class?

2. How would you record and display the data so that it would be easy to read and understand?

One method of recording data is to show it in a table such as the one below.

CLASS BIRTHDAYS

Name	Month	Day	Year	Place

Gather data about the birthdays of the students in your class. Record it in a table.

3. Do any students have the same birthday? If so, what is the date?

4. Which month has the greatest number of birthdays? the least number?

5. How many students were born in the same place?

SHARING IDEAS

6. What other questions could you ask about the data in your table?

7. Why is making a table a useful method of recording data? What makes the data easy to read?

PRACTICE

Use the table at the right to answer Problems 8 and 9.

NUMBER OF STUDENTS OWNING PETS

Fish	Bird	Turtle	Dog	Cat
12	16	9	28	23

8. How many students own birds?

9. What kind of pet is owned by the most students? the fewest students?

Use the table below to answer Problems 10–14.

STUDENTS' HEIGHTS (in inches)

	Short (48 to 52)	Medium (53 to 57)	Tall (58 to 62)
Boys	3	6	3
Girls	2	7	4

10. What heights are shown in the table?

11. What is the greatest height? the least height?

12. How many boys are taller than 57 inches?

13. How many girls are shorter than 53 inches?

14. From the data in the table, would you describe the students in this class as tall, medium, or short?

Gather data to show how many brothers and/or sisters each student in your class has. Record your information in a table.

Student	Number of Sisters	Number of Brothers

Use your table to answer Problems 15–18.

15. How many students have brothers? How many have sisters?

16. Do more students have brothers than have sisters?

17. How many students have neither brothers nor sisters?

18. What other information about your class could you record?

Critical Thinking

19. **What if** you wanted to see how many students were born in January. How would you change the table to show this?

DEVELOPING A CONCEPT

Frequency Tables

Steve took a survey to find out what kinds of hobbies were most popular with others.

WORKING TOGETHER

1. Take a survey to find out what hobbies the students in your class enjoy.

2. Make a table listing the hobbies. How could you find out how many students have each hobby?

3. What might be a better method of recording the information? Compare your method with those of others. Which one seems easiest to use? Why?

Steve recorded the data he gathered in a table like this.

CLASS HOBBIES

Hobby	Tally	Total
Painting	||||| ||	7
Photography	||	2
Tropical fish	|||	3
Stamp collecting	|||||	5
Building models	||||| |||	8
No hobby	|||	3

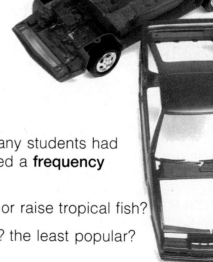

He used tally marks to record how many students had each hobby. This kind of table is called a **frequency table.**

4. Do more students collect stamps or raise tropical fish?

5. Which is the most popular hobby? the least popular? How do you know?

6. Make a frequency table using the data you gathered in Problems 1 and 2.

14   Lesson 1-2

SHARING IDEAS

7. What kind of information does a frequency table give you? How is it useful?

8. How is your frequency table different from Steve's table?

9. Look at your frequency table. Does anyone in your class have more than one hobby? no hobbies?

10. How do you think this information affects your frequency table?

11. What does your frequency table tell you about your class's hobbies?

PRACTICE

Complete the frequency table.

VOTES FOR CLASS PRESIDENT

Name	Tally	Total
José	||||| ||	7
Sarah	||||| |||	12. ■
Tom	13. ■	6
Kris	|||||		14. ■

15. How many votes did José receive?

16. Which two students received the same number of votes?

17. Who won the election? How do you know?

18. Ask your classmates what kinds of pets they have. Make a frequency table of the results.

Critical Thinking

19. When do you think a frequency table would be more useful than a table that lists only information?

UNDERSTANDING A CONCEPT

Numbers to Hundred Thousands

A. Adrienne's favorite sport is long-distance running. She found that the record for the longest nonstop run is held by Bertil Järlåker of Sweden, who ran 617,575 yards.

You can show this number in several ways. In a place-value chart:

Thousands Period			Ones Period		
Hundreds	Tens	Ones	Hundreds	Tens	Ones
6	1	7	5	7	5

Standard form: 617,575

Read: six hundred seventeen thousand, five hundred seventy-five

1. How does the comma between the thousands and ones periods help you read the number?

2. What is the value of the digit 6 in 617,575? How do you know?

3. What is the value of each 5 in the number?

4. In which place is the digit 1? What does this mean?

B. You can use place value to write a number in expanded form.

The expanded form of 617,575 is: 600,000 + 10,000 + 7,000 + 500 + 70 + 5

5. What is the value of each digit in 617,575?

6. How do the values compare with the addends in expanded form?

7. How is the expanded form related to the place-value chart?

8. Here are two ways of writing 703,050 in expanded form:

   700,000 + 0 + 3,000 + 0 + 50 + 0
   700,000 + 3,000 + 50

   How are they the same? How are they different?

TRY OUT

Write the letter of the correct answer. What is the number in standard form?

9. 600,000 + 40,000 + 100 + 5
   a. 6,415   b. 604,105   c. 406,150   d. 640,105

10. six hundred seven
    a. 607   b. 6,007   c. 6,070   d. 60,007

11. three hundred six thousand, five hundred eight
    a. 3,658   b. 36,508   c. 306,580   d. 306,508

PRACTICE

Name the place and the value of the digit 5 in the number.

12. 70,520   13. 485,263   14. 10,050   15. 53,029   16. 506,321

Write the number in expanded form.

17. 6,253   18. 18,041   19. 30,028   20. 250,607   21. 809,006

Write the word name.

22. 85,967   23. 40,293   24. 698,701   25. 130,064   26. 200,002

Write the number in standard form.

27. eight hundred thirty-three thousand, five hundred twelve
28. six hundred four thousand, seventy-four
29. forty-two thousand, one hundred thirty-seven
30. nine hundred six thousand, nine hundred nine
31. 300,000 + 40,000 + 5,000 + 400 + 60 + 8
32. 200,000 + 4,000 + 100 + 9
33. 50,000 + 5,000 + 5

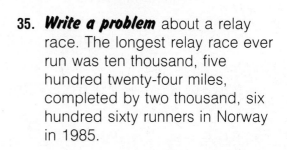

Solve.

34. The marathon is a run of 26 miles 385 yards. Jay Helgerson ran one marathon a week for a year, totaling 1,418 miles. What is the value of the digit 4 in this number? Write the number in expanded form.

35. **Write a problem** about a relay race. The longest relay race ever run was ten thousand, five hundred twenty-four miles, completed by two thousand, six hundred sixty runners in Norway in 1985.

UNDERSTANDING A CONCEPT

Millions and Billions

In a recent year people in the United States spent about $19,725,000,000 on first-class postage. You can show this number in a place-value chart.

Billions Period			Millions Period			Thousands Period			Ones Period		
H	T	O	H	T	O	H	T	O	H	T	O
	1	9	7	2	5	0	0	0	0	0	0

Read: nineteen billion, seven hundred twenty-five million
or
19 billion, 725 million

Write: 19,725,000,000

1. What is the value of the digit 9 in 19,725,000,000? How do you know?

2. Suppose people spent $1,000,000,000 more on first-class postage. How much would they spend? How do you know?

3. Suppose people spent $1,000,000 less. How much would they spend? How do you know?

4. Suppose the place-value chart were extended to include trillions, the next period on the left. How do you think the places would be labeled? Why?

5. What would be the greatest possible number you could write in that expanded place-value chart?

TRY OUT Write the letter of the correct answer.
What is the value of the underlined digit?

6. 378,6<u>4</u>5,904
   a. 4   b. 4,000   c. 40,000   d. 400,000

7. <u>7</u>89,645,321,066
   a. 700,000,000   b. 7,000,000,000   c. 70,000,000,000   d. 700,000,000,000

8. 407,826,<u>9</u>35,602
   a. 9,000   b. 900,000   c. 9,000,000   d. 90,000,000

PRACTICE

Write the digit in the ten millions place and ten billions place.

9. 180,734,509,462　　**10.** 62,310,947,800　　**11.** 907,628,539,140

12. 841,376,007,952　　**13.** 30,654,192,877　　**14.** 428,791,030,561

Write the number in standard form.

15. twelve million, eight hundred five thousand, seventy-one

16. four hundred six billion, two hundred one thousand

17. 340 billion, 111 thousand, 92

18. 280 billion, 39 million, 5

19. 1,000,000 greater than 8,209,745

20. 10,000,000 less than 49,728,073

21. 1,000,000,000 greater than 532,841,029,471

22. 100,000 less than 15,276,399,084

23. 10,000 greater than 999,999,999

Solve.

24. The first stamp ever printed, the 1 Penny Black of Great Britain, was issued in 1840. A total of 68 million, 158 thousand, 80 were printed. Write this number in standard form.

25. The highest prices paid for rare stamps at auction were $935,000 and $864,386. What is the value of the digit 3 in each price?

VISUAL REASONING

Here are three triangular numbers:

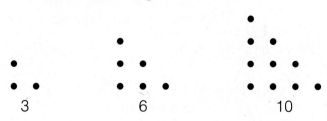

Continue the pattern to find the triangular number between 40 and 50.

PROBLEM SOLVING

✓ UNDERSTAND
✓ PLAN
✓ TRY
✓ CHECK
✓ EXTEND

Using the Five-Step Process

Study how the five-step process is used to solve this problem.

Companies use surveys of TV viewing to help them select programs they wish to sponsor. The CIX Company sponsors programs viewed by more than 145 households out of the 500 that were surveyed. Should CIX sponsor *Pathway to Success*?

Program	Number of Viewing Households in Survey
The Who's Who Show	216
Magic Island	185
90 Minutes	151
Pathway to Success	176
Well-Known Facts	142
Four's a Crowd	134

UNDERSTAND

What do I know? — I know that CIX sponsors TV shows viewed by more than 145 of the households surveyed. The table tells me that 176 households watched *Pathway to Success*.

What do I need to find out? — I need to find out whether 176 is greater than 145. Then I can tell whether CIX should sponsor *Pathway to Success*.

PLAN

What can I do? — I can compare 176 and 145.

TRY

Let me try my plan. — I know that 176 > 145 because 7 tens is greater than 4 tens. So CIX should sponsor *Pathway to Success*.

CHECK

Have I answered the question? — Yes. CIX should sponsor *Pathway to Success*.

EXTEND

What have I learned? — I learned that I can solve some problems by comparing numbers.

How can I apply what I have learned? — By making comparisons I can see that CIX could also sponsor *The Who's Who Show*, *Magic Island*, and *90 Minutes*.

PRACTICE

Apply what you have learned about the five-step process to solve the problem.

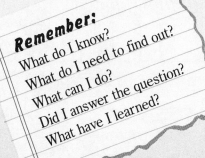

Remember:
What do I know?
What do I need to find out?
What can I do?
Did I answer the question?
What have I learned?

Use the data in the table on page 20 for Problem 1.

1. Which programs should CIX not sponsor?

Use the data in the table below for Problems 2–7.

2. Did a greater number of people watch Channel 2 or 7 during the whole evening?

3. Did a greater number of people watch TV from 7 to 8 or from 10 to 11?

4. Which channel did the greatest number of people watch during the whole evening?

5. Which channel did the least number of people watch during the whole evening?

6. During which hour did the greatest number of people watch TV?

7. During which hour did the least number of people watch TV?

NUMBER OF VIEWERS OUT OF 500 SURVEYED

Channel	Time Slot (P.M.)			
	7–8	8–9	9–10	10–11
2	87	95	80	67
5	101	124	97	75
7	64	76	53	42
9	48	55	39	31

Informal Algebra: Investigating Patterns

A. Marco the Magician built a Marvelous Math Machine. This machine is marvelous because it does interesting things all by itself. For example, one day while Marco is playing with the machine, he puts in a 7. The Marvelous Machine shows a 9. When Marco puts in 16, the machine shows 18.

1. What do you think the machine is doing with each number?

2. What will the machine show if Marco puts in 25? What about 103?

Once you figure out the rule that the Marvelous Machine is using, it starts to do something else.

Guess the rule the machine is using each time. Then complete the Machine Shows column.

3. Rule: _____?_____

Marco Enters	Machine Shows
18	13
103	98
40	35
10	■
72	■
504	■

4. Rule: _____?_____

Marco Enters	Machine Shows
18	3
54	9
36	6
12	■
42	■
0	■

5. Rule: _____?_____

Marco Enters	Machine Shows
3	7
5	11
7	15
4	■
9	■
6	■

6. Make up your own sets of Enters and Shows numbers. Ask a partner to guess each rule.

B. The Marvelous Math Machine also comes in a Deluxe Model. With this model, you can put in two numbers and the machine uses a rule to change the two numbers into one. For example, one day Marco puts in 2 and 4. The machine shows 8. When he enters 3 and 4, the machine shows 12.

7. What is the machine doing with the numbers this time?

8. What will the machine show if you put in 5 and 6?

Guess each rule. Then complete the Machine Shows column.

9. Rule: ____?____

Marco Enters	Machine Shows
4, 5	9
10, 12	22
25, 32	57
16, 4	▪
47, 18	▪
100, 200	▪

10. Rule: ____?____

Marco Enters	Machine Shows
10, 2	5
27, 9	3
48, 6	8
36, 4	▪
72, 8	▪
15, 1	▪

11. Rule: ____?____

Marco Enters	Machine Shows
16, 9	8
13, 7	7
6, 5	2
17, 8	▪
11, 7	▪
10, 10	▪

12. Make up your own sets of Enter and Show numbers for the Deluxe Model. Ask a partner to guess each rule.

UNDERSTANDING A CONCEPT

Comparing and Ordering Numbers

A. For her science project Mei needs to know whether Saturn or Jupiter has the greater diameter.

Saturn
120,053 km
diameter

Planet	Diameter (in kilometers)
Mercury	4,880
Venus	12,104
Earth	12,756
Mars	6,787
Jupiter	142,800
Saturn	120,053

Mei compared 120,053 and 142,800 to find which diameter is greater.

Step 1
Line up the ones. Begin at the left and compare to find the first place where the digits are different.

1 **2** 0, 0 5 3
1 **4** 2, 8 0 0

Step 2
Compare the digits.

2 < 4, or 4 > 2

So 120,053 < 142,800, or 142,800 > 120,053.

Jupiter's diameter is greater than Saturn's.

1. What place value determined which diameter was greater? Why?
2. Which is greater, 987,364 or 985,365? Why?
3. Which is greater, 76,001 or 8,998? How do you know?

B. You can order a group of numbers by comparing them.

Order the diameters of Mercury, Earth, Venus, and Saturn from greatest to least.

Step 1
Line up the ones.

The number with the most digits is the greatest. →
4,880
12,756
12,104
→ 120,053
← The number with the fewest digits is the least.

Step 2
Compare the other numbers.

1 2, **7** 5 6
1 2, **1** 0 4

Think: 7 > 1
So 12,756 > 12,104.

Step 3
Order the numbers from greatest to least.

120,053
12,756
12,104
4,880

4. Order the diameters from least to greatest.

TRY OUT

Compare. Use >, <, or =.

5. 46,927 ● 43,692
6. 103,298 ● 104,298

Order the numbers from least to greatest.

7. 38,726; 9,928; 39,927
8. 98,201; 98,210; 100,438; 9,875

PRACTICE

Compare. Use >, <, or =.

9. 372 ● 357
10. 5,264 ● 998
11. 1,208 ● 1,002
12. 6,824 ● 6,824
13. 127,641 ● 127,461
14. 24,907 ● 249,007
15. 965,079 ● 965,097
16. 100,109 ● 100,109
17. 378 ● 3,780
18. 420 thousand ● 420,000
19. 18 thousand ● 18,020
20. 37,628 ● thirty-seven thousand, six hundred twenty-seven

Write in order from least to greatest.

21. 894; 697; 9,001
22. 325; 2,620; 2,358; 906
23. 301; 3,010; 3,001; 3,100; 30,001
24. 286; 590; 18,137
25. 10,011; 1,011; 10,010; 1,101; 10,001
26. 679; 62,720; 6,738; 60,984

Critical Thinking

Write whether these statements are *always*, *sometimes*, or *never* true. If sometimes true, give examples to support your answer.

27. A 4-digit whole number is greater than a 3-digit whole number.

28. A number with a greater digit in the tens place than another number is greater than the other number.

Solve.

29. Jupiter orbits the Sun at a speed of 47,017 km per hour. Saturn orbits the Sun at 34,705 km per hour. Which planet is faster?

30. **Write a problem** about planets using the information found on page 517 in the Databank. Ask others to solve your problem.

UNDERSTANDING A CONCEPT
Rounding Numbers

A. For a social studies project, Jon made a table showing the populations of some towns in Kansas. He rounded the numbers in his table to make a graph. What is the population of Abilene to the nearest hundred?

POPULATIONS OF SOME TOWNS IN KANSAS

Town	Population
Abilene	6,572
Lawrence	52,738
Manhattan	32,644

You can use a number line to help you round numbers.

Round 6,572 to the nearest hundred.

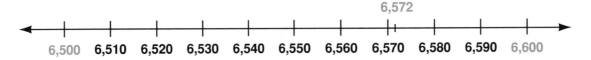

Think: 6,572 is between 6,500 and 6,600. 6,572 is closer to 6,600. Round up.

The population of Abilene rounded to the nearest hundred is 6,600.

When a number is halfway between two numbers, round up.

1. Suppose the population of Abilene were 6,550. What would be the population rounded to the nearest hundred? Why?

2. Round 6,539 to the nearest hundred.

B. You can round numbers without a number line.

Round 6,572 to the nearest hundred.

Step 1	Step 2	Step 3
Find the place to which you are rounding. 6, **5** 7 2	Look at the digit to the right of that place. 6, 5 **7** 2	If it is 5 or greater, round up. If it is less than 5, round down. **Think:** 7 > 5 Round up. 6,572 → 6,600

So 6,572 rounded to the nearest hundred is 6,600.

3. What is the population of Abilene to the nearest ten? Why?

4. What is the population of Manhattan to the nearest thousand?

5. Round $82.57 to the nearest $10; to the nearest $1; to the nearest $.10. Tell how you rounded.

TRY OUT Write the number.

6. Round 882 to the nearest 100.
7. Round 28,299 to the nearest 1,000.
8. Round $37.62 to the nearest $1.00.
9. Round $4.35 to the nearest $.10.

PRACTICE

Round to the nearest 10 or $.10.

10. 53   11. 28   12. $4.92   13. 316   14. 635   15. $.69

Round to the nearest 100 or $1.00.

16. 327   17. 1,298   18. $16.59   19. 21,336   20. 9,268   21. 8,461

Round to the nearest 1,000 or $10.00.

22. 8,275   23. 13,964   24. 95,520   25. $62.70   26. 7,496   27. 57,236

Round to the nearest 10,000 or $100.00.

28. 36,214   29. $145.49   30. 12,692   31. $750.00   32. 43,965   33. $273.01

Critical Thinking

Think about Jon's social studies project.

34. **What if** Jon's table contained 4-, 5-, and 6-digit numbers? What place would you round to? Why?
35. Name some situations in which a rounded number or an estimate of a population would be enough.

Mixed Applications

36. The population of Ottawa, Kansas, is 11,016. What are the values of the digit 1 in this number?
37. Write the populations of the towns in the table on page 26 in order from least to greatest.

UNDERSTANDING A CONCEPT

Coordinate Graphing

A. Jackie's fellow students used a grid to make a map of their town. First they made a list of places to include. Then they named each place with a letter to show it on the map.

A. School   **D.** Post Office
B. Library   **E.** Grocery Store
C. City Hall   **F.** Town Museum

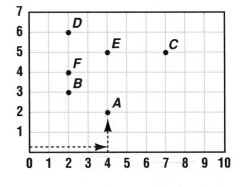

Each point on the map is named by an **ordered pair** of numbers.

To locate point A: Start at 0. Move 4 spaces to the right, then 2 spaces up.
The ordered pair (4, 2) tells you the location of point A. The numbers 4 and 2 are the **coordinates** of point A.

(4, 2)
first coordinate   second coordinate

1. What does point A represent on the map?

2. What would you find on the map if you reversed the coordinates of point A?

3. Why do you think (4, 2) is called an ordered pair?

B. You can locate a point on a grid if you know the ordered pair that names it.

4. What point has the coordinates (2, 6)? What building does it represent?

5. Which three points have the same first coordinate?

TRY OUT Use the grid above.

Name the ordered pair.

6. point B   7. point E   8. point F

Name the point.

9. (7, 5)   10. (2, 4)   11. (4, 5)

PRACTICE

Use the map to answer Exercises 12–27.

A. Train Station
B. Police Station
C. Gas Station
D. Restaurant
E. Pharmacy
F. Paint Store
G. Barber Shop
H. Clothing Store
I. Jeweler
J. Bakery
K. Hardware Store
L. Bookstore

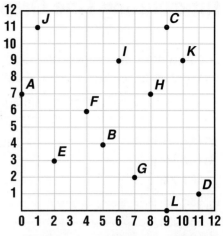

Name the ordered pair that gives the point.

12. A
13. D
14. G
15. E
16. J
17. L
18. C
19. G

Name the point and the place it represents.

20. (5, 4)
21. (9, 11)
22. (6, 9)
23. (10, 9)
24. (4, 6)
25. (8, 7)
26. (0, 7)
27. (11, 1)

Solve.

28. Each square on the grid represents one block. How many blocks are there between the jeweler and the hardware store?

29. Imagine that you cannot take a short cut across any of the blocks. You walk from the pharmacy to the barber shop, then to the clothing store, and then to the restaurant. How many blocks have you walked?

30. Use graph paper to make a map of an area you know well. Make a list of the places and buildings to include. Label each place on the map with a point. Then write an ordered pair for each point.

31. Use graph paper to make a 10 x 10 grid. Number the bottom and side. Connect the following ordered pairs in sequence: (2, 3), (8, 3), (8, 9), (2, 9). Connect (2, 9) and (2, 3). What figure is represented? How do you know?

EXTRA Practice, page 42

Understanding Numbers   29

UNDERSTANDING A CONCEPT

Interpreting Line Graphs

Jessica's class looked at a line graph showing the population of Indianapolis, Indiana, from 1950 to 1980.

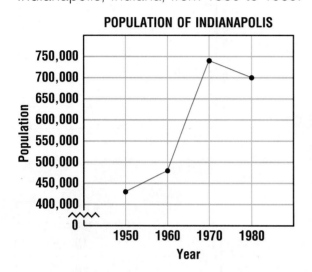

The vertical and horizontal lines that form the graph are called the **vertical axis** and the **horizontal axis.**

1. What information is shown on the vertical axis? the horizontal axis?

2. What does the information on the axes tell you about the purpose of the line graph?

3. Why do you think the populations are given in rounded numbers?

4. Why do you think a broken line is used just below 400,000 on the vertical axis?

5. Use the line graph to find the following information.
   a. the approximate population of Indianapolis in 1960
   b. the exact population of Indianapolis in 1970
   c. the 10-year period when the population decreased

6. Was there some information in Problem 5 that you could not read from the graph? Explain.

TRY OUT Use the line graph above to complete the statement.

7. The population of Indianapolis in 1970 was about ■.

8. The population of Indianapolis increased the most between ■ and ■.

30   Lesson 1-10

PRACTICE

Use the line graph at the right to answer Problems 9–12.

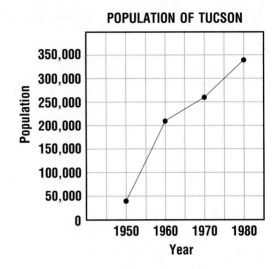

9. About how many people lived in Tucson, Arizona, in 1950? in 1970?

10. In which 10-year period did Tucson's population grow the most? Estimate that growth.

11. About how many more people lived in Tucson in 1980 than in 1950?

12. Write a sentence summarizing what the line graph shows.

Use the line graph at the right to answer Problems 13–15.

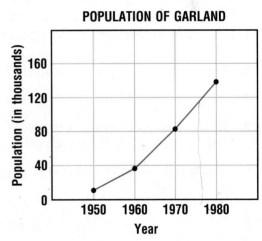

13. About how many people lived in Garland, Texas, in 1960?

14. About how many more people lived there in 1980 than in 1970?

15. Write a sentence summarizing what this graph tells you.

16. What would you predict the population of Tucson to be in 1990? Why?

17. What would you predict the population of Garland to be in 1990? Why?

Mixed Review

Write the number in standard form.

18. 217 million, 30 thousand, 78

19. 802 billion, 9 thousand, 760

20. 360 billion, 275 million, 438 thousand, 9

Write the number in expanded form.

21. 762 million, 400 thousand, 822

22. 5 billion, 367 thousand, 40

DEVELOPING A CONCEPT

Making Line Graphs

The table shows the average number of miles traveled per car in the United States each year from 1975 to 1984.

AVERAGE MILES TRAVELED PER CAR IN THE UNITED STATES

Year	Miles	Year	Miles
1975	9,634	1980	9,135
1976	9,763	1981	9,002
1977	9,839	1982	9,533
1978	10,046	1983	9,654
1979	9,485	1984	9,809

WORKING TOGETHER

Use the data in the table and these steps to make a line graph.

Step 1 Round the data to a convenient place to choose a scale for the vertical axis.
Step 2 Draw and label the horizontal and vertical axes.
Step 3 Place the points on the graph.
Step 4 Connect the points in order with straight lines.
Step 5 Write a title above the graph.

1. To which place did you round the data?
2. What scale did you use for the vertical axis?
3. Which year showed the greatest change in miles traveled? Was this change an increase or a decrease?

Ray graphed the information this way.

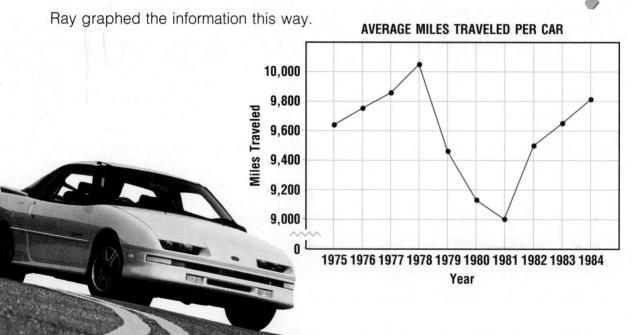

32 Lesson 1-11

SHARING IDEAS

4. Compare your graph with Ray's and those of your classmates. How are they similar? How are they different?

5. **What if** the space between the years on the horizontal scale were doubled? Would the line of the graph rise and fall as sharply? Why or why not?

PRACTICE

Use the data in the table at the right to make a line graph.

6. What scale did you use for the vertical axis?

7. To what place did you round the data?

8. Which 10-year period showed the greatest change in car sales? Was this change an increase or a decrease?

9. Write a sentence to summarize what the graph tells you.

10. Use the information about money spent found on page 517 in the Databank to make a line graph. Write a sentence explaining what the graph shows.

11. Choose a topic of your own. Gather the data you need and use it to make a line graph. Write a sentence summarizing what your graph shows.

CAR SALES

Year	Cars Sold
1920	1,905,560
1930	2,787,456
1940	3,717,385
1950	6,665,863
1960	6,674,796
1970	6,546,817
1980	6,255,340

Mixed Review

Compare. Use >, <, or =.

12. 1,376 ● 1,367   13. 625,123 ● 62,512   14. 90,017 ● 90,071

15. 865,312 ● eight hundred sixty-five thousand, three hundred twelve

Write in order from least to greatest.

16. 28,637; 21,372; 21,732   17. 3,647; 986; 899

Round to the nearest 1,000 or $10.00.

18. 75,248   19. $75.64   20. 398,499   21. 29,901   22. $394.29

EXTRA Practice, page 43

Problem Solving

✓ UNDERSTAND
✓ PLAN
✓ TRY
✓ CHECK
✓ EXTEND

Strategy: Using Number Sense

When Willie looks at something, he uses number sense to think about how much it weighs or about how long or how high it is.

Willie knows that his father is about 6 feet tall. He uses his father's height as a reference when he thinks about the length or the height of an object.

Willie's father can walk through the doorway without bumping his head, so Willie decides that the doorway must be about 7 feet high.

1. Willie's father said the family's two-story house is about 20 feet high. Does that seem reasonable? How can you tell?

2. Willie's little brother said that he is 8 feet tall. How do you know that he cannot be that tall?

Willie's cat weighs about 5 pounds. He uses his cat's weight as a reference when he thinks about how much some things weigh.

3. He thinks that his shoes weigh about 1 or 2 pounds. Does that seem reasonable? How can you tell?

4. Willie estimates that his friend Susie weighs about 15 pounds. Do you think he is right? Why or why not?

5. How can you use number sense to estimate heights and weights?

PRACTICE

Use number sense to solve the problem. Write the letter of the best answer.

6. A pencil is about 6 inches long. About how wide could Jane's school desk be?
   a. 10 inches
   b. 30 inches
   c. 100 inches

7. Cynthia's math book weighs about 1 pound. About how much could her TV weigh?
   a. 1 pound
   b. 2 pounds
   c. 15 pounds

8. A gallon of milk costs about $2. How much could a quart of orange juice cost?
   a. $3
   b. $7
   c. $10

9. About how many students could be in Jim's class?
   a. 70 students
   b. 100 students
   c. 25 students

10. One wall in Jim's classroom is 15 feet long. About how long could the school corridor be?
    a. 10 feet
    b. 500 feet
    c. 1,000 feet

11. A bicycle path is about 4 feet wide. About how wide could a 4-lane highway be?
    a. 10 feet
    b. 15 feet
    c. 40 feet

12. A chicken at the store weighs about 3 pounds. About how much could a turkey weigh?
    a. 1 pound
    b. 10 pounds
    c. 50 pounds

13. **Write a problem** that can be solved by using number sense. Solve the problem. Ask others to solve your problem.

EXTRA Practice, page 43

Understanding Numbers 35

DECISION MAKING

Problem Solving: Joining an After-School Club

SITUATION

Some fifth-grade students plan to join after-school clubs. There are five different clubs to join. The clubs meet on different days and at different times. Each club has a different membership fee and charges extra for field trips that require tickets.

PROBLEM

Which club should the students join?

DATA

COMPUTER CLUB — Become a Computer WIZ. We meet on Mondays, Wednesdays and Fridays from 3:00–5:00. JOIN NOW!

SCIENCE CLUB — JOIN US FOR AN EXCITING LOOK AT THE WORLD AROUND YOU. Meetings every Tuesday and Thursday.

	COMPUTER CLUB	SCIENCE CLUB	SPORTS CLUB	ART CLUB	MUSIC CLUB
Day(s)	Monday Wednesday Friday	Tuesday Thursday	Wednesday	Tuesday Wednesday	Monday Friday
Time	3:00–5:00	3:30–5:00	4:00–6:00	3:00–4:00	3:15–5:00
Membership Fee	none	$20	$10	$15	$5
Field Trips (ticket price)	No trips	Zoo ($1) Botanical Gardens ($1)	Baseball Game ($4) Soccer Match ($3) Track Meet (0)	Art Museum (0) Craft Museum ($1)	Concert ($2) Musical Show ($5) Recital (0)

36 Lesson 1-13

USING THE DATA

1. On which day of the week do the most clubs meet?

How many hours per week does the club meet?

2. Art   3. Computer   4. Science   5. Sports   6. Music

How much will it cost to join the club and attend all the field trips?

7. Art   8. Computer   9. Science   10. Sports   11. Music

MAKING DECISIONS

12. Suppose you could not attend a club on Wednesday or Thursday. Which club should you consider joining?

13. Suppose you wanted to join the most clubs without having two meetings on the same day. Which clubs would you choose?

14. Suppose you could spend only $15 on after-school clubs. Which clubs would you decide to join? Why?

15. **What if** you wanted to attend a club every day of the week? Which clubs would you decide to join?

16. **What if** you wanted to take as many field trips as possible, but you had only $25 to spend on after-school clubs? Tell which club or clubs you would join and which trips you would take.

17. **Write a list** of other things you should consider before joining a club.

18. Which clubs would you join? Why?

Understanding Numbers 37

CURRICULUM CONNECTION

Math and Social Studies

About 50,000 years ago in ancient Babylonia, people found that making marks on the ground was easier than counting on their fingers. Early people then discovered that carving grooves in a board and moving disks along the grooves was a good way of counting.

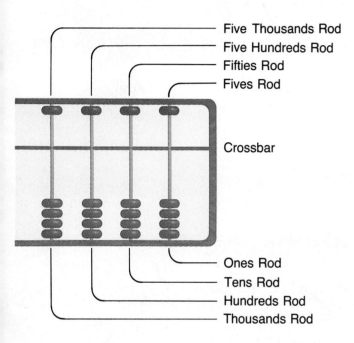

Soon afterward the **abacus** was first used. An abacus, which is made up of a frame that holds rods and movable beads, is still in use today. In fact, some people can figure amounts on an abacus as quickly as people who use modern calculating machines.

Here is part of an abacus. It shows the value of each bead. To show a certain number, the necessary beads are pushed to the crossbar. You can show 9 by moving the 5 bead and the 4 ones beads close to the crossbar. To show 10, simply move 1 tens bead to the crossbar. No ones beads need to be moved.

What number is shown?

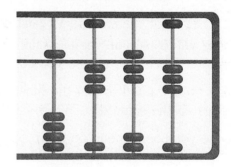

Think: Only look at the beads that have been pushed to the crossbar. There are 3 ones, 1 fifty and 2 tens, 3 hundreds, and 1 five-thousand. The number is 5,373.

ACTIVITIES

1. Do a sketch of an abacus that shows the number 6,519. Remember, each rod above the crossbar holds one bead. Each rod below the crossbar holds four beads.

2. Work in a small group. Make a simple abacus by drawing on paper the frame, crossbar, and rods. Use counters above and below the crossbar to represent the numbers. Take turns giving each other numbers to represent by moving the counters to the crossbar.

Computer Exploration: Powers and Exponents

When the number 10 is used as a factor 3 times, you can write:

$$10 \times 10 \times 10 = 1,000$$

A shorter way of writing this is by using an **exponent** to show how many times 10 is used as a factor. The number 10 is called the **base**. The product 1,000 is a **power** of 10.

$$10 \times 10 \times 10 = 10^3 \leftarrow \text{exponent}$$
3 factors of 10   base

Read: ten to the third power

expanded notation   exponent form

Exponents can also be used with bases other than 10.

You can use the computer program TENPOWER to explore an interesting pattern that occurs when factors of 10 are used. Each time you enter a value for the exponent, the computer will print the factors (expanded notation) and the product.

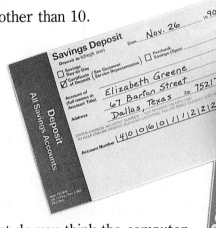

AT THE COMPUTER

1. If you enter 1 for the exponent, what do you think the computer will show for the expanded notation and the product? Check your idea.

2. Now try entering values of 2, 3, 4, 5, 6, 7, and 8 for the exponent. Each time, compare the exponent with the number of zeros shown in the product. (You may wish to write each exponent form and product on a separate sheet of paper so that you can refer to it.)

3. Describe the pattern that you found. Try 9 and 10 for the exponent and check that the pattern continues.

4. Write the exponent form for 10 used as a factor 12 times. How many zeros will the product have? Use the pattern to write the product.

5. How would you write "ten to the zero power" using exponents? What number do you think "ten to the zero power" names?

Understanding Numbers   **39**

EXTRA PRACTICE

Using Tables, page 13

Use the table at the right to answer Questions 1–3.

1. How many students voted for Bird World?
2. Which place was voted for by the most students? the fewest students?
3. Which place is the most popular for a class trip? How do you know?

VOTES FOR PLACES TO VISIT

Petting Zoo	Bird World	Historical Park	Car Museum
12	16	25	28

Frequency Tables, page 15

Copy and complete the frequency table.

1. How many votes did Roberta receive?
2. Which students received the same number of votes?
3. Who won the election? How do you know?

VOTES FOR TEAM CAPTAIN

	Tally	Total
Roberta	ЖТ I	6
Cecelia	ЖТ III	■
Jonathan	ЖТ III	8
Miguel	ЖТ IIII	■

Numbers to Hundred Thousands, page 17

Name the place and value of the digit 9 in the number.

1. 905,638
2. 389,652
3. 80,932
4. 10,090
5. 92,063

Write the number in expanded form.

6. 3,463
7. 87,251
8. 40,019
9. 317,605

Write the word name.

10. 92,487
11. 30,193
12. 452,809
13. 200,050

Write the number in standard form.

14. 400,000 + 30,000 + 2,000 + 700 + 10 + 7
15. 700,000 + 3,000 + 200 + 8
16. nine hundred fifty-three thousand, two hundred thirteen
17. thirty-nine thousand, two hundred forty-nine

EXTRA PRACTICE

Millions and Billions, page 19

Write the digit in the ten millions place and ten billions place.

1. 143,697,421,202
2. 53,685,017,213
3. 809,643,111,017
4. 623,819,407,333
5. 987,137,357,177
6. 78,153,187,179

Write the number in standard form.

7. fifteen million, nine hundred eight thousand, twenty-one
8. 490 billion, 211 thousand, 75
9. 265 billion, 62 million, 6
10. 1,000,000 greater than 7,462,395
11. 10,000,000 less than 27,892,309
12. 1,000,000,000 greater than 618,393,017,325

Problem Solving: Using the Five-Step Process, page 21

Kids News asked 1,000 fifth graders about the type of TV show they liked best. Use the data to answer questions about favorite TV shows.

Comedy Shows	255
Family Shows	245
Talk Shows	200
Sports Shows	300

1. What does the data tell you about sports shows?
2. What two types of shows are about equally favored by fifth graders?
3. How may fifth graders liked talk shows or sports shows?

Comparing and Ordering Numbers, page 25

Compare. Use >, <, or =.

1. 473 ● 437
2. 6,395 ● 987
3. 1,063 ● 1,070
4. 7,347 ● 7,347
5. 119,397 ● 119,973
6. 39,406 ● 399,409
7. 920 thousand ● 920,000
8. 19 thousand ● 19,020

Write in order from least to greatest.

9. 393; 127; 8,075
10. 257; 3,193; 3,939; 850
11. 405; 4,050; 4,005; 4,500; 40,500
12. 6,120; 612; 6,210; 60,201; 6,002

Write in order from greatest to least.

13. 173; 953; 17,687
14. 31,374; 743; 7,374; 70,341

Understanding Numbers 41

Extra Practice

Rounding Numbers, page 27

Round to the nearest 10 or $.10.

1. 24
2. 36
3. $3.57
4. 319
5. $.72

Round to the nearest 100 or $1.00.

6. 417
7. 1,395
8. $15.59
9. 32,609
10. 7,395

Round to the nearest 1,000 or $10.00.

11. 9,134
12. 17,892
13. $72.89
14. 66,530
15. 3,149

Round to the nearest 10,000 or $100.00.

16. 39,116
17. $132.52
18. 11,985
19. $650.00
20. 23,695

Coordinate Graphing, page 29

Use the map to answer Questions 1–12.

- **A.** Laundromat
- **B.** Town Hall
- **C.** Bakery
- **D.** Shopping Mall
- **E.** Shoe Store
- **F.** Transportation Museum
- **G.** Fire House
- **H.** Police Station
- **I.** Historical Park
- **J.** Hardware Store
- **K.** Pool and Patio Store

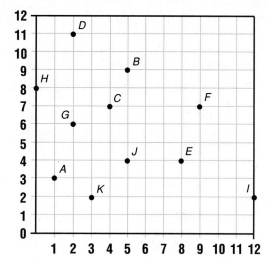

Name the ordered pair that gives the point.

1. A
2. E
3. G
4. K
5. F
6. J
7. B
8. D

Name the point and the place it represents.

9. (12, 2)
10. (0, 8)
11. (3, 2)
12. (9, 7)

Extra Practice

Interpreting Line Graphs, page 31

Use the line graph at the right to answer Questions 1–3.

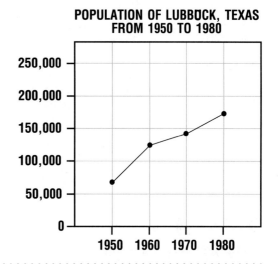

1. About how many people lived in Lubbock, Texas, in 1950? in 1970?

2. In which ten-year period did Lubbock's population grow the most? Estimate the growth.

3. Write a sentence summarizing what the line graph shows.

Making Line Graphs, page 33

Use the data in the table at the right to make a line graph.

REGISTERED MOTORCYCLES

Year	Number Registered
1970	2,824,000
1975	4,964,000
1980	5,694,000
1985	5,444,000

1. What scale did you use for the vertical axis?

2. To what place did you round the data?

3. Which 5-year period showed the greatest change in the number of registered motorcycles? Was this change an increase or a decrease?

4. Write a sentence to summarize what the graph tells you.

Problem-Solving Strategy: Using Number Sense, page 35

Use number sense to solve the problem. Write the letter of the best answer.

1. Angela's portable radio weighs about 1 pound. About how much could her stereo weigh?
   a. 2 pounds **b.** 5 pounds **c.** 20 pounds

2. A quart of juice costs about $1. How much would a gallon of juice cost?
   a. $1 **b.** $3 **c.** $5

3. Gerry's bicycle is 24 inches high. About how high is her younger brother's tricycle?
   a. 36 inches **b.** 20 inches **c.** 12 inches

Understanding Numbers 43

Practice PLUS

KEY SKILL: Comparing and Ordering Numbers (Use after page 25.)

Level A

Compare. Use >, <, or =.

1. 34 ● 32
2. 407 ● 498
3. 100 ● 110
4. 432 ● 432

Write in order from least to greatest.

5. 787; 365; 705
6. 598; 495; 789
7. 321; 341; 211

8. Danielle traveled 687 miles last summer. Brittany traveled 598 miles last summer. Who traveled the most miles?

Level B

Compare. Use >, <, or =.

9. 365 ● 378
10. 8,765 ● 898
11. 7,541 ● 7,541
12. 6,540 ● 6,546
13. 2,001 ● 2,001
14. 6,508 ● 6,580
15. 67,897 ● 98,734
16. 439,710 ● 77,657
17. 23,897 ● 23,987

Write in order from least to greatest.

18. 452; 890; 127
19. 657; 795; 146
20. 570; 678; 568
21. 875; 1,098; 322
22. 543; 907; 111; 1,011
23. 678; 127; 8,909; 2,202

24. Nick moved 2,343 miles from his hometown. Another student, Eve, moved 3,243 miles away. Who moved the farthest?

Level C

Write in order from least to greatest.

25. 876,212; 90,783; 867,212
26. 76,543; 93,750; 81,206
27. 11,919; 11,190; 11,019; 111,900
28. 433,720; 344,720; 44,720; 720,344

Write in order from greatest to least.

29. 439; 678; 1,318
30. 876; 8,760; 8,676; 678
31. 3,011; 3,001; 311; 3,101
32. 7,009; 797; 7,079; 979

33. Leslie has 43,890 miles in her Free-Air Travel program. Rick has a total of 34,879 miles. Who has traveled more miles?

Practice PLUS

KEY SKILL: Rounding Numbers (Use after page 27.)

Level A

Round to the nearest 10.

1. 32 **2.** 78 **3.** 17 **4.** 213 **5.** 432

Round to the nearest 100.

6. 456 **7.** 347 **8.** 42,189 **9.** 6,502 **10.** 3,298

11. The town of Bayfield has a population of 2,289. Round this number to the nearest hundred.

Level B

Round to the nearest 10 or $.10.

12. 87 **13.** $.38 **14.** $3.92 **15.** 745 **16.** $.89

Round to the nearest 100 or $1.00.

17. 636 **18.** $18.90 **19.** 7,543 **20.** 2,298 **21.** 4,098

Round to the nearest 1,000 or $10.00.

22. 6,546 **23.** 17,567 **24.** 87,654 **25.** $56.70 **26.** 24,765

27. Springfield has a population of 35,989. What is the population rounded to the nearest thousand?

Level C

Round to the nearest 100 or $1.00.

28. 654 **29.** $14.39 **30.** 4,390 **31.** 22,109 **32.** 8,017

Round to the nearest 1,000 or $10.00.

33. 6,543 **34.** 12,897 **35.** 89,650 **36.** $35.79 **37.** 95,789

Round to the nearest 10,000 or $100.00.

38. 76,189 **39.** $189.76 **40.** 13,287 **41.** $357.89 **42.** $199.08

43. Rhonda spent $768.90 on her airline ticket. Round this number to the nearest hundred dollars.

Understanding Numbers

Chapter Review

LANGUAGE AND MATHEMATICS

Complete the sentences. Use the words in the chart on the right.

1. You use tally marks to record data in a ■. *(page 14)*
2. 300,000 + 20,000 + 6,000 + 400 + 30 + 9 is written in ■. *(page 16)*
3. You can locate a point on a coordinate grid if you know the ■ that names it. *(page 28)*
4. A ■ is useful in showing changes over a period of time. *(page 32)*
5. **Write a definition** or give an example of the words you did not use from the chart.

VOCABULARY
line graph
round
frequency table
expanded form
ordered pair
vertical axis

CONCEPTS AND SKILLS

Complete the frequency table. *(page 14)*

VOTES FOR CLASS TREASURER

Name	Tally	Total
Lindsay	ЖН I	6
Thomas	ЖН IIII	6. ■
Nathan	7. ■	5
Simone	ЖН II	8. ■

Write the number in expanded form. *(page 16)*

9. 3,425 10. 23,895 11. 102,367 12. 408,907

Write the number in standard form. *(pages 16–19)*

13. 20,000 + 3,000 + 700 + 20 + 8
14. 300,000 + 70,000 + 800 + 5
15. ten million, eight hundred four thousand, seven hundred one

Compare. Use >, <, or =. *(page 24)*

16. 307 ● 381
17. 6,812 ● 889
18. 1,030 ● 1,303
19. 23,817 ● 238,170
20. 300,009 ● 300,009
21. 854,063 ● 854,036

Round each amount to the underlined place. *(page 26)*

22. 6,132   **23.** 14,389   **24.** 88,507   **25.** $73.80   **26.** 1,852

Name the ordered pair that gives the point. *(page 28)*

27. B   **28.** I
29. E   **30.** C
31. A   **32.** D

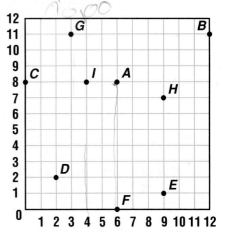

Name the point. *(page 28)*

33. (6, 0)   **34.** (9, 7)   **35.** (3, 11)

36. Use the data in the table below to make a line graph. *(page 32)*

U.S. POPULATION

Year	1930	1940	1950	1960	1970	1980
Number	123,202,624	132,164,569	151,325,798	179,323,175	203,302,031	226,545,518

CRITICAL THINKING

37. When do you think it is better to use a rounded number than an exact number?

38. When making line graphs, how do you decide what scale to use for the vertical axis?

MIXED APPLICATIONS

39. Toby paid $684,509 for a rare antique and $28,769 for a stamp at auction. What is the value of the digit 8 in each price? *(page 16)*

40. *The News at Noon* was watched by 463,685 households, and 435,863 households watched the *Shopping Network*. Which was seen by more households? *(page 24)*

41. The basketball court in the playground is about 25 feet wide. About how long could it be?
a. 20 feet   **b.** 50 feet   **c.** 200 feet

42. Carla's dictionary weighs about 2 pounds. About how much could her set of encyclopedias weigh?
a. 1 pound   **b.** 3 pounds   **c.** 30 pounds

Understanding Numbers **47**

Chapter Test

Use the table at the right to answer Questions 1–3.

1. How many votes did Sasha receive?
2. Which two students received the same number of votes?
3. Who received the fewest number of votes?

VOTES FOR MATH CLUB PRESIDENT

Name	Tally	Total
Willis	IIII	4
Sasha	HHT I	6
Elgin	HHT III	8
Karen	HHT I	6

Write the number in standard form.

4. 200,000 + 50,000 + 2,000 + 300 + 40 + 7
5. four hundred twenty-three thousand, fifty
6. six hundred five billion, four hundred two thousand
7. 783 billion, 49 million, 8
8. 12 million, 708 thousand, 36

Compare. Use >, <, or =.

9. 862,059 ● 862,095
10. 605,819 ● 65,819
11. 346,204 ● 346,204
12. 15,009 ● 150,009

Write in order from greatest to least.

13. 385; 640; 19,167
14. 13,095; 1,095; 13,905; 3,095

Round to the nearest 1,000 or $10.00.

15. 6,489
16. 12,871
17. $53.89
18. $87.09
19. 7,501

Use the grid to answer Questions 20–23. Name the ordered pair that gives the point.

20. C
21. F

Name the point.

22. (3, 1)
23. (6, 3)

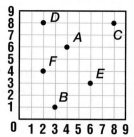

Solve.

24. A watermelon might weigh about 10 pounds. About how much could a grapefruit weigh?
    a. 1 pound b. 5 pounds c. 8 pounds

25. Mrs. Gunther raised $689,205 for the hospital renovation. What is the value of the digit 6 in the dollar amount?

ENRICHMENT FOR ALL

BASE FIVE

A. Our number system is a base-ten system. In a base-ten system you count by ones and group by tens. You use the digits 0 through 9 to name numbers.

In a base-five number system you count by ones and group by fives. You use the digits 0 through 4 to name numbers.

BASE TEN

Hundreds	Tens	Ones

BASE FIVE

Twenty-Fives	Fives	Ones

1. How is the value of each place related to the value of the place to its right in the base-ten chart? in the base-five chart?

2. How would you name the next place to the left in the base-five place-value chart? Why?

B. You can write a number in base ten or base five.

Write 14 in base five.

Think: 14 = 2 fives + 4 ones

Twenty-Fives	Fives	Ones
	2	4

Write: 24_{five}

Write 430_{five} in base ten.

Twenty-Fives	Fives	Ones
4	3	0

Think: 4 twenty-fives + 3 fives + 0 ones
$\quad\quad\quad\quad$ 100 $\quad$ + $\quad$ 15 $\quad$ + $\quad$ 0 $\quad$ = 115

Write: 115

Write the number in base five.

3. 16 **4.** 21 **5.** 10 **6.** 19 **7.** 34 **8.** 26

Write the number in base ten.

9. 11_{five} **10.** 4_{five} **11.** 20_{five} **12.** 22_{five} **13.** 300_{five} **14.** 333_{five}

15. How would $4,000_{five}$ be written in base ten?

Understanding Numbers **49**

Cumulative Review

Choose the letter of the correct answer.

1. Round $38.51 to the nearest $10.00.
   a. $40.00
   b. $30.00
   c. $38.00
   d. not given

2. Which digit is in the ten millions place for 387,403,725,009?
   a. 8
   b. 0
   c. 3
   d. not given

3. Which is the greatest number?
   a. 364
   b. 8,985
   c. 3,075
   d. 8,598

4. Compare: 5,873 ● 8,732
   a. <
   b. >
   c. =
   d. not given

5. Compare: 930 thousand ● 930,000
   a. <
   b. >
   c. =
   d. not given

6. Which digit is in the ten billions place for 678,903,651,898?
   a. 6
   b. 7
   c. 0
   d. not given

7. Round 62 to the nearest ten.
   a. 65
   b. 70
   c. 60
   d. not given

8. What is the standard form for 200,000 + 40,000 + 3,000 + 500 + 7?
   a. 240,357
   b. 243,057
   c. 243,507
   d. not given

9. Round 43,812 to the nearest 10,000.
   a. 43,000
   b. 44,000
   c. 50,000
   d. not given

Use the grid to answer 10–13.

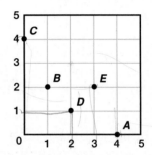

10. Name the ordered pair for point D.
    a. (2, 1)
    b. (1, 2)
    c. (2, 6)
    d. not given

11. Name the ordered pair for point C.
    a. (0, 1)
    b. (4, 0)
    c. (0, 4)
    d. not given

12. Name the point at (3, 2).
    a. D
    b. F
    c. E
    d. not given

13. Name the point at (4, 0).
    a. E
    b. A
    c. B
    d. not given

14. In 1986 the population of El Paso, Texas was 491,800. In 1980 the population was 425,259. In 1970 the population was 322,261. In which year was the population over 500,000?
    a. 1970
    b. 1980
    c. 1986
    d. not given

Using Addition and Subtraction

CHAPTER 2

MATH CONNECTIONS: ALGEBRA • MEASURING LENGTH • TIME • PROBLEM SOLVING

Video Tapes	Price
T30	$1.75
T60	$2.50
T90	$3.25
T120	$4.00

1. What information do you see in this picture?
2. How can you use the information?
3. Which video tape costs the most?
4. Write a problem using the information.

UNDERSTANDING A CONCEPT

Informal Algebra: Meaning of Addition and Subtraction

A. Jerry and Maria packed 7 baseballs in a box that can hold 16. How many more balls can fit in the box?

Jerry found the missing addend.

7 + ■ = 16   7 + 9 = 16
7 + 9 = 16   ↑   ↑   ↑
   ■ = 9   addend sum

Maria subtracted to find the difference.

16 − 7 = ■   16 − 7 = 9
16 − 7 = 9   ↑
   ■ = 9   difference

So 9 more balls can fit in the box.

You can write four related facts using 7, 9, and 16. The facts form a **fact family** that can help you add and subtract.

7 + 9 = 16   9 + 7 = 16   16 − 7 = 9   16 − 9 = 7

1. Write a related fact that can help you find the missing number. Then find the number.
   a. 6 + ■ = 13   **b.** 12 − ■ = 7

B. Addition has three special properties.

Commutative Property
If the order of the addends is changed, the sum remains the same.

5 + 8 = 13
8 + 5 = 13

Identity Property
If one of the addends is zero, the sum is equal to the other addend.

9 + 0 = 9
0 + 9 = 9

Associative Property
If the grouping of the addends is changed, the sum remains the same.

3 + (4 + 5) = 12
(3 + 4) + 5 = 12

2. Since 9 + 6 = 15, which property allows you to write 6 + 9 = 15?

TRY OUT

3. Write the two missing facts in the 6, 9, 15 fact family.
   6 + 9 = 15, 15 − 9 = 6

4. Name the addition property shown.
   4 + (8 + 6) = (4 + 8) + 6

52 Lesson 2-1

PRACTICE

Write a fact family for each group of numbers.

5. 4, 8, 12   **6.** 2, 9, 11   **7.** 7, 3, 10   **8.** 9, 5, 14   **9.** 9, 3, 12

Find the sum or difference.

10. 7 + 6   **11.** 5 + 7   **12.** 7 + 8   **13.** 8 + 6   **14.** 8 + 9   **15.** 4 + 9

16. 12 − 6   **17.** 13 − 5   **18.** 17 − 9   **19.** 13 − 4   **20.** 16 − 8   **21.** 13 − 6

22. 15 − 7   **23.** 17 − 8   **24.** 9 + 9   **25.** 14 − 6   **26.** 9 + 8   **27.** 12 − 3

28. 4 + 7   **29.** 18 − 9   **30.** 11 − 4   **31.** 0 + 8

32. 3 + (5 + 3)   **33.** (4 + 2) + 8   **34.** 7 + (4 + 3)   **35.** (2 + 6) + 8

36. 7 + 2 + 3   **37.** 4 + 6 + 8   **38.** 5 + 2 + 5 + 3   **39.** 3 + 2 + 8 + 4

Complete. What property did you use?

40. 512 + 835 = 835 + ■

41. 201 + (79 + 89) = (201 + 79) + ■

42. 0 + 6,271 = ■

43. 113 + (■ + 599) = (113 + 84) + 599

Critical Thinking

44. What happens when you subtract zero from a number? Give an example to show what happens.

45. What happens when you subtract a number from itself? Does this always happen?

46. Is there a commutative property for subtraction? Give examples to support your answer.

Mixed Applications

47. Maria packed 17 boxes of baseballs. She sent 9 by Ace Delivery Service and the rest by mail. How many boxes did she mail?

48. Ed spent $4 on a sports magazine, $9 on a tee-shirt, $23 on a pair of sneakers, and $18 on a sweater. Which of his purchases cost the most? the least?

UNDERSTANDING A CONCEPT

Estimating Sums, Differences by Rounding

A. Bob and Karen were in charge of selling tickets to the Rodeo and Horse Show. They each estimated the total number of people that attended the show during the three days.

RODEO AND HORSE SHOW

Number of Tickets Sold	
Friday	846
Saturday	3,538
Sunday	2,847

Bob's Method
Round to the greatest place of the greatest number.

```
   846  →   1,000
 3,538  →   4,000
+2,847  → + 3,000
            8,000
```

Think: Round to nearest thousand.

Karen's Method
Round to the greatest place of the least number.

```
   846  →     800
 3,538  →   3,500
+2,847  → + 2,800
            7,100
```

Think: Round to nearest hundred.

1. Which estimate is greater than the exact answer? less than the exact answer? Why?

B. You can also estimate differences by rounding.

Estimate: 5,328 − 864

Method 1
Round to the greatest place of the greater number.

```
 5,328  →   5,000
−  864  → − 1,000
            4,000
```

Method 2
Round to the greatest place of the lesser number.

```
 5,328  →   5,300
−  864  → −   900
            4,400
```

2. Estimate $16.54 − $4.79. Use both methods. Which rounding method results in a more useful estimate? Why?

TRY OUT Write the letter of the correct answer. Estimate by rounding.

3. 8,241 + 754 + 109   a. 7,000   b. 8,000   c. 9,000   d. 12,000
4. 3,406 + 2,679   a. 3,000   b. 4,000   c. 5,000   d. 6,000
5. 9,298 − 2,307   a. 6,000   b. 8,000   c. 7,000   d. 9,000
6. $18.37 − $7.98   a. $10.00   b. $15.00   c. $13.00   d. $18.00

Practice

Estimate by rounding.

7.	1,511 3,745 + 4,203	8.	2,861 1,267 + 3,510	9.	$89.25 22.75 + 9.62	10.	2,903 7,118 + 472	11.	14,252 17,977 + 13,089
12.	5,831 − 3,245	13.	$72.36 − 35.27	14.	8,466 − 757	15.	$64.12 − 8.59	16.	26,487 − 17,398
17.	3,165 4,335 9,060 + 2,703	18.	3,826 722 865 + 191	19.	$23.42 − 7.68	20.	10,789 22,554 + 9,643	21.	7,995 − 799

22. 4,611 + 8,853 + 1,827

23. 5,346 − 289

24. $36.50 + $2.72 + $4.29

25. $91.79 + $4.28 + $3.11

26. 216,829 − 137,596

27. 247,621 − 63,541

Write the letter of the best estimate.

28. 398 + 456 + 767
   - a. > 2,000
   - b. > 1,700
   - c. < 1,700
   - d. < 1,400

29. 42,325 − 8,912
   - a. < 30,000
   - b. > 40,000
   - c. < 33,000
   - d. > 33,000

Critical Thinking

30. Estimate 2,861 − 2,587 by rounding to the nearest thousand. How can you find a more useful estimate?

Mixed Applications

31. Jill rode in three events in the horse show. She received 48, 32, and 61 points. About how many points did she receive?

32. A rodeo ticket cost $7.50 and a ticket to the horse show $5.50. Which ticket cost more?

33. Books and pennants were sold at four booths. Tony listed the sales at each booth: $325.50, $237.25, $195.75, and $438.50. List the amounts from least to greatest.

34. *Write a problem* about estimating ticket sales for an event at school. Solve your problem. Then ask others to solve it.

UNDERSTANDING A CONCEPT

Front-End Estimation: Sums and Differences

A. The director of the town band wants to play three new pieces for the fall concert. The sheet music for the pieces costs $75.00, $135.00, and $208.00. About how much money will he need?

An estimate may help you to answer this question. Here is a method that uses front-end estimation.

Step 1
Add the front digits.
Write zeros for the other digits.

```
   $ 75.00
     135.00
  +  208.00
    $300.00
```

Step 2
Adjust the estimate.

```
   $ 75.00
     135.00
  +  208.00
    $400.00
```
Think: 75 + 35 is about 100.

The director will need about $400.

1. How does adjusting the estimate help you get an estimate that is closer to the exact answer?

2. How would you estimate 2,401 + 7,363 + 6,347 using front-end estimation? What is the estimate?

B. You can also use front-end estimation to estimate differences.

Estimate: 5,609 − 4,512.

Step 1
Subtract the front digits.
Write zeros for the other digits.

```
    5,609
  − 4,512
    1,000
```

Step 2
Adjust the estimate.

```
    5,609
  − 4,512
   > 1,000
```
Think: 6 > 5 The exact answer is greater than 1,000.

3. Estimate 9,375 − 7,624. Is the estimate greater than or less than 2,000? Why or why not?

4. How would you estimate $12.57 − $5.68 using front-end estimation? What is the estimate?

TRY OUT Estimate. Use the front digits and adjust.

5. 307 + 418 + 72

6. 9,534 − 6,203

7. 7,351 − 843

PRACTICE

Estimate by using the front-end digits and adjusting.

8. 131
    437
  + 159

9. 425
    328
  + 84

10. 2,003
    1,198
  + 1,204

11. $6.41
    8.36
  + 2.68

12. 393
    702
  + 27

13.   952
    − 838

14. 5,388
    − 712

15. 4,111
  − 1,372

16. $126.21
  −   80.42

17. 9,446
  − 3,217

18. 6,311
    7,208
  + 5,689

19. 5,061
    4,185
  + 188

20. $43.26
  − 28.15

21. 5,145
    8,120
    3,098
  + 2,167

22. 6,932
     644
  + 533

23. 9,021 + 5,510 + 4,234 + 1,143

24. 6,108 − 851

25. 1,927 − 314

26. 51,102 + 20,211 + 1,146

27. 12,321 − 4,211

28. 34,314 − 1,348

Write the letter of the best estimate.

29. 425 + 1,178 + 250
    a. > 2,000   b. < 2,000
    c. > 1,000   d. > 2,500

30. 7,536 − 2,194
    a. > 4,000   b. < 4,000
    c. < 5,000   d. > 5,000

Critical Thinking

31. When you estimate a sum using front-end estimation without adjusting, do you think the estimate will be more or less than the exact answer? Why?

32. What happens when you adjust the estimate?

Mixed Applications

33. Marilyn played clarinet in the town band. The band performed before audiences of 986 and 1,735 people. Did over 3,000 people attend these concerts?

34. The band decorated the stage for their concert. They used 952 ft of red, white, and blue streamer. If 593 ft were red and blue, about how many feet were white?

UNDERSTANDING A CONCEPT

Mental Math: Adding Whole Numbers

A. David spends $.54 on a notepad and $.37 on a marker. How much does he spend?

First I add the tens. Then, I add on the ones.

To find $.54 + $.37, David adds mentally from left to right.

He adds the tens.

$.54   **Think:** 50 + 30 = 80
+ $.37

He then adds on the ones.

$.54   **Think:** 80 + 4 = $.84
+ $.37   84 + 7 = $.91
$.91

David spends $.91.

1. Can you think of another way to add $.54 + $.37 mentally? How?

2. How could you use the left-to-right method to add 265 + 130 mentally?

B. Margo uses compensation to add mentally.

Add: 64 + 29

$$\begin{array}{c} 64 + 29 \\ -1 \downarrow \quad \downarrow +1 \end{array}$$

Think: 63 + 30 = 93
So 64 + 29 = 93.

Adding numbers with zero in the ones place is easy. 64 + 29 is the same as 63 + 30 and 63 + 30 = 93.

3. Use compensation to add 246 and 198 mentally.

4. Which method would you use to add 2,008 + 3,992 mentally? Why? What is the sum?

TRY OUT Write the letter of the correct answer. Find the sum. Use mental math.

5. 27 + 42   **a.** 15   **b.** 69   **c.** 70   **d.** 79
6. $2.98 + $.72   **a.** $3.70   **b.** $3.96   **c.** $2.26   **d.** $3.00
7. 3,241 + 659   **a.** 5,800   **b.** 2,582   **c.** 3,900   **d.** 3,490
8. 2,207 + 4,341   **a.** 2,206   **b.** 3,134   **c.** 6,548   **d.** 5,688

PRACTICE

Find the sum. Use mental math.

9.	33 + 14	10.	65 + 37	11.	56 + 43	12.	27 + 45	13.	44 + 47	14.	86 + 73
15.	41 + 39	16.	$.35 + .98	17.	18 + 22	18.	289 + 21	19.	301 + 99	20.	429 + 301
21.	103 + 67	22.	186 + 404	23.	$.36 + .98	24.	114 + 46	25.	63 + 44	26.	57 + 63

27. 46 + 31   28. $1.54 + $.36   29. 718 + 328   30. $9.64 + $.36

31. 54 + 62 + 73   32. 4,250 + 125   33. 62 + 15 + 8   34. 12 + 63 + 24

Mixed Applications

35. Regina has $.83. Does she have enough money to buy a notebook that costs $.59 and a pencil that costs $.15? Why or why not?

36. The paper shop is having a sale on partially filled boxes of greeting cards. Carlos buys a box with 15 cards listed on the cover but only 11 inside. How many cards are missing?

37. Mike delivered 47 newspapers on his own route and then helped his brother by delivering 36 more. How many newspapers did Mike deliver?

38. The city newspaper prints 208,700 copies of its magazine section each week. What is the number of copies printed to nearest ten thousand?

Mixed Review

Compare. Use >, <, or =.

39. 3,041 ■ 3,104   40. 10,078 ■ 10,708   41. 729 ■ 402 + 375

42. 350 + 128 ■ 429   43. 517 + 192 ■ 709   44. 610 ■ 230 + 376

45. 602, 375 ■ six hundred two thousand, three hundred seventy-five

46. 4 million, 112 thousand, 893 ■ 4,112,894

UNDERSTANDING A CONCEPT
Adding Whole Numbers

Four grades at school collected metal cans for the Recycling Project. Grade 3 collected 1,521 cans, while Grade 4 collected 982 cans. Grade 5 collected 2,160 cans, while Grade 6 collected 2,407 cans. What was the total number of cans collected by the four grades?

Record your estimate first. Then find the exact answer.

Add: 1,521 + 982 + 2,160 + 2,407

Step 1	Step 2	Step 3	Step 4
Add the ones. Regroup if necessary.	Add the tens. Regroup if necessary.	Add the hundreds. Regroup if necessary.	Add the thousands. Regroup if necessary.
1 1,521 982 2,160 + 2,407 0	1 1 1,521 982 2,160 + 2,407 70	2 1 1 1,521 982 2,160 + 2,407 070	2 1 1 1,521 982 2,160 + 2,407 7,070

The total number of cans collected was 7,070.

1. Compare the exact answer to your estimate. Is the exact answer reasonable? Why?
2. How can you check the answer?

You could have used a calculator to find the total.

1521 (+) **982** (+) **2160** (+) **2407** (=) | *7070.* |

3. Why might using a calculator have been easier than using paper and pencil in solving the problem?
4. **What if** you had to add $3.42 + $1.00 + $2.01? Would you add mentally, use paper and pencil, or use a calculator? Why?
5. **What if** you had to add 277 + 693 + 4,935? Which method would you use? Why?

TRY OUT
Write the letter of the correct answer.

6. 327 + 580 + 23 a. 820 b. 830 c. 920 d. 930
7. 1,420 + 336 + 219 a. 1,965 b. 1,975 c. 2,065 d. 2,075
8. $18.57 + $9.76 a. $27.23 b. $27.33 c. $28.23 d. $28.33

PRACTICE

Add. Use mental math, a calculator, or paper and pencil.

9. 23 + 79
10. $4.27 + 1.82
11. 368 + 275
12. 1,499 + 108
13. 879 + 320

14. 937 + 43
15. 1,238 + 51,698
16. 2,606 + 1,474
17. 46 + 1,798
18. $4,080 + 6,578

19. 2,419 + 3,081
20. 4,275 + 139
21. $60.97 + 5.03
22. 425 + 8,699
23. 15,075 + 6,025

24. 216 + 3,751 + 6,033
25. 4,895 + 16,387 + 723,198
26. $37.65 + 29.89 + 76.45
27. 9,703 + 192 + 805
28. 320,518 + 16,000 + 2,972

29. 517 + 42
30. $.84 + $7.16
31. 6,283 + 8,579
32. 51,264 + 1,075 + 698 + 4,817
33. $3.75 + $18.26 + $68.13 + $3.04

Mixed Applications

Solve. Which method did you use?

34. The fifth grade collected 586 lb of newspaper. The sixth grade collected 414 lb. How many pounds of newspaper were collected?

35. The total number of bottles collected last year was 1,789. This year 2,218 were collected. About how many bottles were collected in all?

36. Kimi kept track of the amount of bottles collected each month. Sept.—35 lb Oct.—41 lb Nov.—37 lb Dec.—63 lb How many pounds were collected in the four months?

ESTIMATION
MENTAL MATH
CALCULATOR
PAPER/PENCIL

UNDERSTANDING A CONCEPT
Mental Math: Subtracting Whole Numbers

A. The book Lenora chose for her history report on the American Revolution has 389 pages. She has read 145 pages. How many pages are left to read?

Sometimes you can subtract mentally by subtracting from left to right.

Subtract: 389 − 145

Think: First subtract the hundreds.	389 − 100 = 289
Next subtract the tens.	289 − 40 = 249
Then subtract the ones.	249 − 5 = 244

Lenora has 244 pages left to read.

1. Use the left-to-right method to subtract 4,865 − 2,432.

B. You can use **equal additions** to subtract mentally. Look to make the number being subtracted end in zero.

Subtract: 572 − 398

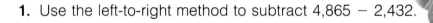
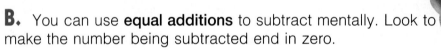

```
         + 2
    572 →   574
         + 2
  − 398 → − 400
           ───
           174
```

Think: Add 2 to each number.

So 572 − 398 = 174.

2. How can you use equal additions to subtract 3,265 − 990 mentally? What is the difference?

TRY OUT
Write the letter of the correct answer.
Subtract. Use mental math.

3. 45 − 23   **a.** 68   **b.** 30   **c.** 28   **d.** 22

4. 575 − 43   **a.** 532   **b.** 618   **c.** 406   **d.** 636

5. 8,721 − 410   **a.** 7,828   **b.** 8,504   **c.** 8,311   **d.** 9,131

6. $3.15 − $1.97   **a.** $1.12   **b.** $1.18   **c.** $5.12   **d.** $4.08

PRACTICE

Find the difference. Use mental math.

7.	65 − 33	8.	279 − 37	9.	513 − 298	10.	$8.36 − .24	11.	2,376 − 324
12.	92 − 49	13.	45 − 19	14.	425 − 99	15.	$3.35 − 1.98	16.	953 − 48
17.	57 − 21	18.	308 − 197	19.	740 − 318	20.	562 − 195	21.	$8.24 − .18
22.	947 − 235	23.	621 − 197	24.	7,862 − 450	25.	5,125 − 998	26.	4,223 − 995

27. 536 − 31
28. $7.35 − $1.25
29. 827 − 45
30. 757 − 48
31. $67.28 − $5.01
32. 8,433 − 429
33. 5,915 − 602
34. 6,590 − 528

Mixed Applications

35. Stephanie is making new curtains for her room. She cuts a piece of fabric 470 cm long from a piece 545 cm long. How much fabric does she have left?

36. Lenora's report is 12 pages long. She also has 8 pages of photographs of battlefields and guns and 7 pages of maps. How many pages long is her project?

Mixed Review

Name the place and the value of the digit 7 in the number.

37. 71,298
38. 31,796,040
39. 57,968,200,634

Write the number in standard form.

40. 20 million, 36 thousand, 1
41. 813 billion, 9 million, 22

UNDERSTANDING A CONCEPT

Subtracting Whole Numbers

A. On a trip up Mt. Washington with the Hiking Club, Kirk and his friends climbed 3,075 ft above sea level. They started their climb from a point 839 ft above sea level. How many feet did they climb?

Record your estimate. Then find the exact answer.

Subtract: 3,075 − 839

Step 1	Step 2	Step 3
Regroup if necessary. Subtract the ones.	Regroup if necessary. Subtract the tens.	Regroup if necessary. Subtract the hundreds. Then subtract the thousands.
$\begin{array}{r} {}^{6\ 15}\\ 3,0\,7\,5 \\ -\quad 8\,3\,9 \\ \hline 6 \end{array}$	$\begin{array}{r} {}^{6\ 15}\\ 3,0\,7\,5 \\ -\quad 8\,3\,9 \\ \hline 3\,6 \end{array}$	$\begin{array}{r} {}^{2\ \ 10\,6\ 15}\\ 3,0\,7\,5 \\ -\quad 8\,3\,9 \\ \hline 2,2\,3\,6 \end{array}$

Kirk and his friends climbed 2,236 ft.

1. Compare the exact answer to your estimate. Is the exact answer reasonable? Why?

2. How can you check the answer?

B. The Hiking Club earned $200 to pay for their trip. They spent $138.65 on supplies. How much money was left?

Sometimes using a calculator to subtract is easier than using paper and pencil.

Subtract: $200 − $138.65

 200 ⊖ 138.65 ⊜ | 61.35 |

3. Why might using a calculator have been easier than using paper and pencil in solving the problem?

4. **What if** you had to subtract 6,278 − 2,062? Would you subtract mentally, use paper and pencil, or use a calculator? Why?

5. **What if** you had to subtract 4,585 − 3,286? Which method would you use? Why?

64 Lesson 2-7

TRY OUT Write the letter of the correct answer.

6. 4,583 − 271   a. 1,873   b. 2,873   c. 3,312   d. 4,312
7. 78,141 − 576   a. 77,565   b. 78,435   c. 78,475   d. 78,675
8. $23.05 − $18.13   a. $4.92   b. $5.12   c. $14.92   d. $15.13

PRACTICE

Subtract. Use mental math, a calculator, or paper and pencil.

9. 82   10. 91   11. 172   12. 563   13. $6.18
  − 56   − 79   − 32   − 258   − 4.19

14. 874   15. 721   16. $9.63   17. 4,585   18. 7,942
  − 56   − 247   − .85   − 3,085   − 851

19. 5,001 20. 3,470 21. $29.55 22. 13,720 23. 18,721
  − 2,312   − 268   − 7.99   − 4,989   − 14,657

24. $651.37 25. 98,901 26. 382,500 27. 531,046 28. 900,000
   − 425.80   − 22,197   − 17,986   − 219,822   − 411,132

29. 325 − 115   30. 946 − 89   31. $32.78 − $12.95
32. 4,165 − 2,100   33. 15,061 − 6,378   34. 127,325 − 18,654

Mixed Applications

Solve. Which method did you use?

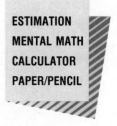

ESTIMATION
MENTAL MATH
CALCULATOR
PAPER/PENCIL

35. Marty climbed from the hikers' camp at 3,075 ft to a rock at 4,140 ft. How many more feet did he climb?

36. Jeanne spent $135.89 on camping equipment. Andrea spent $99.98. How much did they spend together?

- ✓ UNDERSTAND
- ✓ PLAN
- ✓ TRY
- ✓ CHECK
- ✓ EXTEND

PROBLEM SOLVING

Strategy: Choosing the Operation

Aaron and his family are on vacation. They drove 1,608 miles from New York City to Houston. Then they drove 1,067 miles from Houston to Chicago. Aaron wants to know how many miles he and his family drove in all.

1. What information does Aaron know?
2. What does he need to find out?
3. Which operation should he use to solve the problem? Why?
4. How many miles did they drive in all?

Aaron wants to know how much farther it is from New York to Houston than from Houston to Chicago.

5. Which operation should he use? Why?
6. How many more miles is it?

Aaron's family decided to drive from Chicago to Miami, a distance of 1,329 miles. They drove 356 miles on the first day. Aaron wants to know how many more miles they need to drive to reach Miami.

7. Which operation should Aaron use to solve the problem? Why?
8. How many more miles do they still have to drive to reach Miami?

Aaron's family started out with 20 gallons of gas in the car's tank in Chicago. Seven gallons were left in the tank when they stopped. Aaron wants to know how many gallons of gas the car used.

9. Which operation should Aaron use to solve the problem? Why?
10. How much gas did the car use?

PRACTICE

Write which operation you will use. Then solve the problem.

11. Aaron's family had budgeted $1,250 to spend on the trip. By the time they left Chicago, they had spent $837. How much could they spend for the rest of the trip?

12. Aaron's father brought a book to read on the vacation. The book is 837 pages long. By the time he got back home, he had read 692 pages. How many pages did he have left?

13. The George Washington Bridge in New York is 1,067 meters long, and the Verrazano Narrows Bridge is 1,298 meters long. How much longer is the Verrazano Bridge?

14. The Houston Zoological Gardens had 4,613 visitors in June and 5,983 in July. How many people visited the Zoological Gardens in the two months?

Strategies and Skills Review

Solve. Use mental math, estimation, a calculator, or paper and pencil.

15. In Chicago, Aaron went to Wrigley Field, and in Houston he went to the Astrodome to see baseball games. Use the Databank on page 518 to find the difference in seating capacity between the two baseball stadiums.

16. When the United Nations was founded in 1945, there were 51 members. By 1988 there were 159 members. How many nations have joined the United Nations since it was founded?

17. The area of Houston is about 556 square miles. This is about 522 square miles more than the area of Miami. What is the area of Miami?

18. Aaron's car is about 12 feet long. He thinks that a bus is about 35 feet long. Does that seem reasonable? Why or why not?

19. The Empire State Building is 1,250 ft tall. The TV tower on top of the building is 164 ft high. How tall is the building if the TV tower is included?

20. **Write a problem** that can be solved by addition or by subtraction. Solve the problem. Ask others to solve your problem.

What's Your Order?

Informal Algebra: Using Number Concepts

A. Look at this problem: $5 + 3 \times 4 = \blacksquare$

Ann did the problem this way:

$5 + 3 \times 4$
V
$5 + 12 = 17$

Bob did the problem this way:

$5 + 3 \times 4$
V
$8 \times 4 = 32$

It is important that a problem have just one answer.
In mathematics, the established order for doing operations is this:

- First do all multiplications and divisions from left to right.
- Then do all additions and subtractions from left to right.

1. Who had the correct answer, Ann or Bob?

2. Compute $3 \times 2 + 4$. Is the correct answer 10 or 18? Why?

3. Compute $15 - 3 \times 2$. Is the correct answer 24 or 9? Why?

4. Compute $2 \times 3 + 4 \times 5 - 1$. Is the correct answer 69 or 25 or 22? Why?

B. Now try this problem on a calculator. Enter the operations and numbers from left to right.

$$8 + 5 \times 2 - 5 = \blacksquare \quad 21$$

5. What does the display show?

6. Work the problem using paper and pencil. Remember to use the correct order of operations. What is the answer? 13

If your calculator displays 21, the incorrect answer for the problem above, the calculator is performing operations from left to right as entered.

$$8 + 5 \times 2 - 5 = \blacksquare$$
$$\vee$$
$$13 \times 2 - 5 = \blacksquare$$
$$\vee$$
$$26 - 5 = 21$$

This type of calculator is called an *arithmetic* calculator. Some calculators follow the correct order of operations. These are called *scientific* calculators. A scientific calculator will display the correct answer, 13.

To get the correct answer with an arithmetic calculator, you must plan your work. Try the problem this way:

Press: (5) (×) (2) (+) (8) (−) (5) (=)

First try each problem using a calculator. Plan your work. Then compute the answer mentally or with paper and pencil. Do you get the same answer?

7. $4 + 6 - 2 + 9$

8. $9 \times 2 + 3 \times 4$

9. $6 - 2 \times 3 + 4$

10. $36 - 14 \times 2 + 10$

11. $8 + 9 - 2 \times 3$

12. $10 \times 7 + 25 - 5 \times 6$

13. $24 \div 6 + 6 \times 8 - 5$

14. $30 - 12 \div 2 + 4 \times 15$

DEVELOPING A CONCEPT
Measuring and Estimating Length

The **millimeter (mm), centimeter (cm), meter (m),** and **kilometer (km)** are metric units of length.

A is about 1 mm thick.

A is about 1 cm across.

A  is about 1 m long.

A kilometer is about the distance a person can walk in 15 minutes.

WORKING TOGETHER

Use these steps to measure the length of an object.

Step 1 Line up the left side of the object with the left end of the scale on the ruler.

Step 2 Look at the right side of the object. Find the closest measurement mark.

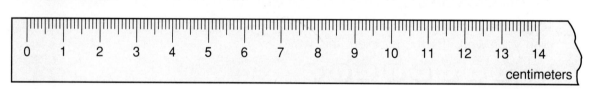

1. What is the length of the pencil to the nearest centimeter?

2. Measure the length of one of your pencils to the nearest centimeter.

3. Choose five or six classroom objects of different lengths to measure. Estimate each length to the nearest centimeter or meter. Then use a centimeter ruler or a meterstick to measure. Record your answers on a table such as this.

Object	Estimate	Measure

SHARING IDEAS

4. Compare your table with those of others. What unit of length did you use for short lengths? for medium lengths? for long lengths?

5. What unit would you use to measure the length and width of a room? Why?

6. What unit would you use to measure the height of a plant?

7. **What if** you had to measure the distance between two cities? What unit would you use?

8. **What if** you wanted to measure the length of an ant? What unit would you use?

PRACTICE

Which unit would you use to measure? Write *millimeter, centimeter, meter,* or *kilometer.*

9. the height of a house
10. the length of a girl's skirt
11. the width of a coin
12. the length of a highway

Which is the most reasonable measure?

13. the distance around a playground   a. 400 cm   b. 400 m   c. 400 km
14. the height of a drinking glass   a. 12 mm   b. 12 cm   c. 12 m
15. the width of a classroom   a. 6 mm   b. 6 cm   c. 6 m
16. the length of a shoe box   a. 20 mm   b. 20 cm   c. 20 m
17. the height of a door   a. 2 cm   b. 2 m   c. 2 km

Name three objects you would measure in each metric unit.

18. millimeter
19. centimeter
20. meter
21. kilometer

Solve.

22. At the hardware store Don asked for an extension cord that was 4 km long. Was this an appropriate length to ask for? If not, what length do you think he should have asked for?

23. **Write a problem** about measurement leaving out all the units of measure. Ask others to fill in the appropriate metric units of measure.

UNDERSTANDING A CONCEPT

Measuring Length

A. Marta measured the entrance to a birdhouse. She measured it to the nearest centimeter. Then she measured it to the nearest millimeter.

The entrance measures 3 cm to the nearest centimeter. It measures 3 cm 3 mm, or 33 mm, to the nearest millimeter.

> 3 cm = 30 mm
> 30 mm + 3 mm = 33 mm

1. Which measurement do you think is closer to the actual measurement, 3 cm or 33 mm? Why?

2. Measure the height of the post to the nearest centimeter. Then measure it to the nearest millimeter.

3. When do you think it is important to have a precise measurement? When would a less precise measurement do just as well? Give examples to show your reasoning.

B. Dan used a centimeter ruler to draw this line segment.

_____

4. What is the length of the line segment to the nearest centimeter? to the nearest millimeter?

5. What is the shortest length in millimeters that is still 5 cm to the nearest centimeter? What is the longest length?

TRY OUT Measure to the nearest centimeter. Then measure to the nearest millimeter.

6. _____ 7. _____

Lesson 2-11

PRACTICE

Measure to the nearest centimeter. Then measure to the nearest millimeter.

8. the width of your math book
9. the length of your pen
10. the width of a calculator
11. the width of a paper clip
12. the length of your thumbnail
13. the length of your shoe
14. the width of a penny
15. the thickness of a button

The **perimeter** of a figure is the distance around the figure. Find the perimeter. Measure to the nearest millimeter.

16.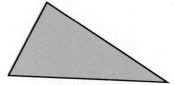
17.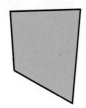
18.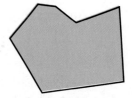

Find the missing length.

19.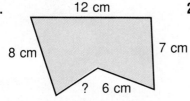

Perimeter = 38 cm

20.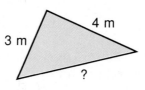

Perimeter = 12 m

21.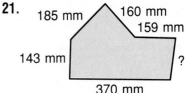

Perimeter = 1,167 mm

Mixed Applications

22. Chris has a bird feeder in her yard. The birds get the birdseed through holes that are 6 mm wide. Draw a line segment 6 mm long without a ruler. Then measure to see how close you were.

23. Chris's bird feeder is 156 cm from the ground. To prevent squirrels from eating the birdseed, Chris attached a shelf 17 cm below the feeder. How far off the ground is the shelf?

24. Chris counted the birds that came to her feeder. She counted 31 starlings, 18 finches, and 12 robins in her yard. How many birds did she count in all?

25. **Write a problem** using three objects you have measured to the nearest centimeter or millimeter. Solve the problem. Then share it with another student.

UNDERSTANDING A CONCEPT

Time and Elapsed Time

A. Paula spent from 11:20 A.M. to 1:35 P.M. playing softball with her brother. How long did they play? To find out how much time has passed, count the hours and then the minutes.

Think: From 11:20 A.M. to 1:20 P.M. is 2 hours.
From 1:20 P.M. to 1:35 P.M. is 15 minutes.

They spent 2 hours and 15 minutes playing softball.

1. Why do we need A.M. and P.M. when referring to the time?

2. Dick took 3 hours and 25 minutes to hike around the lake. If he started at 1:05 P.M., what time did he finish? How did you find the time he finished?

3. Suppose you have been playing tennis for 4 hours and 35 minutes. You finish at 3:10 P.M. What time did you begin? How did you find the starting time?

B. You can add and subtract units of time.

Add: 1 h 35 min + 1 h 50 min

Add the minutes.	Add the hours. Rename if necessary.	
1 h 35 min + 1 h 50 min ─────── 85 min	1 h 35 min + 1 h 50 min ─────── 2 h 85 min or 3 h 25 min	*Think:* 85 min = 60 min + 25 min = 1 h 25 min

Subtract: 3 h 26 min − 1 h 37 min

Subtract the minutes. Rename if necessary.		Subtract the hours.
2 86 3̸ h 2̸6̸ min − 1 h 37 min ─────── 49 min	*Think:* 1 h = 60 min 3 h 26 min = 2 h 86 min	2 86 3̸ h 2̸6̸ min − 1 h 37 min ─────── 1 h 49 min

4. Why was 2 h 85 min renamed as 3 h 25 min in the addition?

5. Why was 3 h 26 min renamed as 2 h 86 min in the subtraction?

Try Out

Find the elapsed time.

6. from 11:13 A.M. to 1:10 P.M.

7. 1 h 18 min after 2:05 P.M.

Add or subtract.

8. 2 h 43 min
  + 5 h 55 min

9. 8 h 11 min
  − 2 h 45 min

Practice

Find the elapsed time.

10. from 7:15 A.M. to 9:35 A.M.

11. from 12:10 P.M. to 2:45 P.M.

12. from 9:45 P.M. to 1:15 A.M.

13. from 8:17 A.M. to 11:36 A.M.

Find the end time.

14. 35 min after 3:45

15. 3 h 26 min after 1:30

16. 17 min after 2:50 P.M.

17. 5 h 17 min after 12 midnight

Find the starting time.

18. 22 min before 1:15 P.M.

19. 2 h 5 min before 10:20 A.M.

20. 3 h 27 min before 1:10 P.M.

21. 6 h 21 min before noon

Add or subtract.

22. 4 h 22 min
  + 6 h 31 min

23. 6 h 17 min
  − 2 h 3 min

24. 9 h 31 min
  − 7 h 45 min

25. 5 h 30 min
  + 2 h 57 min

26. 2 h 41 min
  + 2 h 29 min

27. 8 h 22 min
  − 5 h 25 min

28. 7 h 25 min
  − 1 h 18 min

29. 6 h 47 min
  + 3 h 13 min

Solve.

30. Frances got home from school at 4:25 P.M. On her way she spent 20 minutes in the library, and 15 minutes in the bookstore. It took her 24 minutes to walk home. What time was she dismissed from school?

31. Don has to be at school at 8:15 A.M. He takes 20 minutes to shower and get dressed, 25 minutes to eat breakfast, and 15 minutes to walk to school. What is the latest time he should get up?

PROBLEM SOLVING

Identifying Extra Information

Mr. Scott plans to fly from Seattle to Lake Tahoe on Friday after work. He wants to leave as late in the day as possible. Which flight should he take?

Mr. Scott uses an airline schedule to choose his flight.

FROM SEATTLE TO LAKE TAHOE

Flight	Leaves	Arrives
815	7:05 A.M.	11:03 A.M.
1463	11:55 A.M.	3:25 P.M.
1481	5:57 P.M.	9:20 P.M.
225*	6:35 P.M.	10:15 P.M.

*Saturday only

1. What information does he know?

2. What does he need to find out?

Mr. Scott has more information than he needs.

3. Name all the information he needs to choose his flight.

4. Name the information he does not need.

5. Which flight should he take?

6. **What if** Mr. Scott had planned to fly on Saturday so that he would get to Lake Tahoe as early in the day as possible? Name the information he needs to know to choose his flight. Which flight should he choose?

PRACTICE

List the extra information. Then solve the problem.

7. Mr. Scott took the flight that was scheduled to leave at 5:57 P.M. He left his office at 4:45 P.M. to go to the airport. Because of bad weather the plane did not leave until 6:47 P.M. How many minutes late did the plane take off?

8. It is 663 miles from Seattle to Lake Tahoe, 490 miles from Lake Tahoe to San Diego, and 1,125 miles from San Diego back to Seattle. How many miles will Mr. Scott fly if he travels from Seattle to Lake Tahoe to San Diego?

9. Cloudy Pass has an elevation of 6,478 feet. From there, Mr. Scott could see the top of Glacier Peak, which is 10,541 feet high. Mr. Scott once climbed up Mount Rainier, which is 14,410 feet high. How much higher is Glacier Peak than Cloudy Pass?

10. Mr. Scott's plane was flying at an altitude of 32,683 feet when the plane descended 1,384 feet because of high winds. Then it climbed back 782 feet. What was the plane's altitude before it started climbing back?

Strategies and Skills Review

Solve. Use mental math, estimation, a calculator, or paper and pencil.

11. This year about 365,000 people are expected to visit the South Rim of the Grand Canyon. The railroad will carry 135,000 passengers. How many visitors are expected to use other means of transportation?

12. There are two Niagara Falls. The American Falls are 1,076 feet wide, and the Canadian Falls are 2,200 feet wide. How much wider are the Canadian Falls than the American Falls?

13. Each airplane seat is about 2 feet wide. About how wide could the inside of the plane be? Write the letter of the best answer.
    a. 3 feet   b. 15 feet   c. 50 feet

14. The area of Yosemite National Park is 761,320 acres. The area of Grand Teton National Park is 310,443 acres, and the area of Grand Canyon National Park is 673,575 acres. How much greater is the area of Yosemite than the area of Grand Canyon?

15. Refer to the table on page 76. How long is Flight 1481 scheduled to take from Seattle to Lake Tahoe?

16. **Write a problem** that has extra information. Solve the problem. Ask others to solve your problem.

Using Addition and Subtraction 77

DECISION MAKING

Problem Solving: Making a Class Video

SITUATION

The students in the fifth-grade class of Media Heights Elementary School want to make a class video. They want to tape different class activities. There are several types of videotapes they can purchase.

PROBLEM

Which activities should they tape? Which type of tape or tapes should they buy?

DATA

Class Activities	Estimated Time
Class introduction	10 minutes
Play	30 minutes
Poetry Reading	15 minutes
Spelling Bee	45 minutes
Math Bee	20 minutes
Square Dance	8 minutes each song
Singing Songs	4 minutes each song
Class Skit	10 minutes
Acrobatic Stunts	3 minutes each stunt
Relay Races	10 minutes each race
Gym Games	15 minutes each game
Closing Activity	5 minutes

Video Tapes	Length of Tape	Price
T30	30 minutes	$1.75
T60	60 minutes	$2.50
T90	1 1/2 hours	$3.25
T120	2 hours	$4.00

USING THE DATA

How much time will the activity take?

1. sing 6 songs
2. 3 relay races
3. 2 poetry readings
4. 2 plays
5. class introduction and the closing activity

How much will the items cost?

6. 2 T30 tapes
7. 1 T60 tape and 1 T90 tape
8. 1 T120 tape and 1 T30 tape

How much time can be recorded on the tape?

9. 2 T90 tapes
10. 3 T120 tapes
11. 3 T30 tapes
12. 2 T60 tapes and a T30 tape

MAKING DECISIONS

13. What things should be included in any class video?
14. Which activities should be easy to record on tape? Which are more difficult?
15. The students had $5 to spend on videotape. What is the longest the class video could be?
16. The students wanted to tape one acrobatic stunt and one gym game. Which length tape should they buy? Why?
17. The students wanted to mix 2 musical activities with 2 athletic activities on a 1-hour tape. Which activities should they choose? Why?
18. The students had a T120 tape and wanted to fill it with poetry and plays. How many of each can they tape?
19. **What if** you planned 90 minutes of activity? What type of tape would you use? Why?
20. **Write a list** of other things the class should think about when planning their class video.
21. Which tape would you buy? Which activities would you tape? Why?

Using Addition and Subtraction

CURRICULUM CONNECTION

Math and Art

The copyright law protects works produced by writers, painters, musicians, and other artists. Without the copyright law, anyone could copy an artist's work and sell it or use it as though it were his or her own. The law says that only the holder of the copyright has the right to reproduce, sell, or rent the work. Anyone else who wants to use the work must get permission from the copyright owner. The person seeking permission must often pay a fee.

In 1978, a new copyright law went into effect. Any work copyrighted before January 1, 1978, is protected for a first term of 28 years. It can be renewed for a second term of 47 years for a total of 75 years. Works first copyrighted on or after January 1, 1978, are protected for the author's lifetime plus 50 years after his or her death.

What if an author publishes and copyrights a book in 1993? The author lives until 2003. What year will the copyright end? How long is it valid?

Think: From 1993 when the book is copyrighted until the author's death in 2003 is 10 years. The book is copyrighted another 50 years. The total is 60 years. The copyright will end in 2053, 50 years from the author's death.

ACTIVITIES

1. Find out when your favorite books were copyrighted. Make a time line showing the dates, authors, and titles.

2. Many books that were published before 1914 are said to be in "the public domain." Find out what this term means.

Computer Spreadsheet: Travel Costs

Mr. Zelaya is planning a two-day trip to Washington, D.C., to do some sightseeing. He wants to know how much money the trip is likely to cost him, depending on what choices he makes among different items.

The computer spreadsheet program TRAVEL COSTS makes it easy to explore the cost of Mr. Zelaya's trip. Once you enter the cost of each item, the computer will calculate item totals, daily totals, and a grand total for the whole trip. If you change the cost of one or more items, the computer will quickly recalculate all the totals.

AT THE COMPUTER

Mr. Zelaya will spend $89.50 each day for his hotel and $37.75 each day for a rental car (transportation). He estimates meals will cost $25.00 each day, and he budgets $18.00 a day for admissions.

1. Enter these costs in the spreadsheet. How much will Mr. Zelaya spend in all for his hotel? How much will he spend in all each day? How much will the whole trip cost?

2. Suppose instead of a rental car, Mr. Zelaya uses buses and taxis, which only cost $21.00 each day. How much will the whole trip cost?

3. Since he saved money on transportation, Mr. Zelaya decides he would like to spend $50.00 on souvenirs on Day 2. How will you enter this on the spreadsheet? How much will he spend on Day 2? How much will the whole trip cost?

4. The round-trip airfare is $198.66. Enter this under Other for Day 1. Now how much will the whole trip cost? If Mr. Zelaya does not want to spend more than $525 in all, what changes can he make? Experiment on the computer.

Using Addition and Subtraction

Extra Practice

Meaning of Addition and Subtraction, page 53

Write a fact family for each group of numbers.

1. 5, 7, 12
2. 8, 5, 13
3. 6, 9, 15
4. 4, 7, 11

Find the sum or difference.

5. 3 + 9
6. 8 + 4
7. 9 + 7
8. 13 − 8
9. 16 − 7
10. 18 − 9

11. 7 + 5
12. 16 − 8
13. 12 − 6
14. 0 + 7
15. 7 + (3 + 2)
16. (3 + 4) + 9
17. 6 + (2 + 7)
18. (3 + 3) + 9
19. 4 + 3 + 5
20. 6 + 4 + 7
21. 3 + 2 + 5 + 4
22. 1 + 0 + 3 + 4

Complete. What property did you use?

23. 516 + 325 = 325 + ■
24. 172 + (38 + 49) = (172 + 38) + ■
25. 0 + 3,129 = ■
26. (192 + ■) + 562 = 192 + (24 + 562)

Estimating Sums, Differences by Rounding, page 55

Estimate by rounding.

1. 2,352 + 1,607 + 3,985
2. 3,105 + 1,389 + 4,873
3. 3,871 − 802
4. 62,255 − 18,498

5. 3,012 + 8,352 + 1,985
6. 6,872 − 1,655
7. $32.19 + $3.79 + $4.19
8. 317,652 − 157,892
9. 872,187 − 79,900
10. $82.17 + $9.89 + $2.10

Front-End Estimation: Sums, Differences, page 57

Estimate by using the front-end digits and adjusting.

1. 112 + 529 + 262
2. 315 + 281 + 75
3. 1,030 + 2,175 + 1,309
4. $139.17 − 65.87
5. 8,456 − 6,397

6. 8,072 + 1,895 + 3,073
7. 5,875 − 602
8. 1,873 − 412
9. 28,193 + 11,811 + 1,796
10. 36,489 − 4,376
11. 23,685 − 2,362

EXTRA PRACTICE

Mental Math: Adding Whole Numbers, page 59
Find the sum. Use mental math.

1. 24 + 17
2. 38 + 27
3. 57 + 33
4. 25 + 39
5. 82 + 17
6. 95 + 27
7. $.35 + .89
8. 117 + 38
9. 401 + 89
10. 52 + 37
11. $.17 + .87
12. 104 + 26
13. 33 + 27
14. $1.89 + $.32
15. 612 + 325
16. $8.20 + $.99
17. 53 + 24 + 32
18. 3,125 + 109
19. 35 + 12 + 7

Adding Whole Numbers, page 61
Add.

1. 35 + 87
2. $3.25 + 1.72
3. 462 + 182
4. 1,099 + 209
5. 5,457 + 350
6. 822 + 187
7. 2,139 + 37,182
8. 1,409 + 2,373
9. 26 + 2,897
10. $1,070 + 3,679
11. 117 + 2,607 + 3,095
12. 5,872 + 13,109 + 323,176
13. $47.82 + 33.89 + 72.49
14. 8,709 + 185 + 807
15. 319,852 + 17,825 + 1,873
16. 212 + 35
17. $.83 + $6.12
18. 3,482 + 8,195
19. 23,819 + 1,073 + 395 + 4,093
20. $3.65 + $17.85 + $3.18 + $2.09

Mental Math: Subtracting Whole Numbers, page 63
Find the difference. Use mental math.

1. 68 − 16
2. 45 − 23
3. 82 − 59
4. $.67 − $.42
5. 39 − 11
6. 322 − 107
7. $3.27 − $1.11
8. 412 − 109
9. $8.19 − $4.16
10. 409 − 187
11. 350 − 102
12. 753 − 199
13. 643 − 212
14. 419 − 181
15. 777 − 86
16. 414 − 389
17. $8.20 − $5.10
18. 4,099 − 176
19. $63.85 − $4.01

Using Addition and Subtraction

EXTRA PRACTICE

Subtracting Whole Numbers, page 65
Subtract.

1. 67 − 29
2. 81 − 62
3. 163 − 26
4. 452 − 243
5. $8.16 − 6.17

6. 987 − 28
7. 673 − 347
8. $8.65 − .78
9. 3,498 − 2,079
10. 9,761 − 692

11. 3,002 − 1,789
12. 2,370 − 496
13. $28.95 − 8.99
14. 12,890 − 5,898
15. 17,523 − 14,564

16. 436 − 112
17. 879 − 54
18. $29.89 − $11.10
19. 8,965 − 2,200
20. 17,854 − 7,091
21. 136,934 − 20,007

Problem-Solving Strategy: Choosing the Operation, page 67
Write the operation to use. Then solve the problem.

1. There are 458 runners in the City Park race. Only 393 finish the race. How many runners do not finish the race?

2. There are 1,600 people watching the race at the start line. There are 1,500 people at the finish line. How many people are watching at the two lines?

Measuring and Estimating Length, page 71
Which unit would you use to measure? Write *millimeter, centimeter, meter,* or *kilometer.*

1. the height of a skyscraper
2. the length of a window curtain
3. the width of a thumbtack
4. the length of a runway

Which is the most reasonable measure?

5. the distance across a river   a. 300 cm   b. 300 m   c. 300 km
6. the height of a mug   a. 10 mm   b. 10 cm   c. 10 m
7. the width of a bedroom   a. 6 mm   b. 6 cm   c. 6 m

Name three objects you would measure in each metric unit.

8. millimeter
9. centimeter
10. meter
11. kilometer

84 Chapter 2

EXTRA PRACTICE

Measuring Length, page 73

Measure to the nearest centimeter. Then measure to the nearest millimeter.

1. the width of your notebook
2. the length of your foot
3. the width of your hand
4. the length of a pencil
5. the width of a quarter
6. the thickness of a book

The perimeter of a figure is the distance around the figure. Find the perimeter. Measure to the nearest millimeter.

7.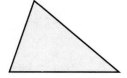
8.
9.

Time and Elapsed Time, page 75

Find the elapsed time.

1. from 8:45 A.M. to 1:05 P.M.
2. from 6:17 A.M. to 10:46 A.M.

Find the end time.

3. 18 min after 3:50 P.M.
4. 4 h 8 min after 12 midnight

Find the starting time.

5. 23 min before 2:19 P.M.
6. 3 h 19 min before 12 noon

Problem Solving: Identifying Extra Information, page 77

List the extra information. Then solve the problem.

Al uses the rush-hour trains to get to the city in the morning. Use this schedule to solve the problems.

Rush Hour Train Schedule
Lv 7:05—7:50 Ar
Lv 7:35—8:20 Ar
Lv 8:05—8:50 Ar
Lv 9:05—9:50 Ar

1. Al usually takes the 8:05 train. Today he wants to be in his office in the city by 8:30. What is the latest train he can take to be on time?
2. The 7:05 was delayed by 15 minutes on Tuesday and 10 minutes on Wednesday. When did it arrive on Tuesday?

Using Addition and Subtraction

Practice PLUS

KEY SKILL: Adding Whole Numbers (Use after page 61.)

Level A

Add.

1. 32 + 89
2. $3.67 + 2.72
3. 543 + 177
4. 654 + 287
5. 1,298 + 399
6. $8.39 + 2.65
7. 857 + 33
8. 1,406 + 32,109
9. 367 + 459
10. 3,405 + 2,198

11. The recycling committee collected 876 pounds of newspaper last month and 978 pounds this month. How many pounds of newspaper did it collect in all?

Level B

Add.

12. 3,219 + 3,892
13. 7,895 + 139
14. 55 + 3,498
15. 13,087 + 5,689
16. 11,098 + 2,986
17. $70.89 + 28.98
18. 543 + 2,987
19. 14,086 + 6,743
20. 4,054 + 2,769
21. 23,209 + 86,421

22. Hannah and her friends collected 27,654 aluminum cans for recycling during the summer. In the winter, they collected 10,009 cans. How many cans did they collect in all?

Level C

Add.

23. 403 + 3,490 + 8,045
24. 4,896 + 13,678 + 432,109
25. $56.76 + 39.04 + 43.10
26. 8,654 + 282 + 302
27. 420,018 + 18,430 + 3,987

28. 32,198 + 43,048 + 13,032 + 789
29. $3.87 + $18.65 + $32.23 + $56.98

30. Bart received $56.98, $32.21, $10.87, and $4.65 for recycling paper. How much did he receive in all?

86 Chapter 2

Practice PLUS

KEY SKILL: Subtracting Whole Numbers (Use after page 65.)

Level A
Subtract.

1. 75 − 26
2. 81 − 65
3. 45 − 18
4. $3.18 − 1.79
5. 131 − 54

6. 654 − 89
7. 652 − 378
8. $8.65 − .76
9. 3,115 − 2,226
10. 6,905 − 4,086

11. Zack climbed from a hikers' camp at 2,078 feet to a mountain at 5,019 feet. How many feet did he climb?

Level B
Subtract.

12. 987 − 89
13. 3,098 − 569
14. $38.74 − 6.54
15. 15,650 − 4,388
16. 17,542 − 13,753

17. 6,227 − 3,188
18. $678.43 − 269.80
19. 32,098 − 17,940
20. 60,781 − 29,536
21. 87,325 − 12,876

22. Jeanine had $570.50 to spend on camping gear. She spent $289.75. How much money did she have left?

Level C
Subtract.

23. 7,010 − 3,987
24. 78,543 − 22,762
25. 652,891 − 345,765
26. $389.54 − 207.65
27. 759,501 − 547,654

28. $876.41 − 254.87
29. 453,123 − 227,431
30. 800,700 − 688,998
31. 432,981 − 132,890
32. 900,000 − 311,078

33. 5,987 − 4,980
34. 17,850 − 3,897
35. 154,875 − 12,087

36. Rebecca had $670.00 to spend on a tent. She bought one on sale for $569.69. How much money did she have left?

Using Addition and Subtraction

CHAPTER REVIEW

LANGUAGE AND MATHEMATICS

Complete the sentences. Use the words in the chart on the right.

1. The ■ property states that if the grouping of the addends is changed, the sum remains the same. *(page 52)*

2. A ■ is a group of related facts that can help you add and subtract. *(page 52)*

3. A relay race is a distance of about 10 ■. *(page 70)*

4. ■ is a method of adding mentally. *(page 58)*

5. **Write a definition** or give an example of the words you did not use from the chart.

VOCABULARY
fact family
compensation
equal additions
associative
commutative
kilometers

CONCEPTS AND SKILLS

Complete. What property did you use? *(page 52)*

6. 318 + 452 = 452 + ■

7. 0 + 4,673 = ■

Estimate by rounding. *(page 54)*

8.   1,473
     2,874
   + 3,163

9. $36.42
   − 11.13

10. $87.25
      32.12
   + 40.17

11. 8,997
   −   899

Estimate by using the front-end digits and adjusting. *(page 56)*

12. $3.87
      2.09
   + 4.12

13. $142.17
   −   30.55

14. 463
     341
   + 38

15. 33,812
   − 4,463

Find the sum. Use mental math. *(page 58)*

16. 107
   + 33

17. $.35
   + .25

18. 304
   + 26

19. $8.63
   + 3.88

Add. *(page 60)*

20. 873
   + 49

21.   1,345
   + 32,987

22. $60.52
   +   3.95

23.   4,683
      32,485
   + 347,810

88 Chapter 2

Find the difference. Use mental math. *(page 62)*

| **24.** 38 − 21 | **25.** 407 − 298 | **26.** $8.28 − 3.50 | **27.** 6,816 − 403 |

Subtract. *(page 64)*

| **28.** 91 − 38 | **29.** 453 − 148 | **30.** $28.64 − 8.92 | **31.** 800,743 − 318,484 |

Which unit would you use to measure? Write *millimeter, centimeter, meter,* or *kilometer.* *(page 70)*

32. the length of a football field

33. the width of a box

34. the height of a flagpole

35. the distance from New York to Pittsburgh

Add or subtract. *(page 74)*

| **36.** 3 h 38 min + 2 h 27 min | **37.** 9 h 13 min − 4 h 42 min | **38.** 6 h 24 min + 3 h 38 min |

CRITICAL THINKING

39. Is there an associative property for subtraction? Give examples to support your answer.

40. How would you estimate $13.65 − $13.17 using rounding? What is the estimate?

MIXED APPLICATIONS

41. Gary spent $329.17 on airfare to San Francisco. David paid $289.25 for his tickets. How much did they spend together? *(page 60)*

42. Katie has $50. She spends $38.65 on souvenirs at the airport. How much money does she have left? *(page 66)*

43. Adam earned $17.50 watching people's pets, $25 mowing lawns, and $15.50 baby-sitting. How much did he earn baby-sitting and mowing lawns? *(page 60)*

44. Laura earns $2.50 an hour baby-sitting. She sat from 7:00 P.M. to 10:00 P.M. How long did Laura baby-sit? *(page 74)*

Chapter Test

Estimate by using front-end estimation and rounding.

1. 2,517
   4,683
   + 5,181

2. $35.27
   − 17.53

3. 30,627
   12,543
   + 18,734

4. 897,354
   − 407,216

5. $34.12
   15.28
   + 1.16

Find the sum or difference.

6. 13
   + 29

7. 216
   − 98

8. $6.95
   + 3.07

9. 2,168
   + 63,095

10. 38,603
    − 17,974

Which unit would you use to measure? Write *millimeter, centimeter, meter,* or *kilometer.*

11. the height of a building

12. the width of a paper clip

Which is the most reasonable measure?

13. the length of a bathrobe   a. 120 cm   b. 120 mm   c. 120 m

14. the height of a ladder   a. 2 mm   b. 2 cm   c. 2 m

Measure the line segment to the nearest centimeter.

15. _____   16. _____   17. _____

Find the end time.

18. 18 minutes after 3:40 P.M.

19. 4 hours 12 minutes after 1:13 A.M.

Find the starting time.

20. 25 minutes before 3:12 P.M.

21. 19 minutes before 12:18 P.M.

22. 4 hours 27 minutes before 12:10 P.M.

23. 2 hours 10 minutes before 9:20 A.M.

Solve.

24. Alison spent $137.42 on ski equipment. Robin spent $89.73. How much did they spend together? Which operation did you use?

25. It takes Raymond 25 minutes to walk to school. It takes 10 minutes if he rides his bicycle. Raymond left his house at 8:05 A.M. and walked to school. What time did Raymond get to school? What is the extra information?

Enrichment For All

OTHER NUMBER SYSTEMS

A. The ancient Romans did not use a place-value system of numeration. They used letters as symbols to name numbers.

1	5	10	50	100	500	1,000
I	V	X	L	C	D	M

The value of a Roman numeral was either added or subtracted.

When the symbol for a greater or equal value comes first, add.
MDLXI is 1,000 + 500 + 50 + 10 + 1, or 1,561.

When the symbol for a smaller value comes first, subtract the smaller value from the larger value.
DCCXLIX is 500 + 100 + 100 + (50 − 10) + (10 − 1), or 749.

B. The Egyptians used these symbols to name numbers.

The number 438 could be written as:

1. Is the Egyptian number system a place-value system? How do you know?
2. How are the Roman and Egyptian number systems similar?

Write the number.

3. DCCCLXX　　　**4.** MCDLXXXVI　　　**5.** MDI

6. ∧∧||||||　　**7.** ⟟⟟⟟∧|||　　**8.** ||||∧∧⟟⟟⟟
　　∧∧　　　　　　　　　　　　　　　　∧∧⟟⟟⟟

Write the Roman numeral and the Egyptian number for each.

9. 298　　**10.** 927　　**11.** 1,054　　**12.** 3,136　　**13.** 3,890

Cumulative Review

Choose the letter of the correct answer.

1. Estimate by rounding to the greatest place of the lesser number.
   $29.89
   − 8.97

   a. $15   c. $21
   b. $20   d. not given

2. Which digit is in the hundred millions place of 302,719,354,073?
   a. 1   c. 7
   b. 0   d. not given

3. Round $485.27 to the nearest $100.
   a. $480   c. $490
   b. $500   d. not given

4. Compare: 325,809 ● 353,908
   a. <   c. =
   b. >   d. not given

5. Round 1,389 to the nearest 100.
   a. 2,000   c. 1,400
   b. 1,300   d. not given

6. Which digit is in the hundred billions place of 123,489,605,897?
   a. 1   c. 6
   b. 8   d. not given

7. Estimate by rounding.
   2,312
   4,819
   + 3,205

   a. 10,000   c. 9,000
   b. 8,000   d. not given

8. Which is the standard form for twenty-one thousand, thirty-two?
   a. 21,320   c. 2,132
   b. 21,032   d. not given

9. 389
   + 437

   a. 716   c. 826
   b. 816   d. not given

10. $40.83
    + 6.27

    a. $47.06   c. $47.10
    b. $46.00   d. not given

11. 3,002
    − 1,728

    a. 1,383   c. 1,284
    b. 1,274   d. not given

12. 800,013
    − 311,894

    a. 588,217   c. 487,118
    b. 488,119   d. not given

13. 3 h 13 min
    + 4 h 29 min

    a. 8 h 2 min   c. 7 h 32 min
    b. 7 h 42 min d. not given

14. Mandy's last class ended at 3:07 P.M. She left school at 3:18 P.M. She arrived at home at 4:12 P.M. How long did it take her to walk home from school?
    a. 44 min   c. 54 min
    b. 1 h   d. not given

92 Chapter 2

Adding and Subtracting Decimals

CHAPTER 3

MATH CONNECTIONS: GRAPHING • STATISTICS • PROBLEM SOLVING

Girl's 10-speed bike
no accessories
$105.00

All-terrain bike
(wide tires)
19-speed,
girl's model
$250.00

Boy's 12-speed bike
some accessories
$185.00

1. What information do you see in this picture?
2. How can you use the information?
3. What is the least amount of money you can spend on a new bike?
4. Write a problem using the information.

UNDERSTANDING A CONCEPT
Tenths and Hundredths

A. In a section of the auditorium, 2 of the 10 rows are reserved for the families of the performers. What number can you write to show this?

You can write a fraction or a decimal for a number less than 1.

Fraction: $\frac{2}{10}$
Decimal: 0.2 ← Write 0 before the decimal point to show that there are no ones.
Read: two tenths
So 0.2 of the section is reserved for families.

Fraction: $\frac{15}{100}$
Decimal: 0.15
Read: fifteen hundredths

Fraction: $\frac{8}{100}$
Decimal: 0.08 ← This 0 means that there are no tenths.
Read: eight hundredths

B. You can write a decimal for a number greater than 1.

Mixed number: $1\frac{7}{10}$
Decimal: 1.7
Read: one and seven tenths

Mixed number: $1\frac{23}{100}$
Decimal: 1.23
Read: one and twenty-three hundredths

1. Write two and six hundredths as a decimal.

C. Decimals that show the same amount are **equivalent decimals.**

  0.4 of the region is shaded.

  0.40 of the region is shaded.

0.4 = 0.40 ← equivalent decimals

2. Which pairs of decimals are equivalent? Draw diagrams, if needed.
   a. 1.6 and 1.60   **b.** 1.6 and 1.06   **c.** 1 and 1.00

TRY OUT

Write the decimal.

3.   4. $\frac{7}{10}$

Write an equivalent decimal.

5. 0.20   6. 5.1

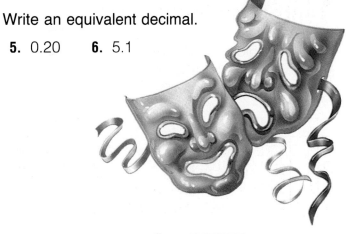

PRACTICE

Write the decimal.

7.   8.   9.

10.   11.   12.

13. eleven hundredths
14. one and seven tenths
15. eight tenths
16. two and thirty hundredths
17. five hundredths
18. eight and one hundredth

Write the word name.

19. 5.4   20. 7.21   21. 0.3   22. 18.02   23. 1.6
24. 0.07 25. 12.1   26. 4.10 27. 0.5   28. 2.04

Write an equivalent decimal.

29. 0.2   30. 1.30   31. 1.7   32. 3.9   33. 0.50

Solve.

34. Ms. Reed used $\frac{5}{100}$ of a roll of streamers to decorate the stage. Write this number as a decimal.

35. Mr. Ross said that $\frac{90}{100}$ of the fifth graders participated in the show. Write a pair of equivalent decimals to show this.

UNDERSTANDING A CONCEPT
Thousandths

A. This region is divided into hundredths. Think of dividing each hundredth into 10 equal parts. Then the region would be divided into thousandths.

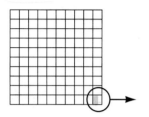 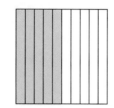

Fraction: $\frac{5}{1,000}$
Decimal: 0.005
Read: five thousandths

1. What do the zeros mean in 0.005?

B. A place-value chart can help you read decimals. If there is a whole number, read the whole number first. Then read the decimal point as *and*. Next, read the decimal just as you would a whole number before reading the place value of the last digit.

Ones	Tenths	Hundredths	Thousandths
7 .	4	1	5
2 .	3	7	8

Read:
7 and 415 thousandths
2 and 378 thousandths

2. How would you read 4.607?

You can also use a place-value chart to show the value of a digit.

3. What is the value of the digit 7 in 2.378? the digit 5 in 7.415? How do you know?

4. **What if** the place-value chart were extended another place to the right? What would be the name of that place?

TRY OUT   Write the letter of the correct answer.

5. one hundred eighteen thousandths   **a.** 0.018   **b.** 0.118   **c.** 1.18

6. three and fourteen thousandths   **a.** 0.314   **b.** 3.014   **c.** 3.14

7. $\frac{9}{1,000}$ **a.** 0.009   **b.** 0.09   **c.** 0.9

PRACTICE

Write the decimal.

8. seven hundred fifteen thousandths
9. thirty-four thousandths
10. six and eighty-one thousandths
11. twelve and two thousandths
12. ten and forty-seven thousandths
13. eight thousandths
14. twenty and five hundred nine thousandths
15. four hundred twenty-eight thousandths

Write the word name.

16. 1.375
17. 6.029
18. 0.103
19. 0.004
20. 0.095
21. 16.020
22. 0.093
23. 0.726
24. 7.001
25. 4.108

Write the value of the underlined digit.

26. 4.2
27. 6.85
28. 0.79
29. 9.632
30. 0.007
31. 5.371
32. 25.070
33. 13.026
34. 0.054
35. 10.295

Write an equivalent decimal.

36. 0.7
37. 1.3
38. 2.08
39. 0.41
40. 7.50
41. 4.9
42. 0.800
43. 1.550
44. 6.0
45. 15

Solve.

46. The fastest-moving garden snail can travel a distance of 0.83 m in a minute. Write this distance as thousandths of a meter.

47. Erik walked a distance of thirty-three thousandths km to the library. Write this distance as a decimal.

48. Find three decimals in a newspaper or magazine. List them. Then write the word names for these numbers.

49. Jaime rode 0.5 km on his bike. Write this distance as hundredths of a km.

Mixed Review

Find the answer. Which method did you use?

MENTAL MATH
CALCULATOR
PAPER/PENCIL

50. 37,561 + 4,982
51. 90,350 − 89,724
52. $100.00 − 29.96
53. 639,287 − 20,087

UNDERSTANDING A CONCEPT

Comparing and Ordering Decimals

A. In the cross-country skiing world championships, the fastest speed for the first race is 14.25 miles per hour (mph). In the second race, the fastest speed is 14.38 mph. Which speed is faster?

Compare: 14.25 and 14.38

Step 1	Step 2	Step 3
Line up the decimal points.	Begin at the left. Compare to find the first place where the digits are different.	Compare the digits.
14.25 14.38	14.**2**5 14.**3**8	2 < 3 or 3 > 2 So 14.25 < 14.38 or 14.38 > 14.25.

Since 14.38 > 14.25, the faster speed is 14.38 mph.

1. How is comparing decimals similar to comparing whole numbers?

2. Marika compared 17.39 and 17.6. She said that 17.39 is greater because 39 > 6. Do you agree or disagree? Why?

B. You can order the decimals 1.33, 1.306, and 0.453 from least to greatest by comparing.

Step 1	Step 2	Step 3
Line up the decimal points. Write equivalent decimals if necessary.	Compare the other decimals.	Order the decimals.
1.330 ← *Think:* 1.33 = 1.330 1.306 0.453 ↑ The number without ones is the least.	1.3**3**0 1.3**0**6 *Think:* 0 < 3 So 1.306 < 1.330.	From least to greatest: 0.453 1.306 1.33

3. Is the decimal with the greatest number of decimal places always the greatest decimal? Why?

98 Lesson 3-3

Try Out

Compare. Use >, <, or =.

4. 0.11 ● 0.29 **5.** 4.37 ● 4.307 **6.** 10.60 ● 10.6

7. Order from greatest to least: 4.95, 5.17, 5.2

8. Order from least to greatest: 19.196, 19.107, 18.99

Practice

Compare. Use >, <, or =.

9. 1.3 ● 2.7   **10.** 0.4 ● 0.39   **11.** 1.5 ● 0.9   **12.** 3.15 ● 3.015

13. 2.41 ● 2.14   **14.** 8.16 ● 8.160   **15.** 7.015 ● 7.105   **16.** 18.01 ● 17.29

17. 1 thousandth ● 0.101   **18.** 9.058 ● 9 and 6 thousandths

19. 7.02 ● 72 hundredths   **20.** 4 and 6 hundredths ● 4.060

What digits can ■ be if:

21. 1.4 > 1.■?   **22.** 7.■ < 7.8?   **23.** 0.■ > 0.5?

Write in order from least to greatest.

24. 6.5, 2.31, 9.860   **25.** 0.9, 8.0, 0.08   **26.** 8.7, 8.06, 7.8, 7.68

Write in order from greatest to least.

27. 0.80, 0.82, 0.81   **28.** 1.7, 1.307, 1.037   **29.** 6.53, 6.35, 6.5, 6.305

Critical Thinking

30. What decimal is exactly between 2.5 and 2.6? How do you know?

Mixed Applications

31. The fastest downhill ski speeds are 139.030 mph and 133.234 mph. Which is faster?

32. In 1978 two skiers skied down Mt. Everest from 26,900 ft to 20,340 ft. How far did they ski?

33. Bill skied 30 mi in 1 h 59 min 47 s. The official record time for this distance was 2 h 10 min 50 s. How much faster was Bill than the record?

34. In a cross-country race, four skiers had speeds of 15.57 mph, 14.38 mph, 14.25 mph, and 15.95 mph. What was the fastest speed? the slowest speed?

EXTRA Practice, page 126

Adding and Subtracting Decimals

UNDERSTANDING A CONCEPT
Rounding Decimals

A. The fastest running speed recorded for a male is 26.32 miles per hour. What is this speed to the nearest mile per hour?

You can round using a number line.

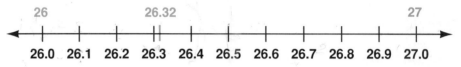

Think: 26.32 is between 26 and 27.
26.32 is closer to 26 than to 27.
Round down.

So 26.32 rounded to the nearest whole number is 26.

The fastest running speed for a male to the nearest mile per hour is 26.

When a number is halfway between two numbers, round up to the greater number.

Since 26.5 is halfway between 26 and 27, round up to 27.

1. How would you round 26.72 to the nearest whole number? Why?

B. You can round decimals without using a number line.

Round 8.418 to the nearest hundredth.

Step 1	Step 2	Step 3
Find the place to which you are rounding.	Look at the digit to the right of that place.	If it is 5 or greater, round up. If it is less than 5, round down.
8.4**1**8	8.41**8**	*Think:* 8 > 5 Round up. 8.418 → 8.42

So 8.418 rounded to the nearest hundredth is 8.42.

2. What is 8.741 rounded to the nearest tenth? to the nearest hundredth? How did you round? Why?

100 Lesson 3-4

TRY OUT Write the number.

3. Round 7.365 to the nearest whole number.
4. Round 2.349 to the nearest hundredth.
5. Round $5.38 to the nearest $.10

PRACTICE

Round to the nearest whole number.

6. 3.1
7. 4.8
8. 10.36
9. 12.69
10. 3.872
11. 4.05
12. 10.82
13. 9.15
14. 0.98
15. 5.096

Round to the nearest tenth.

16. 23.86
17. 1.07
18. 2.547
19. 0.809
20. 5.87
21. 0.42
22. 20.55
23. 32.84
24. 10.97
25. 0.981

Round to the nearest hundredth.

26. 4.908
27. 3.251
28. 0.609
29. 1.908
30. 15.072
31. 0.375
32. 6.059
33. 18.301
34. 12.998
35. 5.006

To what place has the number been rounded?

36. 17.456 rounds to 17.46
37. 0.706 rounds to 1
38. 8.061 rounds to 8.1
39. 2.346 rounds to 2.35
40. 5.635 rounds to 5.64
41. 13.025 rounds to 13.0

Mixed Applications

42. The fastest running speed recorded for a female is 22.5 mph. Round this speed to the nearest whole number.

43. On May 25, 1936, athlete Jesse Owens won a race of 100 yd and two races of 220 yd each. How many yards did he run?

44. In the 1936 Summer Olympic Games, Marjorie Gestring scored 89.27 points in the Springboard Diving event. In 1972 Micki King scored 450.03 points in the same event. How many years passed between these two Olympics?

45. In 1964 Robert Webster scored 148.58 points in the Platform Diving event of the Summer Olympic Games. In 1956 Joaquin Capilla scored 152.44 points in the same event. Who scored the greater number of points?

EXTRA Practice, page 126

UNDERSTANDING A CONCEPT
Estimating Decimal Sums and Differences

A. The Parks Department is building a bicycle trail in Truman Park. The trail will have three sections that will measure 7.55 km, 8.52 km, and 9.6 km. About how long will the trail be?

You can estimate to solve the problem. Here are two methods.

Rounding

Round to the nearest whole number.

```
7.55  →    8
8.52  →    9
+ 9.6 → + 10
           27
```

Front-End

Add the front digits. Then adjust the estimate.

```
 7.55      7.55  ⎫ Think:
 8.52      8.52  ⎬ about 1
+ 9.6     + 9.6  ⎭
   24        25
```

The bicycle trail will be about 25 to 27 miles long.

1. How does the exact answer compare with the estimate found by rounding? by using front-end estimation?

B. You can also use rounding and front-end estimation to estimate differences.

Rounding

Round to the nearest whole number.

```
 7.23 →   7
- 1.84 → - 2
          5
```

Front-End

Subtract the front digits. Then adjust the estimate.

```
 7.23      7.23     Think: 8 > 2
- 1.84    - 1.84    The exact answer is less than 6.
    6        < 6
```

TRY OUT

Estimate by rounding to the nearest whole number.

2. 7.43 + 0.96 + 2.08
3. 11.2 + 7.54 + 3.6
4. 14.5 − 6.7

Estimate. Use the front digits and adjust.

5. 15.9 + 2.34 + 3.71
6. 8.76 − 0.43
7. 9.2 − 5.67

PRACTICE

Estimate by rounding to the nearest whole number.

8. 6.78
   4.32
   + 7.09

9. 13.5
   8.24
   + 4.69

10. 4.94
    23.83
    + 15.6

11. 5.08
    3.17
    6.22
    + 2.47

12. 1.63
    17.4
    8.3
    + 12.58

13. 40.5
    − 18.9

14. 6.4
    − 0.57

15. 7.26
    − 5.8

16. 14.86
    − 2.5

17. 7.28
    − 0.831

18. 7.6 + 3.37 + 5.4

19. 18.8 − 9.93

20. 8.26 − 6.63

Estimate. Use the front digits and adjust.

21. 2.6
    5.4
    + 6.1

22. 4.25
    15.7
    + 11.3

23. 0.2
    6.8
    + 14.3

24. 2.07
    3.96
    5.07
    + 8.4

25. 12.6
    5.73
    3.38
    + 9.4

26. 25.3
    − 8.7

27. 6.5
    − 4.1

28. 7.38
    − 0.8

29. 9.06
    − 6.7

30. 9.95
    − 5.72

31. 4.03 + 12.8 + 5.71

32. 8.6 − 2.74

33. 6.21 + 8.7 + 5.04

Write the letter of the better estimate.

34. 13.83 + 4.01 + 1.9   a. less than 20   b. greater than 20
35. 14.06 − 5.28   a. less than 10   b. greater than 10
36. 7.8 − 2.9   a. less than 5   b. greater than 5
37. 4.32 + 17.81 + 2.05   a. less than 20   b. greater than 20

Mixed Applications

Solve. Which method did you use?

38. The Parks Department estimated that it would cost $120,500 to build a bike trail. The actual cost was $181,250. By how much was their estimate off?

39. Erin rode her bike 3.35 km from home to the bike trail. Then she rode along the trail and back home again, a distance of 16.8 km. About how far did she ride?

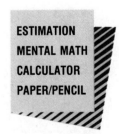

ESTIMATION
MENTAL MATH
CALCULATOR
PAPER/PENCIL

UNDERSTANDING A CONCEPT

Adding Tenths and Hundredths

A. Alicia jogged 1.8 km. Then she walked another 0.65 km. What was the total distance she covered?

You can add to find the total distance.

Add: 1.8 + 0.65

Step 1	Step 2	Step 3
Line up the decimal points. Write an equivalent decimal if necessary.	Add the hundredths. Add the tenths. Regroup if necessary.	Add the ones. Regroup if necessary. Write the decimal point.
1.8**0** + 0.65	¹ 1.80 + 0.65 ‾‾‾45	¹ 1.80 + 0.65 ‾‾‾2.45

Think: 1.8 = 1.80

Alicia covered 2.45 km.

1. Is 2.45 km a reasonable answer? Why?

B. Sometimes you can add decimals mentally. Here are two methods.

Add 1.6 + 2.3 mentally.

Left to Right

Add the ones.   Add on the tenths.

```
  1.6              1.6    Think:
+ 2.3            + 2.3    3 + 0.6 = 3.6
───              ───      3.6 + 0.3 = 3.9
  3                3.9
```

Add $.98 + $10.35 mentally.

Compensation

Change one of the addends to end in zero.

$.98 + $10.35
+ $0.02 ↓   ↓ − $0.02
$1.00 + $10.33 = $11.33

2. Add mentally. Which method did you choose? Why?
   a. 16.35 + 4.98   **b.** 6.2 + 8.5 + 3.6

TRY OUT Add. Do as many as you can mentally.

3. 3.78 + 1.9   **4.** 0.21 + 5.6 + 1.52   **5.** $7.83 + $4.79   **6.** 8.98 + 5.42

Lesson 3-6

PRACTICE

Add. Do as many as you can mentally.

| 7. | 4.7
+ 3.9 | 8. | 8.01
+ 0.97 | 9. | 9.3
+ 6.2 | 10. | $5.78
+ .16 | 11. | $.57
+ .92 |

| 12. | 3.7
+ 2.56 | 13. | 0.93
+ 6.2 | 14. | 7.5
+ 10.3 | 15. | 15.46
+ 2.5 | 16. | 8.4
+ 10.79 |

| 17. | 5.3
2.7
+ 4.1 | 18. | 6.56
2.31
+ 0.4 | 19. | $20.63
1.75
+ 12.98 | 20. | 12.5
3.63
+ 5.4 | 21. | $1.69
.83
+ .27 |

22. 4.7 + 3.6
23. $15.76 + $38.65
24. 0.6 + 0.49
25. 0.29 + 7.83
26. $18.06 + $10.92
27. 25.6 + 9.8 + 0.47

28. Find the sum of 3.6 and 19.51.

29. The addends are 0.64, 5.92, and 19.3. What is the sum?

Find the perimeter.

30.

31.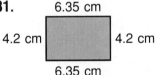

32.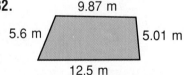

Critical Thinking

Do you agree or disagree with the following statements?
Give examples to support your answer.

33. When you add any two decimals less than 1, the sum is always less than 1.

34. When you add two decimals each greater than 0.5, the sum will always be greater than 1.

Mixed Applications

35. During the week Eva ran a total distance of 6.92 km. On Saturday Eva ran 1.3 km. On Sunday she ran 2.75 km. About how far did she run that week?

36. Jay earned $17 raking leaves, $9.50 baby-sitting, and $10 painting a fence. How much money did he earn?

Word Wealth

Experimenting and Predicting

A. Suppose each letter of the alphabet is worth a certain number of dollars: A is worth $1, B is worth $2, C is worth $3, and so on, so that Y is worth $25 and Z is worth $26.

1. How much is your first name worth?

2. Find out who has the most valuable name in your class. Who has the least valuable name?

3. Make a graph like the one below. Use Xs. Show all the first-name values for your class. Discuss what the graph shows. Does each part of the graph have about the same number of values, or are some parts much more crowded than others? Why do you think this is so?

20–39	40–59	60–79	80–99	100–119	120–139	140–159	160 or more
		X					
		X			X		
	X	X			X		

4. Next find the total value of your first and last names. If you have a middle name, add its value to the total.

   Which of your names is worth the most?

5. What is the total value of your teacher's first and last names? your principal's names?

6. Work with a partner. Pick a group such as MOVIE STARS or BASEBALL PLAYERS. How much are the names of your favorites worth?

B. You can compute the value of word names for numbers, too.

7. How much is the word name of each number from one to ten worth? Which number has the greatest value?

8. Can you find a number between ten and ninety-nine whose value is the same as its name?

   First, try to think of some shortcuts for doing this problem. For example, what letters are used in every name from thirteen to nineteen? from twenty to ninety-nine?

   How does this help you estimate some names?

9. Work with a partner. Can you find some words that are worth exactly $100? For example: QUARTER = $17 + $21 + $1 + $18 + $20 + $5 + $18 = $100.

10. **Write a problem** of your own. For example, find even-number word names whose values are odd.

UNDERSTANDING A CONCEPT

Subtracting Tenths and Hundredths

A. Pat is building a fence around his garden. The fence posts are 1.3 m high and the wire fencing is 0.85 m high. How much higher than the fencing are the posts?

Subtract: 1.3 − 0.85

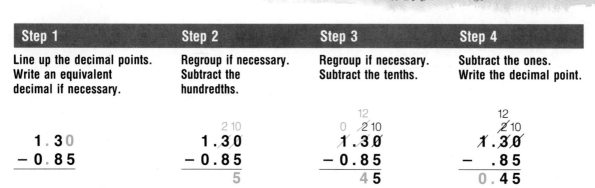

Step 1	Step 2	Step 3	Step 4
Line up the decimal points. Write an equivalent decimal if necessary.	Regroup if necessary. Subtract the hundredths.	Regroup if necessary. Subtract the tenths.	Subtract the ones. Write the decimal point.
1.30 − 0.85	2 10 1.3̸0̸ − 0.85 5	12 0 2̸ 10 1̸.3̸0̸ − 0.85 45	12 2̸ 10 1̸.3̸0̸ − .85 0.45

Think: 1.3 = 1.30

The posts are 0.45 m higher than the fencing.

1. Subtract: 7 − 1.36. What equivalent decimal did you need? Why?

B. Sometimes you can subtract decimals mentally. Here are two methods.

Subtract 8.7 − 5.2 mentally.

Left to Right

Subtract the ones.   Subtract the tenths.

    8.7   3.7
 − 5   − 0.2
    3.7   3.5

Subtract 15.75 − 13.9 mentally.

Equal Additions

Look to make the number being subtracted end in zero.

          + 0.1
  15.75 ⟶   15.85
          + 0.1
− 13.9 ⟶ − 14.00
                       1.85

2. Subtract mentally. Which method did you use? Why?
   a. 9.5 − 7.3   **b.** 17.35 − 12.8

TRY OUT   Subtract. Do as many as you can mentally.

3. 36.4 − 25.83   **4.** 17.9 − 0.46   **5.** $5.75 − $2.90   **6.** 8.71 − 5.99

PRACTICE

Subtract. Do as many as you can mentally.

| 7. 9.6 − 2.4 | 8. 8.3 − 3.7 | 9. $5.46 − 2.37 | 10. $11.97 − 9.70 | 11. 16.9 − 10.5 |

| 12. 12.3 − 6.2 | 13. 37.9 − 0.37 | 14. $.73 − .56 | 15. 3.78 − 2.9 | 16. 67.1 − 40.93 |

| 17. 8.65 − 4.7 | 18. $15.21 − 13.75 | 19. 0.65 − 0.39 | 20. 9 − 4.72 | 21. 18 − 6.9 |

22. 14.7 − 10.7
23. 6.28 − 0.3
24. 7.38 − 4.57
25. 11.06 − 5.33
26. 3.6 − 1.48
27. 25 − 8.14
28. How much less is 7.54 than 16.1?
29. How much more is 0.72 than 0.08?

Find the missing length.

30.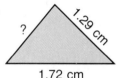

Perimeter: 4.22 cm

31.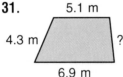

Perimeter: 20.2 m

32.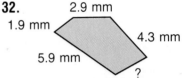

Perimeter: 17.9 mm

Mixed Applications

33. Marian grew pumpkins and sold them at Thanksgiving. She borrowed $6.50 to buy seeds and fertilizer, and she made $21.25 selling the pumpkins. How much money did she have after she paid back the money she borrowed?

34. **Write a problem** in which you must find the sum or difference of two decimals. Solve the problem. Then ask others to solve it.

Mixed Review

Find the answer. Which method did you use?

| 35. 2.56 + 1.89 | 36. 9,380 + 4,255 | 37. 1.27 − 0.43 | 38. 15,237 − 8,785 |

MENTAL MATH
CALCULATOR
PAPER/PENCIL

PROBLEM SOLVING

✓ UNDERSTAND
✓ PLAN
✓ TRY
  CHECK
✓ EXTEND

Strategy: Choosing the Operation

Lowell School is having a track and field day. Nancy and Brian run as a relay team. Nancy runs her leg of the race in 31.28 seconds. Brian runs his leg in 33.19 seconds. Nancy wants to know what their total time is.

To solve the problem Nancy plans to write a **number sentence.** Then she will solve the number sentence.

She knows that she is combining her time and Brian's time for the race. So she writes an addition sentence.

$$31.28 + 33.19 = \blacksquare$$

1. What is their total time?

Brian wants to know how much faster Nancy ran her leg of the race than he did. Since he is finding the difference between the two times, he writes a subtraction sentence.

$$33.19 - 31.28 = \blacksquare$$

2. How much faster than Brian did Nancy run her leg of the race?

3. **What if** another relay team had run the race in 1.39 seconds less time than Nancy and Brian? What number sentence would you write to find their time for the race? How fast did the other team run?

PRACTICE

Write a number sentence. Then solve the problem.

4. The women's indoor record for the longest throw of a Frisbee is 229.6 feet. The women's outdoor record is 179.6 feet greater than the indoor record. What is the outdoor record?

5. In 1985 Lou Scripa did 100,003 sit-ups in 50 hours. He did 60,405 sit-ups during the first 24 hours. How many sit-ups did he do during the rest of the time?

6. The first automobile race took place in 1895. The winner averaged 15.01 miles per hour. In 1987 a car was timed at 212.81 miles per hour. How much faster was the 1987 car than the 1895 car?

7. Josh ran the hurdles in 58.7 seconds. Mark's time was 0.8 second longer. What was Mark's time for the race?

Strategies and Skills Review

Solve. Use mental math, estimation, a calculator, or paper and pencil.

8. The longest distance anyone ever rode a bicycle in 24 hours is 516.2 miles. This was done on an indoor track. The outdoor record is 515.8 miles. How much greater is the indoor than the outdoor record?

9. The great basketball player Wilt Chamberlain scored 25,362 points on field goals and 6,057 points on free throws in his career. How many points did he score in his career?

10. Fran said that she once did a high jump of 10 feet. Does that seem reasonable? How can you tell?

11. Mary entered the run-and-skip race. She ran 100 yards in 13.8 seconds. Then she skipped 20 yards in 18.4 seconds. How long did it take her to finish the race?

12. The shortest major league baseball game on record lasted 51 minutes. The longest game lasted 8 hours and 6 minutes and was played over 2 days. How much longer did it take to play the longest game than the shortest?

13. **Write a problem** that can be solved by writing and solving a number sentence. Solve the problem. Ask others to solve your problem.

UNDERSTANDING A CONCEPT

More Estimating: Sums and Differences

A. Delia and Jay are helping their teacher cover three bookshelves with adhesive paper. The shelves measure 1.75 m, 1.45 m, and 0.59 m in length. About how much paper do they need?

You can estimate to find the answer. Here are two methods you can use.

Rounding

Round to greatest place of greatest number.

```
 1.75  →   2
 1.45  →   1
+0.59  → + 1
           4
```

Round to greatest place of least number.

```
 1.75  →   1.8
 1.45  →   1.5
+0.59  → + 0.6
           3.9
```

Front-End

Add the front digits. Adjust the estimate.

```
 1.75
 1.45
+0.59
   3
```
Think: 0.45 + 0.59 is about 1. 2 + 1 = 3

Delia and Jay need between 3 m and 4 m of paper to cover the bookshelves.

1. Which estimate is closest to the exact answer? Why?

B. You can also use rounding and front-end estimation to estimate differences.

Estimate: 5.482 − 0.83

Rounding

Round to greatest place of greater number.

```
 5.486 →   5
−0.83  → −1
           4
```

Round to greatest place of lesser number.

```
 5.486 →   5.5
−0.83  → −0.8
           4.7
```

Front-End

Subtract the front digits. Adjust the estimate.

```
 5.486
−0.83
  <5
```
Think: 8 > 4 The exact answer is less than 5.

TRY OUT Estimate.

Use rounding.

2. 5.6
    0.87
　 +0.354

3. 0.925
　 −0.49

Use the front digits and adjust.

4. 0.12
    0.89
　 +0.60

5. 0.63
　 −0.185

PRACTICE

Estimate by rounding.

6.	0.43 0.7 + 0.12	**7.**	0.88 0.675 + 0.36	**8.**	0.1 0.29 + 0.74	**9.**	6.3 0.55 + 0.462	**10.**	4.56 0.71 + 0.39
11.	0.97 − 0.74	**12.**	0.62 − 0.27	**13.**	0.445 − 0.18	**14.**	7.2 − 0.56	**15.**	8.17 − 0.529

16. 0.56 + 0.201 + 0.917   **17.** 0.61 − 0.47   **18.** 3.8 + 0.11 + 0.86

19. 2.4 − 0.89   **20.** 6.713 + 2.94 + 0.47   **21.** 0.908 − 0.53

Estimate. Use the front digits and adjust.

22.	0.16 0.52 + 0.73	**23.**	0.43 0.14 + 0.92	**24.**	0.225 0.38 + 0.596	**25.**	7.3 0.28 + 0.51	**26.**	2.65 0.4 + 3.91
27.	0.51 − 0.17	**28.**	0.609 − 0.23	**29.**	0.85 − 0.146	**30.**	9.3 − 0.75	**31.**	4.48 − 0.29

32. 0.75 + 0.13 + 0.52   **33.** 0.37 − 0.19   **34.** 8.71 − 0.53

35. 5.1 + 0.83 + 6.14   **36.** 3.29 + 8.3 + 0.5   **37.** 7.29 − 5.8

Write the letter of the better estimate.

38. 0.74 − 0.47   **a.** less than 0.3   **b.** greater than 0.3

39. 0.652 + 0.981   **a.** less than 1.5   **b.** greater than 1.5

40. 0.62 + 0.536 + 0.431   **a.** less than 1.5   **b.** greater than 1.5

41. 0.986 − 0.35   **a.** less than 0.6   **b.** greater than 0.6

Mixed Applications

42. Josh is making a banner to hang in the gym. The banner is 20.5 m wide, and the gym is 25.4 m wide. About how much rope must he attach to the banner in order to hang it from wall to wall?

43. Mr. Spencer's fifth-grade class decorated their classroom for a party. They began at 1:30 and finished at 2:45. How long did it take them to decorate the classroom?

EXTRA Practice, page 128   Adding and Subtracting Decimals

UNDERSTANDING A CONCEPT

Adding and Subtracting Thousandths

The world record for the long jump is 8.903 m. The next longest jump is 8.795 m. How much longer is the world record jump?

You can subtract to find how much longer the world record jump is.

Calculator

8.903 ⊖ 8.795 ⊜

The world record jump is 0.108 m longer than the next longest jump.

Paper and Pencil

```
    8 9 13
  8.9 0 3
- 8.7 9 5
  0.1 0 8
```

1. Is 0.108 a reasonable answer? Why?

2. **What if** you have to subtract 7.293 − 3.04? Would you use mental math, paper and pencil, or a calculator? Why?

3. **What if** you have to find these sums? Would you use mental math, paper and pencil, or a calculator?
   a. 4.8 + 3.67 + 9.056   b. 9.3 + 0.5   c. 325.4 + 45.37

4. When do you think using a calculator is the best way to find a sum or difference? When might mental math be better?

TRY OUT Write the letter of the correct answer. Use mental math, a calculator, or paper and pencil.

5. 42 + 9.317   a. 9.359   b. 13.517   c. 51.317

6. 0.92 − 0.862   a. 0.058   b. 0.062   c. 0.162

7. 2.98 + 16.109   a. 4.5909   b. 16.407   c. 19.089

8. 36 − 8.635   a. 5.035   b. 27.365   c. 28.635

PRACTICE

Add or subtract. Use mental math, a calculator, or paper and pencil.

9.	3.572 + 2.838	**10.**	9.498 − 3.407	**11.**	0.669 + 7.583	**12.**	15.381 + 10.999	**13.**	12.047 − 8.169
14.	23.06 − 14.751	**15.**	3.981 + 7.244	**16.**	23 + 18.687	**17.**	5.725 − 0.925	**18.**	17.5 − 6.511
19.	7.29 0.683 + 1.945	**20.**	5.2 12.603 + 10.192	**21.**	37.21 − 19.845	**22.**	6.34 7.298 + 0.706	**23.**	49.998 − 24.298

24. 63.56 + 2.013 + 5.3

25. 7.962 − 2.8

26. 4.7 + 3.09 + 12.312

27. 18.04 − 9.723

28. Find the sum of 16.08 and 12.403.

29. Find the sum of 4.706 and 19.281.

30. Find the difference between 20 and 5.728.

31. Find the difference between 12.38 and 9.574.

Compare. Use >, <, or =.

32. 3.79 + 4.657 ● 8.45

33. 10.62 − 5.281 ● 5.339

34. 0.25 ● 0.72 − 0.473

35. 6.35 ● 1.079 + 5.28

36. 12.6 − 9.041 ● 2.3 + 1.256

37. 21.3 + 18.359 ● 52 − 12.493

Mixed Applications

Solve. Which method did you use?

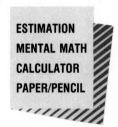

38. Joe is 137.8 cm tall. He marked this length on the ground, then did a running long jump to see how far he could jump. He jumped a distance of 145.5 cm. How much longer was his jump than his height?

39. In the snail race, Linda's snail crawled 3.72 m, stopped for a rest, and then crawled 2.5 m farther. How far did her snail go?

40. Sharon found that the length of her running shoe is 25.6 cm. The length of her brother's running shoe is 33.8 cm. About how much longer is her brother's running shoe?

PROBLEM SOLVING

✓ UNDERSTAND
✓ PLAN
✓ TRY
✓ CHECK
✓ EXTEND

Strategy: Using Estimation

A village library wants to collect at least 1,000 books to sell at the book fair. Volunteers collected 345 books the first week, 456 books the second week, and 312 books the third week. They want to know if they collected enough books.

Mrs. Lock, the librarian, plans to use estimation to help her solve the problem. To make sure there will be enough books, she plans to *underestimate*.

Mrs. Lock tries her plan. She rounds each number *down* and adds to find the total.

$$
\begin{array}{r}
345 \rightarrow 300 \\
456 \rightarrow 400 \\
+\,312 \rightarrow +\,300 \\
\hline
1{,}000
\end{array}
$$

1. Will the volunteers have enough books? How can you tell?

2. What might happen if she overestimates the number of books?

Mrs. Lock has budgeted $350 to buy equipment for the new community room at the library. She wants to buy a TV for $219.23, a cassette recorder for $67.25, and a TV stand for $49.50. Can she buy all three items for the amount she budgeted?

To make sure she has enough money, Mrs. Lock plans to *overestimate* the cost of the items. So she rounds each number *up*.

$$
\begin{array}{r}
\$219.23 \rightarrow \$220 \\
67.25 \rightarrow 70 \\
+\,49.50 \rightarrow +\,50 \\
\hline
\$340
\end{array}
$$

3. Does she have enough money to buy the items? How can you tell?

4. What might happen if Mrs. Lock underestimates the cost of the items?

116   Lesson 3-12

PRACTICE

Use estimation to solve the problem. Did you overestimate or underestimate? Why?

5. A dictionary publisher does not charge for the postage on packages that weigh under 10 pounds. Is there a charge for a package containing dictionaries that weigh 1.8 pounds, 0.8 pounds, 1.7 pounds, 1.5 pounds, and 2 pounds?

6. An author on marathon running recommends that a person hike at least 10 hours a week. Oscar hiked 4.2 hours on Monday, 3.6 hours on Wednesday, and 4.2 hours on Friday. Has he hiked for the recommended number of hours?

7. The library hopes to collect $2,100 from the book sale. Last weekend the workers collected $1,030. This weekend they collected $428 on Friday, $573 on Saturday, and $380 on Sunday. Have they collected at least $2,100?

8. At the book fair Alma wants to buy books for $35.50, bookends for $12.89, and a videotape for $5.75. She has $50. Is that enough money to pay for these items?

Strategies and Skills Review

Solve. Use mental math, estimation, a calculator, or paper and pencil.

9. There are 213 books on 8 shelves in one section of the library and 192 books on 6 shelves in another section. How many books are there in the two sections?

10. Marlowe bought a book for $15.75. He gave the cashier a $20 bill and got $5.25 back in change. Did he get the correct change? How can you tell by estimation?

11. The large dictionary in the library weighs about 10 pounds. About how many pounds does a set of encyclopedias weigh? Write the letter of the best answer.
    a. 60 lb **b.** 300 lb **c.** 600 lb

12. The library dictionary is 2,015 pages long. The almanac has 1,087 fewer pages than the dictionary. How many pages does the almanac have?

13. Amy has painted 24 pictures in her picture book. She has 18 pictures left to paint. How many pictures are there in the book?

14. ***Write a problem*** that can be solved by using estimation. Solve the problem. Ask others to solve your problem.

UNDERSTANDING A CONCEPT

Interpreting Bar Graphs

A. In a survey conducted by a sports magazine, fans rated the sports they enjoyed watching most. The results of the survey are shown in the bar graph to the right.

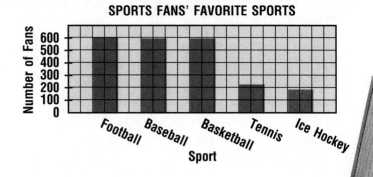

1. What information is shown on the horizontal axis? on the vertical axis?
2. What is the most preferred sport? How can you tell?

B. You can use a double-bar graph to compare more than one set of data.

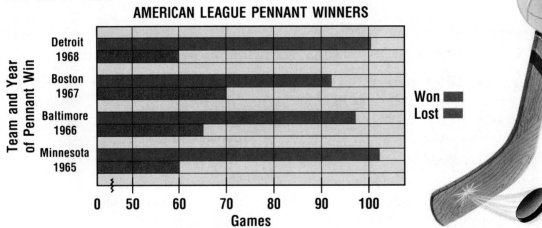

3. What information is compared on the double-bar graph?
4. Why is the horizontal scale broken between 0 and 50?
5. What does ■ represent? ■ represent?

TRY OUT   Solve. Use the graphs above.

6. What is the least preferred sport?
7. Which two sports are preferred by the same number of fans?
8. Which team won the fewest games? lost the fewest games?

PRACTICE

Use the bar graph at the right to answer Problems 9–11.

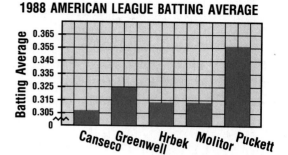

9. Who has the highest batting average?

10. Who has the lowest batting average?

11. Which two players have the same batting average?

Use the double-bar graph below to answer Problems 12–17.

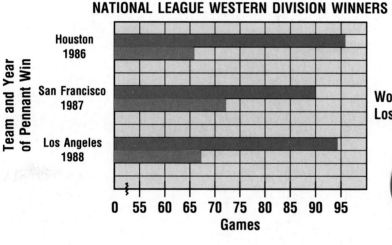

12. What information is compared on this graph?

13. Which bar represents games won? games lost?

14. Which team won the fewest games? the most games?

15. Which two teams lost about the same number of games?

16. About how many games did Los Angeles win in 1988?

17. About how many games did San Francisco lose in 1987?

Critical Thinking

18. **What if** you want to show the favorite month of the students in your class? Would you show it on a bar graph or on a line graph? Why?

19. **What if** you want to show the change in the number of students who participated in sports over the last five years? Would you show it on a bar graph or on a line graph? Why?

DEVELOPING A CONCEPT

Making Bar Graphs

The table at the right shows the number of home runs scored by baseball's all-time home run leaders.

ALL-TIME HOME RUN LEADERS

Player	Number of Home Runs
Hank Aaron	755
Harmon Killibrew	573
Willie Mays	660
Frank Robinson	586
Babe Ruth	714

WORKING TOGETHER

Use the data in the table and these steps to make a bar graph.

Step 1 Round the data to a convenient place to choose a scale for the vertical axis.

Step 2 Draw and label the horizontal and vertical axes.

Step 3 Draw the bars on the graph.

Step 4 Write a title above the graph.

1. To which place did you round the data? Why?
2. What scale did you use for the vertical axis? How did you decide?
3. Does the vertical axis begin at zero? If not, how did you show the axis?
4. What is the title of your graph?

Here is the way Elena graphed the data:

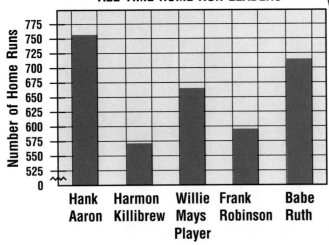

120 Lesson 3-14

SHARING IDEAS

5. Compare the vertical axis of your graph to Elena's and to those of others. How are they similar? How are they different?

6. How do differences in the vertical scale change the appearance of the graph?

7. Which player had the most home runs? the fewest?

8. From reading the graph can you tell the exact difference between the greatest and the least number of home runs? What would be a reasonable estimate?

PRACTICE

Use the data in the table at the right to make a bar graph.

9. To which place did you round the data?

10. What scale did you use for the vertical axis?

LEADERS IN STOLEN BASES

Player	Bases Stolen
Lou Brock	938
Max Carey	738
Eddie Collins	742
Rickey Henderson	794

Use your bar graph to answer the question.

11. Which player has the highest number of stolen bases?

12. Which player has the second highest number?

13. Which two players have about the same number?

14. Include the following players and number of stolen bases on the bar graph you made: Honus Wagner, 703 Ty Cobb, 892

15. Which player has the second highest number of stolen bases now? the lowest number?

16. Write some conclusions you can draw from a bar graph comparing the seating capacities of six different baseball stadiums. First use the information found on page 518 in the Databank to make this graph.

17. Write three questions that can be answered by reading a bar graph of sports favorites. Answer the questions. Then ask others to solve them. To make the graph, take a survey of the other students to find their favorite sport.

DECISION MAKING

Problem Solving: Buying a New or Used Item

SITUATION

Julie wants to buy a bicycle. She plans to use the bicycle to ride to school each day and to ride for fun.

PROBLEM

Should Julie buy a new or a used bicycle?

For Sale: Girl's 10-speed bike, yellow, excellent cond., 6 mo. old, $68. Call 333-4256

DATA

BICYCLE BILL'S NEW AND USED BICYCLES

$105
10-speed bicycle
All frame sizes and styles
Red, blue, white
1-year warranty

$185
12-speed bicycle
All frame sizes and styles
Blue, black, silver
2-year warranty

$250
18-speed all-terrain bicycle
All frame sizes and styles. Red, blue
2-year warranty

All used bicycles guaranteed for 30 days.

Used Girls' Bicycles
26" red all-terrain fair condition 12-speed 5 years old .... $ 82
20" green 10-speed good condition 10 years old .... $ 59
28" pink 10-speed excellent condition 2 years old $115

Used Boys' Bicycles

For Sale: Girl's 10-speed bike, orange, 24", good cond., 4-yr. old, $35. Call days: 852-6721

122 Lesson 3-15

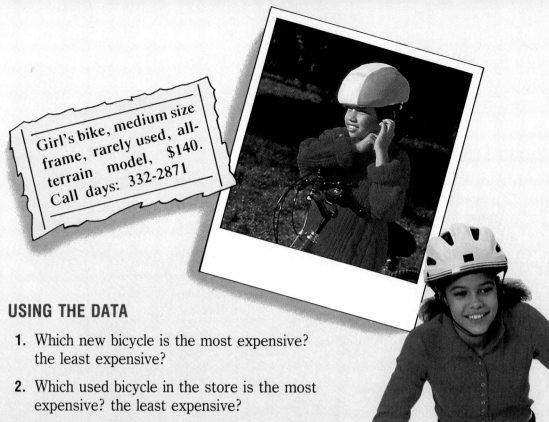

Girl's bike, medium size frame, rarely used, all-terrain model, $140. Call days: 332-2871

USING THE DATA

1. Which new bicycle is the most expensive? the least expensive?

2. Which used bicycle in the store is the most expensive? the least expensive?

3. Which used bicycle advertised in the want ads is the most expensive? the least expensive?

4. Which newspaper want ad gives you the least information? Explain.

MAKING DECISIONS

5. What are some of the advantages of buying a used bicycle from a bicycle shop?

6. Julie decided to buy a used bicycle but none was her size. What might she do?

7. **What if** you could not have a new bicycle and the used bicycle in your size was not the color you wanted? What might you do?

8. Suppose you needed a rugged bike to ride to school over dirt roads. Which bike would you choose? Why?

9. **What are** some of the things that might happen when you call about a used bike listed in the want ads?

10. **Write a list** of other things you should think about when deciding whether to buy a new or used bicycle.

11. **What if** you are going to buy a bicycle? Would you choose a new or a used bicycle? Why?

Adding and Subtracting Decimals

CURRICULUM CONNECTION

Math and Science

The strength of an earthquake is measured with a seismograph and reported as a number from 1 to 8 on the Richter scale. Decimals are used to show strength between one number and the next. For example, an earthquake might register 3.5.

Look at the scale. It shows how the strength increases tenfold with each level.

What if you want to compare an earthquake that measures 4 with one that measures 2?

Think: From 2 to 3 is a tenfold increase—from 10 to 100.
From 3 to 4 is another tenfold increase—from 100 to 1,000.

A quake that measures 4 on the Richter scale is 100 times greater in strength than one measuring 2.

THE RICHTER SCALE

Richter Number	Increase in Magnitude
8	10,000,000
7	1,000,000
6	100,000
5	10,000
4	1,000
3	100
2	10
1	1

ACTIVITIES

1. Find out about seismology, the science that deals with earthquakes. How does a seismograph work? Write a paragraph that a younger student could understand.

2. Use an almanac to find out about 5 major earthquakes in the last 50 years. Make a bar graph showing the strength of each earthquake. Be sure to show the date and location of each earthquake.

Computer Graphing: Reaction Time

If you are bicycling down the road and someone walks in front of your bicycle, how long does it take your foot to hit the brake? If your hand accidentally touches a stove, how long does it take for you to pull it away? The time it takes you to respond is known as the **reaction time**. On the average, a person can react in about one-fifth of a second.

To gather data on reaction time, work with a partner and follow these steps:

Step 1 Take a twelve-inch ruler and hold it vertically against the wall. The 0-inch mark should be at the bottom.

Step 2 Have your partner place the middle of his or her thumb at the 0-inch mark without touching the ruler.

Step 3 Release the ruler. Have your partner use his or her thumb to stop the ruler against the wall. Notice at what marking the middle of your partner's thumb is now. Round this distance to the nearest half inch and write the amount as a decimal.

Step 4 Do four more trials by repeating Steps 1–3.

Step 5 Now reverse roles with your partner and do five more trials.

Now use the computer graphing program REACT to calculate your reaction times.

AT THE COMPUTER

Enter your initials and your partner's initials at the top of the columns. Then enter the number of inches the ruler dropped for each trial. The computer will now create two line graphs showing your reaction time to the nearest hundredth of a second.

1. What was your fastest reaction time? What was your partner's fastest reaction time?

2. Did you improve with practice? Did your partner improve? If you kept practicing, how fast do you think you could stop the ruler?

Adding and Subtracting Decimals

Extra Practice

Tenths and Hundredths, page 95
Write the decimal.

1. six hundredths
2. five and two hundredths

Write the word name.

3. 3.7
4. 4.25
5. 0.7
6. 17.5

Write an equivalent decimal.

7. 0.3
8. 2.10
9. 0.40
10. 3.7

Thousandths, page 97
Write the decimal.

1. eight hundred twelve thousandths
2. fifty-three thousandths
3. four and twenty-two thousandths
4. thirteen and four thousandths

Write the value of the underlined digit.

5. 3.4
6. 3.67
7. 0.62
8. 12.395

Comparing and Ordering Decimals, page 99
Compare. Use >, <, or =.

1. 0.72 ● 0.71
2. 6.015 ● 6.105
3. 19.07 ● 18.12
4. 2 thousandths ● 0.202
5. 5 and 4 hundredths ● 5.040

Write in order from least to greatest.

6. 5.3, 2.72, 8.625
7. 0.7, 7.0, 0.06
8. 4.6, 4.03, 6.4, 6.48

Rounding Decimals, page 101
Round to the nearest whole number.

1. 3.7
2. 2.1
3. 13.52
4. 3.073
5. 10.52

Round to the nearest tenth.

6. 24.79
7. 2.09
8. 3.85
9. 33.82
10. 5.99

Round to the nearest hundredth.

11. 3.809
12. 0.507
13. 13.897
14. 5.067
15. 3.007

EXTRA PRACTICE

Estimating Decimal Sums and Differences, page 103

Estimate by rounding to the nearest whole number.

1.	15.8 + 3.5	2.	13.52 − 5.8	3.	3.8 + 12.7	4.	50.3 + 4.7	5.	17.5 − 3.8
6.	7.25 2.73 + 5.89	7.	61.098 − 10.6	8.	6.879 − 2.068	9.	16.80 − 3.485	10.	13.89 6.7 + 12.075

Estimate. Use the front digits and adjust.

11.	3.6 + 6.5	12.	5.35 + 16.8	13.	0.3 7.9 + 15.4	14.	3.08 4.79 + 6.08	15.	13.7 4.97 4.49 + 9.5
16.	26.4 − 9.8	17.	7.6 − 5.2	18.	8.79 − 0.3	19.	10.07 − 7.8	20.	20.05 − 6.82

Adding Tenths and Hundredths, page 105

Add. Do as many as you can mentally.

1.	3.6 + 2.9	2.	7.09 + 0.65	3.	8.2 + 3.7	4.	$3.67 + .15	5.	$.38 + .85
6.	4.2 3.8 + 2.1	7.	6.73 2.21 + 0.3	8.	$10.53 2.85 + 13.89	9.	11.5 2.64 + 3.7	10.	$1.28 .74 + .12

11. 3.4 + 2.3 **12.** $12.79 + $48.62 **13.** 15.6 + 8.8 + 0.49

Subtracting Tenths and Hundredths, page 109

Subtract. Do as many as you can mentally.

1.	8.7 − 1.3	2.	7.4 − 2.7	3.	$3.37 − 1.28	4.	$12.89 − 7.80	5.	19.8 − 10.7
6.	13.2 − 6.1	7.	35.8 − 0.36	8.	$.72 − .55	9.	4.89 − 2.9	10.	57.2 − 40.83

Adding and Subtracting Decimals

EXTRA PRACTICE

Problem-Solving Strategy: Choosing the Operation, page 111

Write a number sentence. Then solve the problem.

1. Heidi saved $35.00 earned by babysitting. She spent $21.95 on a new leotard. How much did she have left over?

2. Ivan's grandmother lives 257.8 miles away. His Uncle Peter lives 23.2 miles farther. How far away does Ivan live from his uncle?

More Estimating: Sums and Differences, page 113

Estimate by rounding.

1. 0.46 + 0.39
2. 0.72 − 0.38
3. $0.88 − 0.16
4. $0.25 + 1.47
5. 4.375 − 0.187

6. 2.897 − 0.516
7. 0.17 + 0.891
8. 0.473 + 3.289 + 0.893
9. $ .38 + .27 + 19.85
10. 0.785 + 0.876 + 0.15

Estimate. Use front digits and adjust.

11. 0.22 + 0.76 + 0.17
12. 1.65 + 1.35 + 0.39
13. 5.32 − 0.76
14. 0.73 − 0.208
15. 0.48 − 0.23

16. 0.348 + 0.584 + 0.28
17. 0.639 − 0.22
18. 3.131 + 1.71 + 0.4
19. 7.88 + 0.45 + 0.79
20. 8.36 − 2.74

Adding and Subtracting Thousandths, page 115

Add or subtract.

1. 4.589 + 3.895
2. 8.375 − 2.309
3. 0.558 + 6.385
4. 12.281 + 10.829
5. 16.048 − 3.125

6. 8.29 + 0.385 + 1.872
7. 3.3 + 11.609 + 10.185
8. 47.31 − 18.685
9. 24 − 17.365
10. 13.4 − 6.811

11. 35.26 + 3.017 + 3.3
12. 6.873 − 2.5
13. 16.07 − 8.893

128 Chapter 3

EXTRA PRACTICE

Problem-Solving Strategy: Using Estimation, page 117

Use estimation to solve the problem. Did you overestimate or underestimate?

1. Mrs. Lock drives 8.8 miles from work to her evening class. She drives 4.7 miles from her class to her home. Does she drive at least 15 miles?

2. A local charity wants to raise $50,000. It raised the following amounts: $23,119, $27,019, and $12,015. Did it reach its goal?

Interpreting Bar Graphs, page 119

Use the bar graph at the right to answer Questions 1–3.

1. Which club has the highest batting average?

2. Which club has the lowest batting average?

3. Which two clubs have the same batting average?

Use the double-bar graph to answer Questions 4–7.

4. What information is compared on this graph?

5. Which symbol represents games won? games lost?

6. Which team won the fewest games? the most games?

7. Which team lost the most games?

Making Bar Graphs, page 121

Use the data in the table at the right to make a bar graph.

1. What scale did you use for the vertical axis?

2. To which place did you round the data?

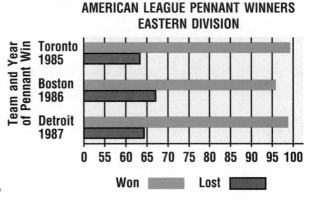

LEADERS IN STRIKEOUTS

Player	Strikeouts
Nolan Ryan	5,007
Gaylord Perry	3,534
Don Sutton	3,574
Tom Seaver	3,640

Adding and Subtracting Decimals

Practice PLUS

KEY SKILL: Adding Tenths and Hundredths (Use after page 105.)

Level A
Add.

1. 0.8
  + 0.7

2. 7.0
  + 0.6

3. 0.4
  + 0.2

4. $3.98
  + 0.12

5. $.78
  + .34

6. 2.9
  + 3.5

7. 0.5
  + 0.4

8. 0.45
  + 3.2

9. 3.9
  + 10.2

10. 7.8
  + 11.5

11. Melissa walked 2.8 km on Monday and 1.7 km on Wednesday. What was the total distance she walked?

Level B
Add.

12. 6.07
  + 4.12

13. $3.65
  + 2.89

14. $.98
  + .32

15. 4.3
  + 4.89

16. 6.8
  + 3.9

17. 0.76
  + 4.38

18. 5.7
  + 12.6

19. 13.65
  + 3.48

20. $21.98
  + 3.08

21. 8.9
  + 14.55

22. Martin jogged 4.5 miles on Saturday and 6.87 miles on Monday. How many miles did he jog in all?

Level C
Add.

23. 6.58
  + 44.7

24. $76.34
  + 12.40

25. 0.56
  + 5.77

26. 3.8
  + 4.89

27. 6.8
  + 23.85

28. 4.9
  3.5
  + 2.9

29. 7.74
  2.08
  + 4.01

30. $38.45
  22.12
  + 8.54

31. 18.9
  6.76
  + 3.87

32. $3.89
  1.37
  + .93

33. Kesha spent $87.65 on running shoes, $32.98 on sweatpants, and $10.98 on wristbands. How much money did he spend in all?

Practice Plus

KEY SKILL: Subtracting Tenths and Hundredths (Use after page 109.)

Level A
Subtract.

1. 8.6 − 2.1	2. 6.5 − 1.4	3. $3.48 − 2.17	4. $10.76 − 8.40	5. 15.7 − 11.4
6. 12.4 − 4.2	7. $11.67 − 5.43	8. 23.8 − 0.4	9. 3.8 − 1.2	10. 43.7 − 20.6

11. Bonnie has $5.46. She spends $2.74 on petunias at the garden shop. How much money does she have left?

Level B
Subtract.

12. 8.2 − 3.4	13. $.98 − .39	14. $16.82 − 14.53	15. 0.72 − 0.31	16. 9.8 − 3.9
17. 5.4 − 0.6	18. $30.08 − 11.54	19. 4.3 − 2.7	20. $23.04 − 11.96	21. 14.9 − 3.7

22. Quentin is planting lettuce in 15.3 square meters and carrots in 8.74 square meters. How much more space is he allowing for the lettuce?

Level C
Subtract.

23. 0.76 − 0.38	24. 45 − 6.54	25. $76.87 − 43.98	26. 43.9 − 0.87	27. 18.15 − 0.87
28. 74 − 3.98	29. 8.43 − 3.6	30. $75.04 − 32.10	31. 16.54 − 3.9	32. 65 − 23.7

33. Sally has $100. She spends $75.87 on gardening tools. How much money does she have left?

Adding and Subtracting Decimals

Chapter Review

LANGUAGE AND MATHEMATICS
Complete the sentences. Use the words in the chart on the right.

1. A fraction or a ■ can be written for a number less than 1. *(page 94)*
2. 6.807 rounded to the nearest ■ is 6.81. *(page 100)*
3. ■ is a method of subtracting mentally. *(page 108)*
4. A ■ can be used to compare more than one set of data. *(page 118)*
5. ***Write a definition*** or give an example of the words you did not use from the chart.

VOCABULARY
bar graph
decimal
hundredth
equal additions
double-bar graph
thousandth

CONCEPTS AND SKILLS
Write the decimal. *(pages 94–97)*

6. four tenths
7. three and forty-three hundredths
8. nine hundred twelve thousandths
9. thirty and two hundred eight thousandths

Compare. Use >, <, or =. *(page 98)*

10. 1.7 ● 2.9
11. 3.16 ● 3.016
12. 16.09 ● 18.17
13. 3 and 8 hundredths ● 3.080
14. 3.09 ● 39 hundredths

Write in order from least to greatest. *(page 98)*

15. 5.3; 4.73; 2.073
16. 0.7; 0.007; 8.0

To what place has the number been rounded? *(page 100)*

17. 13.378 rounds to 13.38
18. 0.504 rounds to 1
19. 7.052 rounds to 7.1
20. 2.457 rounds to 2.46

Write the letter of the best estimate. *(page 102)*

21. 12.52 + 3.07 + 2.8   **a.** less than 20   **b.** greater than 20
22. 16.09 − 7.13   **a.** less than 10   **b.** greater than 10

Add or subtract. *(pages 104, 108, 114)*

23. 2.8 + 3.64

24. $30.63 + 2.09 + 11.85

25. 16.3 + 4.7 + 3.85

26. $1.38 + .65 + .87

27. 11.7 − 3.5

28. $.64 − .36

29. 38.2 − 20.89

30. $16.11 − 11.75

31. 2.875 + 3.893

32. 4.093 + 6.179

33. 89.897 − 34.297

34. 8.763 − 0.874

35. 6.8 + 12.7
36. $19.95 − $4.89
37. 0.5 + 0.76
38. 2.3 − 1.64
39. 14 − 8.39
40. 0.75 + 12.4 + 3.28

CRITICAL THINKING

41. Without calculating, determine which number sentence has the greatest sum and which number sentence has the least sum. Why?

    a. 3.097 + 4
    b. 3.097 + 0.4
    c. 3 + 0.04

42. Which method would you use to subtract? Why?
    a. 8.5 − 6.3   b. 13.85 − 7.5

MIXED APPLICATIONS

43. Brooke has $10.75. She buys a wallet for $8.19. How much change does she have left? *(page 108)*

44. Don runs 1.8 km on Monday, 3.7 km on Tuesday, and 5.2 km on Wednesday. What is a reasonable estimate of the number of kilometers he ran? *(page 116)*
    a. 8 km   b. 11 km   c. 15 km

45. The fifth grade raised money by organizing a comedy show. They earned $45.75 in ticket receipts, $26.09 for programs, and $10.20 for refreshments. What is a reasonable estimate of the amount of money earned? *(page 116)*
    a. $55   b. $75   c. $80

46. Marlo is putting a decorative fence around her flower garden. The posts are 1.2 m high and the fencing is 0.65 m high. How much higher are the posts than the fencing? *(page 108)*

Chapter Test

Write the decimal.

1. twelve hundredths
2. eight and fifty-four thousandths

Write an equivalent decimal.

3. 3.07
4. 16

Compare. Use >, <, or =.

5. 5.17 ● 5.170
6. 8 and 4 hundredths ● 8.4

Round to the nearest hundredth.

7. 3.885
8. 13.998

9. Order 6.734, 5.734, and 5.729 from least to greatest.

Estimate by rounding or using front-end estimation.

10. 4.67 + 3.04 + 8.19
11. 8.04 − 3.6
12. 5.13 − 0.734

Add or subtract.

13. 4.6 + 3.75
14. 18.8 − 11.43
15. $12.85 − $6.50

Use the bar graph at the right to answer Questions 16–18.

16. Which team has the lowest batting average?
17. Which two teams have the same batting average?
18. Houston had a team average of 0.244. How would you show this information on the graph?

1988 NATIONAL LEAGUE CLUB BATTING AVERAGES

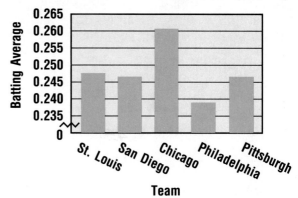

Solve.

19. Danny earned $19.50 mowing the lawn, $8.77 for recycling cans, and $17.05 in tips from his paper route. About how much money did Danny earn?

20. Barney rode his bicycle 4.48 km to the playground. Then he rode a distance of 12.7 km around the track. How far did he ride? Write the number sentence you used to solve this problem.

ENRICHMENT FOR ALL

NUMBER SEQUENCES

A **sequence** is a set of numbers that follows a certain pattern.

Study the pattern for each sequence.

A. 53, 49, 45, 41, ...

B. 3, 6, 5, 8, 7, ...

C. 0.5, 1, 1.25, 1.75, 2, ...

The pattern for sequence A is subtract 4.

1. What is the pattern for sequence B? for sequence C?
2. Write the next four numbers in sequences A, B, and C.

Find the pattern for the sequence. Then write the next four numbers in the sequence.

3. 3, 10, 17, 24, ...
4. 6, 5, 10, 9, 14, ...
5. 18, 19.5, 21, 22.5, 24, ...
6. 45, 37, 40, 32, 35, ...
7. 9, 12, 11.75, 14.75, 14.5, ...
8. 8.9, 7.8, 6.7, 5.6, ...
9. 6.5, 6.9, 6.7, 7.1, 6.9, ...
10. 16, 23.5, 22, 29.5, 28, ...
11. Make up a sequence involving addition and subtraction. Ask others to find the pattern and write the next four numbers in the sequence.

Adding and Subtracting Decimals

Cumulative Review

Choose the letter of the correct answer.

1. Which is the best estimate?
   225 + 1,489 + 350
   a. 3,000
   b. 2,000
   c. 2,500
   d. not given

2. Which digit is in the ten billions place of 328,507,392,481?
   a. 2
   b. 0
   c. 5
   d. not given

3. 38.874 − 12.312
   a. 26.56
   b. 26.502
   c. 26.652
   d. not given

4. Which is the standard form for 400,000 + 30,000 + 3,000 + 5?
   a. 433,050
   b. 433,500
   c. 433,005
   d. not given

5. $53.89 + $17.45 + $89.25
   a. $140.59
   b. $160.49
   c. $160.59
   d. not given

6. 375,400 − 18,563
   a. 356,837
   b. 355,726
   c. 355,847
   d. not given

7. Which is an equivalent decimal for 2.6?
   a. 0.26
   b. 0.260
   c. 2.60
   d. not given

8. 0.85 + 3.3
   a. 4.15
   b. 3.13
   c. 4.11
   d. not given

9. Compare: 3.019 ● 3.109
   a. >
   b. <
   c. =
   d. not given

10. Round 6.7 to the nearest whole number.
    a. 6
    b. 7
    c. 8
    d. not given

11. Which is the best estimate?
    13.06 − 1.9
    a. < 5
    b. < 10
    c. > 10
    d. > 15

12. Which is the decimal for two and eight hundred sixteen thousandths?
    a. 0.2816
    b. 2.816
    c. 208.16
    d. not given

13. Compare: 895,063 ● 859,045
    a. >
    b. <
    c. =
    d. not given

14. Pat is 147.5 cm tall. Her shadow is 175.3 cm long. How much longer is Pat's shadow than he actual height?
    a. 20 cm
    b. 27.8 cm
    c. 322.8 cm
    d. not given

136 Chapter 3

Multiplying Whole Numbers

MATH CONNECTIONS: ALGEBRA • PERIMETER • AREA • CIRCUMFERENCE • PROBLEM SOLVING

CHAPTER 4

1. What information do you see in this picture?
2. How can you use the information?
3. What shapes do you see in this picture?
4. Write a problem using the information.

UNDERSTANDING A CONCEPT

Informal Algebra: Meaning of Multiplication

A. Alberto wants to frame his teammates' pictures and hang them on his wall. He can make 3 rows with 7 pictures in a row. How many pictures can he hang in all?

You can add to find how many in all.

$7 + 7 + 7 = 21$

You can also multiply.

$3 \times 7 = 21$ ← factors, product

$\begin{array}{r} 7 \\ \times 3 \\ \hline 21 \end{array}$ ← factors, product

Alberto can hang 21 pictures in all.

1. How are the addition and multiplication sentences similar?

2. Why can you multiply as well as add?

B. Multiplication properties can help you multiply.

Commutative Property
If the order of the factors is changed, the product remains the same.

$9 \times 4 = 36 \quad 4 \times 9 = 36$

Identity Property
If one factor is 1, the product is equal to the other factor.

$7 \times 1 = 7 \quad 1 \times 7 = 7$

Property of Zero
If one factor is 0, the product is 0.

$8 \times 0 = 0 \quad 0 \times 8 = 0$

Associative Property
If the grouping of the factors is changed, then the product remains the same.

$(4 \times 2) \times 3 = \blacksquare \quad 4 \times (2 \times 3) = \blacksquare$
$\quad\; 8 \times 3 = 24 \quad\quad 4 \times\; 6 = 24$

Distributive Property
If one factor is a sum, multiplying each addend by the other factor before adding does not change the product.

$3 \times (5 + 4) = (3 \times 5) + (3 \times 4)$
$3 \times\quad 9\quad =\quad 15\quad +\quad 12$
$\quad\quad 27\quad =\quad 27$

3. Do you think the distributive property works with subtraction? Give an example to support your answer.

TRY OUT Find the missing number.

4. 6 × 7 = 7 × ■
5. (4 × 2) × 2 = 4 × (2 × ■)
6. 3 × (7 + 2) = (3 × ■) + (3 × 2)
7. 8 × 7 = ■

PRACTICE

Multiply.

8. 4 × 5
9. 8 × 2
10. 3 × 9
11. 6 × 2
12. 5 × 9
13. 7 × 4
14. 3 × 4

15. 5 × 7
16. 3 × 8
17. 9 × 4
18. 1 × 7
19. 6 × 6
20. 4 × 3
21. 2 × 9

22. 2 × 8
23. 9 × 6
24. 0 × 4
25. 5 × 3
26. 6 × 7
27. 7 × 9
28. 3 × 6

29. 0 × 7
30. 4 × 4
31. 6 × 8
32. 8 × 1
33. 9 × 3
34. 5 × 5
35. 8 × 6

36. 5 × 6
37. 3 × 7
38. 9 × 0
39. 8 × 4
40. 7 × 2
41. 4 × 1
42. 8 × 3
43. 3 × 6
44. 4 × 5 × 1
45. 3 × 8 × 0
46. 2 × 3 × 4
47. 4 × 2 × 7

Solve.

48. The factors are 5 and 2. What is the product?
49. One factor is 4. The product is 28. What is the other factor?

Complete. Which property did you use?

50. 6 × 2 = 2 × ■
51. ■ × 108 = 108
52. 4 × (2 + 6) = (■ × 2) + (4 × 6)
53. 67 × 1 = ■
54. ■ × 131 = 0
55. 3 × (2 × 4) = (3 × ■) × 4

Mixed Applications

56. The Wong family bought 4 tickets to the game. Each ticket cost $6. How much did they spend on tickets?
57. The Wongs left for the ballpark at 11:30 A.M. They were back home at 5:15 P.M. How long had they been gone?

UNDERSTANDING A CONCEPT

Basic Facts and Strategies

Ryan and two teammates are ordering T-shirts for the baseball teams in their league. They plan to order 9 packages. Each package has 8 T-shirts. How many T-shirts will they receive?

You can multiply to find how many in all.

Multiply: 9 × 8

Ryan and his teammates do not remember the multiplication fact. But each one uses an easier fact that he or she does know.

Ryan writes 8 as a sum and then uses the distributive property. He thinks:

9 × 8 = 9 × (5 + 3)
       = (9 × 5) + (9 × 3)
       = 45 + 27
       = 72

Belinda thinks:

9 × 8 = (10 × 8) − 8
       = 80 − 8
       = 72

Kevin uses doubles. He thinks:

9 × 8 = (8 × 8) + 8
       = 64 + 8
       = 72

They will receive 72 T-shirts.

1. In what other ways can you find the product?

TRY OUT Write the letter of the correct answer. Which is the correct number sentence?

2. 6 × 7 = ■   **a.** (6 × 6) + 7 = ■   **c.** (6 + 6) × 6 = ■
                b. (5 × 7) + 7 = ■   **d.** (5 × 7) + 6) = ■

3. 8 × 6 = ■   **a.** (8 × 6) + (1 × 6) = ■   **c.** (4 × 6) + (4 × 6) = ■
                b. (3 × 6) + (3 × 6) = ■   **d.** (7 × 6) + (6 × 6) = ■

PRACTICE

Find the product.

4. 6 ×5	5. 7 ×3	6. 4 ×8	7. 3 ×5	8. 8 ×0	9. 7 ×5	10. 4 ×6
11. 8 ×3	12. 3 ×3	13. 2 ×9	14. 7 ×7	15. 1 ×9	16. 5 ×4	17. 4 ×7
18. 8 ×8	19. 6 ×9	20. 7 ×6	21. 5 ×8	22. 9 ×7	23. 9 ×9	24. 8 ×7
25. 3 ×9	26. 6 ×8	27. 5 ×5	28. 6 ×6	29. 8 ×6	30. 9 ×5	31. 8 ×9

32. 4 × 9   33. 8 × 5   34. 3 × 9   35. 2 × 7

36. 2 × 9 × 3   37. 4 × 3 × 2   38. 7 × 2 × 2   39. 3 × 5 × 2

Compare. Use >, <, or =.

40. 9 × 5 ● 14 + 28   41. 6 × 4 ● 12 + 6   42. 7 × 9 ● 52 + 10

43. 24 + 32 ● 7 × 8   44. 45 + 18 ● 8 × 9   45. 24 + 18 ● 6 × 7

Complete.

46. Rule: × 7

Input	Output
3	■
7	■
6	■
8	■

47. Rule: × 9

Input	Output
5	■
4	■
9	■
2	■

48. Rule: ■

Input	Output
6	18
4	12
7	21
3	9

49. Rule: ■

Input	Output
4	12
8	16
5	13
9	17

Mixed Applications

50. Tanya has $10. She buys a softball for $6. How much money does she have left?

51. Mark spends $7 for a bat and $9 for a glove. How much does he spend in all?

52. Ellen buys 3 baseball caps for her brothers. Each cap costs $4. How much does she spend for the caps?

53. **Write a problem** that can be solved using the multiplication sentence 6 × 7 = ■.

UNDERSTANDING A CONCEPT
Mental Math: Using Multiplication Patterns

A. Amy, Ben, and Cathy used patterns to multiply large numbers. They found the products below.

Amy	Ben	Cathy
8 × 1 = 8	3 × 2 = 6	5 × 4 = 20
8 × 10 = 80	3 × 20 = 60	5 × 40 = 200
8 × 100 = 800	3 × 200 = 600	5 × 400 = 2,000
8 × 1,000 = 8,000	3 × 2,000 = 6,000	5 × 4,000 = 20,000
8 × 10,000 = 80,000	3 × 20,000 = 60,000	5 × 40,000 = 200,000

B. You can use a similar pattern to multiply when both factors are multiples of 10, 100, and 1,000.

4 × 7 = 28
40 × 70 = 2,800
40 × 700 = 28,000
40 × 7,000 = 280,000
400 × 700 = 280,000

1. What pattern could you have used to find the products mentally?

C. Sometimes you can use patterns and multiplication properties to help you multiply mentally.

Ewen and Fran are trying to find 5 × 9 × 2.

Ewen starts to multiply mentally. Then he uses his calculator.

5 × 9 × 2
45 [×] 2 [=] | 90. |

Fran uses multiplication properties to rearrange the factors. Then she multiplies mentally.

5 × 9 × 2
5 × 2 × 9 ← commutative property
(5 × 2) × 9 ← associative property
10 × 9 = 90

2. How could you find the product of 5 × 3 × 8 mentally?

TRY OUT Write the letter of the correct answer.

3. 7 × 10,000   **a.** 7,000   **b.** 70,000   **c.** 700,000   **d.** 7,000,000

4. 6 × 500   **a.** 300   **b.** 3,000   **c.** 30,000   **d.** 300,000

5. 50 × 4,000   **a.** 2,000   **b.** 20,000   **c.** 200,000   **d.** 2,000,000

6. 6 × 7 × 5   **a.** 210   **b.** 2,100   **c.** 2,110   **d.** 21,000

Practice

Find the product mentally.

7. 3 × 100
8. 9 × 10
9. 5 × 1,000
10. 5 × 200
11. 6 × 400
12. 8 × 70
13. 9 × 10,000
14. 3 × 7,000
15. 8 × 5,000
16. 30 × 60
17. 40 × 80
18. 50 × 800
19. 400 × 50
20. 80 × 6,000
21. 3,000 × 90
22. 600 × 900
23. 700 × 800
24. 600 × 500
25. 6 × 8 × 5
26. 5 × 9 × 8
27. 4 × 6 × 5
28. 6 × 5 × 4 × 2
29. 4 × 5 × 5 × 8
30. 2 × 5 × 7 × 5
31. 20 × 30 × 50
32. 40 × 60 × 50
33. 50 × 70 × 60

Compare. Use >, <, or =.

34. 20 × 60 ● 2 × 600
35. 20 × 7,000 ● 700 × 200
36. 300 × 50 ● 1,500

Mixed Applications

37. The Melendez Trio gave two concerts at the city stadium last weekend. On Saturday 3,176 people attended the concert and on Sunday 5,378 people attended. How many people attended last weekend?

38. Gary Wade's new movie was seen by 2,000 people in the giant Century Theater on opening night. If each person paid $8, how much money did the theater take in?

Mixed Review

Find the answer. Which method did you use?

39. 7,500
    × 800

40. $40.05
    − 12.98

41. 7.392
    + 4.005

42. 643
    × 121

43. 72
    × 50

MENTAL MATH
CALCULATOR
PAPER/PENCIL

UNDERSTANDING A CONCEPT
Estimating Products

A. The Mountain Middle School has 368 students. Each student is given 8 pens to use during the school year. About how many pens should the principal order?

Estimate: 8 × 368

Here are two ways to estimate the product:

Front-End Estimation
Use the front digits.

$$\begin{array}{r} 368 \\ \times\ \ 8 \\ \hline \end{array}$$ ***Think:*** $$\begin{array}{r} 300 \\ \times\ \ \ 8 \\ \hline 2{,}400 \end{array}$$

The estimate is 2,400.

Rounding
Round 368 to the nearest hundred.

$$\begin{array}{r} 368 \\ \times\ \ 8 \\ \hline \end{array}$$ ***Think:*** $$\begin{array}{r} 400 \\ \times\ \ \ 8 \\ \hline 3{,}200 \end{array}$$

The estimate is 3,200.

1. Which method results in an estimate that is less than the exact answer? greater than the exact answer? Why?

2. Should the principal order 2,400 or 3,200 pens? Why?

B. You can use both front-end estimation and rounding to estimate products when both factors have two or more digits.

Estimate: 18 × 375

Front-End Estimation
Use the front digits.

18 × 375
↓   ↓
Think: 10 × 300 = 3,000

Rounding
Round each factor to its greatest place.

18 × 375
↓   ↓
Think: 20 × 400 = 8,000

3. Estimate 83 × 2,187 by front-end estimation and rounding. Compare the estimates. What did you find?

TRY OUT

Estimate. Use front-end estimation.

4. 8 × 26   **5.** 32 × 421

Estimate. Use rounding.

6. 33 × $5.15   **7.** 729 × 863

PRACTICE

Estimate. Use front-end estimation.

8.	612 × 7	9.	909 × 5	10.	3,419 × 3	11.	2,310 × 6	12.	$5.23 × 4
13.	$7.43 × 3	14.	$42.50 × 9	15.	$18.45 × 8	16.	31 ×65	17.	87 ×11

18. 93 × 85   19. 54 × 62   20. 57 × 439   21. 84 × 526
22. 24 × 686   23. 75 × 823   24. 5 × $27.90   25. 31 × 792

Estimate. Use rounding.

26.	471 × 7	27.	553 × 9	28.	4,225 × 8	29.	9,034 × 2	30.	$3.76 × 6
31.	$2.78 × 9	32.	$97.89 × 4	33.	$38.66 × 5	34.	78 ×63	35.	56 ×72

36. 37 × 68   37. 73 × 19   38. 47 × 439   39. 98 × 204
40. 82 × 475   41. 21 × 685   42. 8 × $38.60   43. 46 × 875

Mixed Applications

44. The smallest school in the Free City School District has 198 students. The largest school has 632 students. About how many more students are in the largest school?

45. Ari wants to buy a crafts kit that costs $35. He earns $6.75 per hour baby-sitting. He estimates he will have enough money for the kit after working 5 hours. Is he right? Why?

46. The Valley School's library has 2,469 books. The Hilltop School's library has 3,769 books. When the schools are combined, about how many books will be in the new school's library?

47. An art supply kit costs $67.75 per student per year. Is $2,200 enough to supply an art class of 28. How do you know?

UNDERSTANDING A CONCEPT
Multiplying by a 1-Digit Number

A. There are 235 students in the Fourth Street School. The cafeteria manager orders 1 pint of milk for each student each day. How many pints of milk does he order for each 5-day school week?

Multiply: 5 × 235

Step 1	Step 2	Step 3
Multiply the ones. Regroup if necessary.	Multiply the tens. Add any new tens. Regroup if necessary.	Multiply the hundreds. Add any new hundreds. Regroup if necessary.
2 235 × 5 ――― 5	1 2 235 × 5 ――― 75	1 2 235 × 5 ――― 1,175
Think: 25 ones = 2 tens 5 ones	***Think:*** 5 × 3 tens = 15 tens 15 tens + 2 tens = 17 tens	***Think:*** 10 hundreds + 1 hundred = 11 hundreds

The cafeteria manager orders 1,175 pints of milk for each school week.

1. Is the exact answer reasonable? How do you know?

B. Be especially careful when there is a 0 in the factor.

Multiply: 7 × 806

Step 1	Step 2	Step 3
Multiply the ones. Regroup if necessary.	Multiply the tens. Add any new tens. Regroup if necessary.	Multiply the hundreds. Add any new hundreds. Regroup if necessary.
4 806 × 7 ――― 2	4 806 × 7 ――― 42	4 806 × 7 ――― 5,642
Think: 42 ones = 4 tens 2 ones	***Think:*** 7 × 0 tens = 0 tens 0 tens + 4 tens = 4 tens	***Think:*** 7 × 8 hundreds = 56 hundreds

2. How would you multiply a 4-digit number by a 1-digit number?

3. How does knowing that 5 × 235 = 1,175 help you find 5 × $2.35?

TRY OUT
Multiply.

4. 4 × 318
5. 3 × 2,107
6. 7 × $40.35
7. 6 × 247

PRACTICE
Multiply.

8. 29 × 7
9. 36 × 2
10. 741 × 8
11. 463 × 5
12. 985 × 9

13. 150 × 6
14. 905 × 8
15. 204 × 7
16. $6.09 × 8
17. 5,020 × 4

18. 2,573 × 3
19. 246 × 8
20. $10.62 × 5
21. 7,186 × 9
22. 4,205 × 6

23. 43,120 × 5
24. 17,088 × 3
25. 60,825 × 7
26. $251.43 × 9
27. 94,587 × 8

28. 3 × 76
29. 7 × 5,109
30. 8 × 29,460
31. 4 × $12.56

Do only those exercises with products less than 5,000.

32. 6 × 798
33. 4 × 1,709
34. 3 × 949
35. 9 × 589
36. 5 × 1,066
37. 7 × 822
38. 8 × 599
39. 8 × 498

Mixed Applications

40. Last year there were 4,235 students in New City. This year there are 5,189 students. How many more students are there this year?

41. A cook in the school cafeteria makes 250 gallons of soup each week. How many gallons of soup does he make in 8 weeks?

42. Bread for students' sandwiches costs $215.50 per month. How much is spent on bread in 6 months?

43. **Write a problem** that can be solved using the multiplication sentence 4 × $25.95 = ■. Ask others to solve your problem.

UNDERSTANDING A CONCEPT
Multiplying by a 2-Digit Number

A. Simms Sound Store orders 30 cartons of tapes. Each carton contains 48 tapes. How many tapes does the store order in all?

Multiply: 30 × 48

Step 1	Step 2
Multiply by the ones.	**Multiply by the tens.**
48 × 30 —— 0	2 48 × 30 —— 1,440
Think: 0 × 48 = 0	*Think:* 3 tens × 48 = 144 tens

The store orders 1,440 tapes in all.

1. Is 1,440 a reasonable answer? How do you know?
2. How would you multiply 40 × 72?

B. You can use these steps to multiply by 2-digit numbers.

Multiply: 25 × 36

Step 1	Step 2	Step 3
Multiply by the ones.	**Multiply by the tens.**	**Add the products.**
3 36 × 25 —— 180 ← 5 × 36	1 3̶ 36 × 25 —— 180 720 ← 20 × 36 Cross out before multiplying.	1 3̶ 36 × 25 —— 180 + 720 —— 900

TRY OUT   Write the letter of the correct answer.

3. 80 × 36   **a.** 288   **b.** 2,448   **c.** 2,480   **d.** 2,880
4. 24 × 19   **a.** 114   **b.** 240   **c.** 356   **d.** 456
5. 49 × $63   **a.** $819   **b.** $3,087   **c.** $3,187   **d.** $3,287

PRACTICE

Multiply.

6. 28 × 30	**7.** 45 × 70	**8.** 37 × 50	**9.** 39 × 25	**10.** 61 × 52	**11.** 21 × 14
12. 42 × 90	**13.** 37 × 24	**14.** 94 × 36	**15.** $26 × 72	**16.** 59 × 29	**17.** 61 × 40
18. 98 × 80	**19.** 64 × 72	**20.** 75 × 27	**21.** 39 × 65	**22.** $48 × 79	**23.** 53 × 30

24. 18 × 42   **25.** 85 × 64   **26.** 60 × 59   **27.** $72 × 98

28. 3 × 9 × 46   **29.** 4 × 6 × 18   **30.** 7 × 6 × 21   **31.** 8 × 6 × 33

Critical Thinking Do you agree or disagree with the following statement? Give examples to support your answer.

32. The product of a 2-digit number and a 2-digit number may be a 2-digit number.

33. The product of a 2-digit number and a 2-digit number may be a 5-digit number.

Mixed Applications

34. A clock radio is priced at $32. Simms puts it on sale for $28. How much will you save by buying the clock radio on sale?

35. Adarsh unpacks 15 cartons of cassette players. Each carton has 45 players. How many players does Adarsh unpack?

MENTAL MATH

Breaking factors apart can help you multiply mentally.

12 × 25 = (10 × 25) + (2 × 25)   or   12 × 25
       = 250 + 50   = 3 × 4 × 25
       = 300   = 3 × 100
                                                          = 300

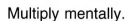

Multiply mentally.

1. 16 × 15   **2.** 35 × 14   **3.** 25 × 28   **4.** 18 × 45

PROBLEM SOLVING

✓ UNDERSTAND
  PLAN
  TRY
  CHECK
✓ EXTEND

Finding Needed Information

Leah reads in a newspaper that last year about 30,192,000 people used New York's John F. Kennedy Airport. She knows that New York's other airport, La Guardia Airport, is not as busy as Kennedy. She wonders how many more people used Kennedy Airport than La Guardia Airport.

1. What information does Leah know?

2. What does she want to find out?

Sometimes there is not enough information given to solve the problem. You need to look up additional information.

3. What additional information does Leah need? Where might she find this information?

Leah looked in an almanac and found that last year about 24,226,000 people used La Guardia Airport.

4. Does she have enough information to solve the problem now?

5. About how many more people used John F. Kennedy Airport than La Guardia Airport?

6. **What if** Leah wanted to fly to New York City? How would this information help her decide which airport to fly to? What other information might she want to know?

PRACTICE

If there is not enough information to solve the problem, list what additional information is needed. If there is enough information, solve the problem.

7. The first crossing of the Atlantic Ocean in an airplane took place in 1919. The plane was in the air for 54 hours. It flew at an average speed of 87 miles per hour. What was the distance of the flight?

8. Orville Wright, the first man to fly an airplane, was born in 1871. How old was he at his death?

9. The Boeing 747-SP can seat 331 passengers. The Boeing 707-320B is a smaller aircraft. How many fewer passengers can it seat?

10. The highest altitude attained by an unpiloted balloon was about 170,000 feet and by a piloted balloon about 123,800 feet. How much higher was the altitude of the unpiloted balloon?

Strategies and Skills Review

Solve. Use mental math, estimation, a calculator, or paper and pencil.

11. Ruth is flying from Dallas to Boston. She pays $8.95 for a van to take her to the airport in Dallas and $18.50 for a cab to get from the Boston airport to her hotel. The plane fare is $159. She estimates that she is not spending more than $190 on transportation. Is she correct? Why?

12. An airline makes three flights a day from Houston to Chicago. Today there were 369 passengers on the first flight, 431 on the second flight, and 298 on the third. How many people flew from Houston to Chicago today on that airline?

13. The air distance from New York to London is 3,469 miles. The distance from New York to Hong Kong is 8,060 miles. How much greater is the distance from New York to Hong Kong than from New York to London?

14. The Boeing 747 and the McDonnell-Douglas DC8 are both four-engine jets. The 747 has a wing span of 195.7 feet. The wing span of the DC8 is 53.5 feet less. What is the wing span of the DC8?

15. Dan builds model airplanes. He sells them for $16 each. How much money will he make if he sells 9 model planes?

16. **Write a problem** that has a missing fact needed to solve the problem. Ask another student to tell what additional information is needed.

Investigating Patterns

A. Suppose you decide on the following savings plan. On the first day of the month you will save 1¢, on the second day 2¢, on the third day 4¢, on the fourth day 8¢, and so on. Each day you double the amount you save.

How much will you save on the twelfth day?

You can solve this problem two ways with a calculator. The first way:

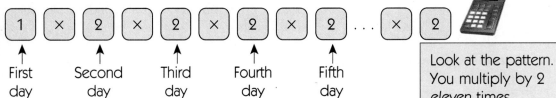

Look at the pattern. You multiply by 2 *eleven* times.

1. Try it. What is the answer?

2. How many times would you multiply by 2 to find the amount saved on the twentieth day?

   The second way to do the problem is by using the constant feature of your calculator for repeated multiplication. Experiment with your calculator. See if it makes the first number the constant,

   ② × ① = = = = = ... =

   or the second number the constant.

   ① × ② = = = = = ... =

3. How many times will you press the = key to find the amount saved on the twelfth day?

4. How many times will you press the = key to find the amount saved on the twenty-fifth day?

5. Which of the two ways do you prefer? Why?

6. Suppose you decide to save 1¢ on the first day and then triple the amount each day thereafter (1¢, 3¢, 9¢ . . .). How much will you save on the tenth day? How did you find the answer?

Thinking MATHEMATICALLY

B. Here is a game to play with a partner. You need one calculator.

The only move a player can make is to enter 1 + or 2 +. Take turns. The winner is the first player to get 20 in the display.

For example, a game might start off like this:

	Move	Display
First player	1+	1
Second player	2+	3
First player	2+	5
Second player	1+	6
First player	2+	8
.	.	.
.	.	.

7. Play the game a few times. Think about how to win.

8. In order for you to win, what should the display show after your next-to-last move? Why?

9. In order to get the calculator to display the number 17, what should you get on the display one turn earlier? Why?

10. Suppose you are the first player. What number should you start with to be sure of winning?

11. Suppose the only moves allowed are 1 +, 2 +, 3 +, and 4 +. The winner is the first person to get 30. Can you find a way to win if you are the first player? Why or why not? What if you are the second player?

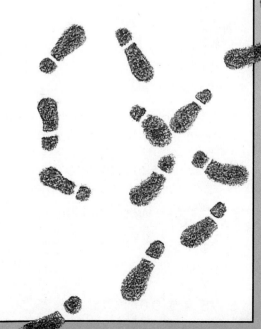

153

UNDERSTANDING A CONCEPT
More Multiplying by a 2-Digit Number

A. An astronomy museum gives special shows in its auditorium. One show was sold out for 28 days in a row. How many tickets did the museum sell if the auditorium seats 645 people?

Multiply: 28 × 645

Step 1	Step 2	Step 3
Multiply by the ones.	**Multiply by the tens.**	**Add the products.**
3 4 645 × 28 ――― 5,160 ← 8 × 645	1 3̸ 4̸ 645 × 28 ――― 5160 12,900 ← 20 × 645	1 3̸ 4̸ 645 × 28 ――― 5160 12900 ――― 18,060

The museum sold 18,060 tickets.

1. If 28 × 645 = 18,060, what does 28 × $6.45 equal?

B. Be especially careful when there is a 0 in the factor.

Multiply: 72 × 304

```
   2
  304
×  72
 ―――
  608
21280
 ―――
21,888
```

2. How would you multiply a 4-digit number by a 2-digit number?

3. Multiply 28 × 5,020.

TRY OUT   Write the letter of the correct answer.

4. 47 × 618   **a.** 6,798   **b.** 29,046   **c.** 30,246   **d.** 31,046

5. 56 × 7,034   **a.** 77,374   **b.** 393,904   **c.** 394,904   **d.** 396,104

6. 75 × $10.07 **a.** $80.25   **b.** $120.48   **c.** $755.25   **d.** $1,180.95

PRACTICE

Multiply.

| 7. 147 × 23 | 8. 276 × 19 | 9. 353 × 64 | 10. 595 × 28 | 11. $862 × 86 |

| 12. 205 × 12 | 13. 504 × 17 | 14. 770 × 36 | 15. 930 × 50 | 16. $8.06 × 42 |

| 17. 5,241 × 36 | 18. 4,058 × 24 | 19. 2,806 × 75 | 20. $74.45 × 25 | 21. 9,050 × 19 |

22. 83 × 456 **23.** 27 × 903 **24.** 52 × 638 **25.** 12 × 863

26. 18 × $4.30 **27.** 64 × $12.98 **28.** 32 × 6,008 **29.** 27 × 6,425

Do only those exercises with products greater than 10,000.

30. 98 × 99 **31.** 35 × 509 **32.** 26 × 298 **33.** 217 × 625

34. 27 × 450 **35.** 42 × 370 **36.** 47 × 183 **37.** 52 × 175

Compare. Use >, <, or =.

38. 52 × 80 ● 64 × 64 **39.** 72 × 48 ● 36 × 96

40. 18 × 512 ● 7,708 + 4,508 **41.** 41 × 605 ● 18,796 + 5,768

42. 23 × 432 ● 15,430 − 2,386 **43.** 58 × 298 ● 14,372 − 2,106

Mixed Applications

Solve. You may need to use the Databank on page 518.

44. The museum store sells telescopes for $175. During the special show it sells 32 telescopes. How much money does the museum make from telescope sales?

45. Last Saturday 1,587 people visited the museum. On Sunday there were 1,938 visitors. How many more visitors came on Sunday?

46. Wanda buys an astronomy book for $10.79 and a set of postcards for $3.75. How much money does Wanda spend in all?

47. Ms. Graves buys 18 copies of *Beyond Our Galaxy* for the Astronomy Club. How much do the books cost in all?

UNDERSTANDING A CONCEPT

Multiplying Large Numbers

A. Mr. Selig's class visited a bicycle factory. They found out that the factory had produced 435 bicycles each day for 305 days. What was the total number of bicycles the factory produced?

Multiply: 305 × 435

Two students did this multiplication in different ways.

Juanita	Josh
435	435
× 305	× 305
2 175 ← 5 × 435	2 175 ← 5 × 435
0 000 ← 0 × 435	130 500 ← 300 × 435
130 500 ← 300 × 435	132,675
132,675	

The factory produced a total of 132,675 bicycles.

1. How are the students' methods similar? How are they different?

2. Why could Josh omit multiplying by 0 tens?

B. *What if* the factory had produced 425 bicycles in 328 days? How many bicycles would the factory have produced then?

Multiply: 328 × 425

You can use a calculator or paper and pencil to solve.

Calculator

328 ⊗ 425 ⊜ | *139400.* |

Paper and Pencil

```
    425
  × 328
  3 400
  8 500
127 500
139,400
```

The factory would have produced 139,400 bicycles.

3. *What if* you entered 328 ⊗ 452 for the problem above and saw | *148256.* |? How would you know that you made an error?

156 Lesson 4-10

TRY OUT   Multiply. Use a calculator or paper and pencil.

4. 546 × 738
5. 202 × 140
6. 450 × 4,608
7. 671 × 8,325

PRACTICE

Multiply. Use a calculator or paper and pencil.

8. 600 × 300
9. 364 × 219
10. 240 × 800
11. 749 × 503
12. 428 × 646

13. 340 × 900
14. 817 × 225
15. 642 × 357
16. $4.05 × 210
17. $8.00 × 600

18. 3,562 × 475
19. 5,080 × 120
20. 1,592 × 314
21. $40.79 × 500
22. $39.95 × 325

23. 546 × 738
24. 202 × 140
25. 450 × 4,608
26. 754 × 5,627

27. 300 × 219
28. 623 × 145
29. 300 × 500
30. 850 × 917

Mixed Applications   Solve. Which method did you use?

31. Ben's Bike Shop sold 498 five-speed bicycles last year for $128 each. How much money did it receive for these bicycles?

32. Ben ordered 350 three-speed bicycles and 275 ten-speed bicycles. How many more three-speed bicycles than ten-speed bicycles did Ben order?

33. Last year chain locks cost $25.79. This year they cost $32.50. How much has the price increased?

MENTAL MATH
CALCULATOR
PAPER/PENCIL

Mixed Review

Write the decimal.

34. three and nine tenths
35. six hundredths
36. two and eight thousandths
37. forty-one thousandths

Write the value of the underlined digit.

38. 6.<u>8</u>
39. 1.07<u>9</u>
40. 0.1<u>3</u>2
41. 3.<u>0</u>04
42. 0.08<u>5</u>

EXTRA Practice, page 172   Multiplying Whole Numbers   157

PROBLEM SOLVING

✓ UNDERSTAND
✓ PLAN
✓ TRY
✓ CHECK
✓ EXTEND

Strategy: Solving a Multistep Problem

The fifth-grade students at Turner School plan to have a Thanksgiving party for the whole school. Joyce is in charge of napkins. She will need 775 napkins. There are 280 napkins left from last year's party, and she has bought 13 packages of napkins with 25 napkins in each package. Joyce needs to know how many more napkins she must buy.

1. What information does Joyce know?

2. What does she need to find out?

Sometimes it takes several steps to solve a problem. When there is more than one step, it helps to make a plan.

3. What does Joyce need to find out before she can decide how many more napkins to buy?

4. What can she do to find out how many napkins she already has?

Joyce tries the following plan.

PLAN

Step 1 $13 \times 25 = 325$

Step 2 $325 + 280 = 605$

Step 3 $775 - 605 = 170$

5. What did she find first?

   What did she find next?

   What did she find last?

6. How many more napkins should she buy?

158 Lesson 4-11

PRACTICE

Solve the problem. Show the steps you used.

7. Claire buys 3 packages of paper plates for $1.98 a package. She also buys cups for $6.25 and forks for $4.90. She gives the cashier $20. How much change should she receive?

8. Jay buys 18 grape juice packs with 6 cans in each pack and 15 orange juice packs with 8 cans in each pack. How many juice cans does he buy in all?

9. A group of 7 students went trick-or-treating for 45 minutes. The students stopped at one house every 5 minutes. Each student received 3 treats at each house. How many treats did the students receive in all?

10. Students at the Turner School collected money for charity. Grade 3 collected $334.96. Grade 4 collected $75.63 less than grade 3, and grade 5 collected $51.49 more than grade 4. How much money did the students collect in all?

Strategies and Skills Review

Solve. Use mental math, estimation, a calculator, or paper and pencil.

11. The PTA hopes to sell at least 250 tickets to the party. The members sold 56 tickets the first week, 68 the second week, and 75 the third week. Will they meet their goal if they sell 83 tickets the fourth week? How would you use estimation to find out?

12. The students rehearsed for the Thanksgiving play for 45 minutes each day on Monday, Wednesday, and Friday and for 30 minutes on Tuesday and Thursday. How many minutes did they rehearse in all?

13. Laurie bought 16 packages of pipe cleaners to use in making party favors. There are 32 pipe cleaners in each package. She made 119 favors. How many pipe cleaners did she buy?

14. Jamie is making 15 quarts of punch for the party. She mixes grape juice and apple juice. What other information do you need to find how many quarts of apple juice she uses?

15. Sam inflated 153 balloons for the party. If 27 balloons broke during the party, how many balloons were left after the party?

16. *Write a problem* that takes several steps to solve. Solve the problem. Ask others to solve your problem.

EXTRA Practice, page 172

UNDERSTANDING A CONCEPT
Perimeter

A. Rima is making a quilt for the Folk Art Fair. She wants to sew trimming around the edge of the quilt. The quilt is a rectangle with a length of 3 m and a width of 2 m. How much trimming does Rima need?

The **perimeter** of a figure is the distance around the figure.

You can add to find the perimeter.

Perimeter = 3 + 2 + 3 + 2
          = 10

You can also multiply and then add.

Perimeter = 2 × length + 2 × width
$P = (2 \times \ell) + (2 \times w)$
    = (2 × 3) + (2 × 2)
    =   6   +   4
    =   10

Rima needs 10 m of trimming.

1. What do you know about the length of the opposite sides of a rectangle that allows you to write the formula $P = (2 \times \ell) + (2 \times w)$?

B. You can use this formula to find the perimeter of a square with sides of length s.

What if Rima's quilt were a square with sides 3 m long? What would be the perimeter of the quilt?

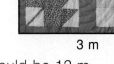

Perimeter = 4 × length of side
          = 4 × s
          = 4 × 3
          = 12   The perimeter of the square quilt would be 12 m.

2. What do you know about the length of the sides of a square that allows you to write the formula $P = 4 \times s$?

TRY OUT Write the letter of the correct answer. Find the perimeter.

3.
16 mm square, 16 mm
  a. 16 mm
  b. 32 mm
  c. 48 mm
  d. 64 mm

4. Rectangle, 14 cm by 22 cm
  a. 36 cm
  b. 72 cm
  c. 50 cm
  d. 58 cm

PRACTICE

Find the perimeter.

5.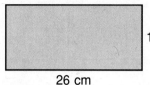
6.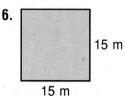
7.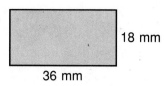

8. a rectangle with length of 10 m, width of 8 m
9. a square with sides of 12 cm
10. a rectangle with length of 24 mm, width of 18 mm

Measure the sides to the nearest millimeter. Then find the perimeter.

11.
12.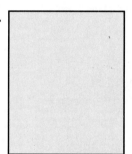
13.

Find the missing length in each rectangle.

14.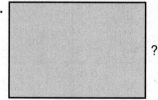
Perimeter = 28 m

15.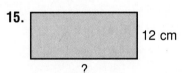
Perimeter = 74 cm

16.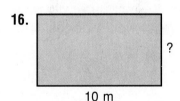
Perimeter = 30 m

Mixed Applications

17. Carla is knitting a baby blanket that is a rectangle 90 cm by 120 cm. How much ribbon will she need to trim the edge?

18. Vera buys 12 m of quilting material. The material costs $6.98 a meter. How much money does Vera spend for the material?

19. Sam is making a frame for one of Rima's quilts. The quilt is a square with sides of 260 cm. How much framing does Sam need?

20. There are 45 quilts in the Folk Art Fair this year. This is 18 more quilts than last year. How many quilts were shown last year?

UNDERSTANDING A CONCEPT
Area

A. The supply room for the art classes is a 5-m by 6-m rectangle. What is the area of the floor?

The **area** of a figure is the number of square units enclosed by the figure. Area is often measured in **square centimeters** (cm^2), **square meters** (m^2), or **square millimeters** (mm^2).

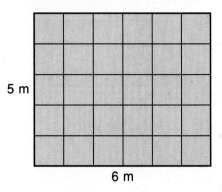

You can count the square meters to find the area of the floor. You can also use the formula at the right.

Area = length × width
= ℓ × w
= 6 × 5
= 30

The area of the floor is 30 m^2.

1. What do the length and width tell about the rows of the rectangular floor?

B. You can use this formula to find the area of a square with sides of length s.

What if the floor of the supply room were a square with sides 5 m long? What would be the area of the floor?

Area = s × s
= 5 × 5
= 25

The area of the floor would be 25 m^2.

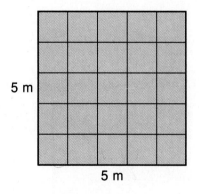

2. How is the formula for finding the area of a square similar to the formula for finding the area of a rectangle? How is it different?

TRY OUT
Write the letter of the correct answer. Find the area.

3. 12 cm × 12 cm
   a. 24 cm^2
   b. 48 cm^2
   c. 144 cm^2
   d. 240 cm^2

4. 5 mm × 15 mm
   a. 20 mm^2
   b. 40 mm^2
   c. 60 mm^2
   d. 75 mm^2

PRACTICE

Find the area.

5.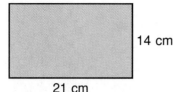
14 cm
21 cm

6.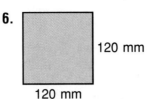
120 mm
120 mm

7.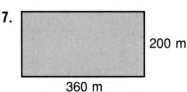
200 m
360 m

8. a rectangle with length of 12 m, width of 10 m

9. a square with sides of 15 cm

10. a rectangle with length of 122 mm, width of 70 mm

Measure the sides to the nearest millimeter. Then find the area.

11.

12.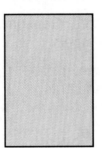

13.

Critical Thinking Use graph paper to solve the problem.

14. If the length of a rectangle increased from 5 cm to 6 cm, by how much would the area change?

15. Draw all the different rectangles that have a perimeter of 20 units. Which one has the greatest area?

16. Repeat the activity in Problem 15 for rectangles that have a perimeter of 28 units and 36 units. What can you conclude?

Mixed Applications

Solve. You may need to use the Databank on page 518.

17. Thomas wove a rug that is a 120-cm by 250-cm rectangle. What is the area of the rug?

18. Paul wants to frame a poster that is an 80-cm by 125-cm rectangle. How much framing does he need?

19. Su made a square painting that has 125-cm sides. How many square centimeters of canvas did she paint?

20. Ms. Lennon ordered 48 copies of *Art in Early America* for her art classes. What is the total cost of the books?

DEVELOPING A CONCEPT

Circumference

All points on a circle are the same distance from the **center** of the circle.

A **chord** is a line segment with endpoints on the circle.

A **diameter** is a chord that passes through the center of the circle.

A **radius** is a line segment with one endpoint at the center of the circle and the other endpoint on the circle.

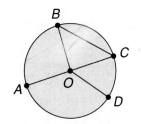

WORKING TOGETHER

Collect circular objects of different sizes. Get some string and a centimeter ruler or measuring tape.

1. Measure the radius and the diameter of each object. Record the measurements in a table like this:

Object	Radius (r)	Diameter (d)

2. Look for a pattern in the table. How would you describe the relationship between a radius and diameter of a circle?

The distance around a circle is called the **circumference.** Add these columns to your table.

Circumference (C)	C ÷ d

3. Measure the circumference of each object in your table. Record the circumference in the table.

4. Use a calculator to find $C \div d$ for each object. Record the number in the table.

SHARING IDEAS

5. Compare the numbers you get for C ÷ d with those of others. What do you notice about the numbers?

6. About how many times greater is the circumference than the diameter?

7. Write a formula to find the circumference of a circle. Can you use an equal sign? Why?

PRACTICE

Use your formula to find the approximate circumference.

8. 10 m

9. 75 mm

10. 110 cm

11. 39 cm

12. 309 mm

13. 9 m

14. a circle with diameter of 120 cm

15. a circle with diameter of 14 m

16. a circle with radius of 75 mm

Measure the diameter or radius to the nearest millimeter. Then find the approximate circumference.

17.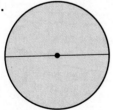

18.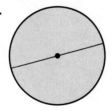

19.

Mixed Applications

20. There is a circular running path in New City Park. The diameter of the circle is 350 m. What is the approximate length of one lap around the path?

21. The entire park is a rectangle with length of 972 m and width of 788 m. If you jogged once around the park, how far would you jog?

DECISION MAKING

COOPERATIVE LEARNING

Problem Solving: Planning a Day Trip

SITUATION

The students in Mrs. Green's class are planning a day trip. The students must decide whether to visit City Hall or to go see a play. They have collected some information about each trip.

PROBLEM

Where should the students go?

DATA

Day Trip to City Hall

Travel time -- 45 minutes each way

Walking tour of historic district around City Hall -- 1½ hours

Tour of City Hall -- every hour on the hour from 10:00 a.m. to 4:00 p.m.

Lunch in park -- 30 minutes

CITY HALL TOURS

Day Trip to the Theater

PLAY TO SEE

"The Girl Who Wanted To Sing"
Musical. Story about a young girl with musical talent who became a famous singer. Wonderful cast!

Starts at 12 noon. Ends at 2:30 p.m.

"Young George Washington"
Drama. Story based on events in the life of the first president when he was a boy. Well done! Don't miss!

Starts at 1:00 p.m. Ends at 3:00 p.m.

"How the Old West Was"
Historical Play. A history of how the western part of the United States was settled. Newest hit!

Starts at 2:00 p.m. Ends at 4:00 p.m.

Travel time - 1½ hours each way
Lunch at local restaurant - 45 minutes

USING THE DATA

1. What is the round-trip travel time to:
   a. City Hall?   b. the theaters?

2. How long is the:
   a. musical?   b. historical play?   c. drama?

3. How long is the day trip to City Hall, including time for lunch?

4. How long is the day trip to see the drama including lunch?

5. If the students eat lunch before attending the play, what time should they leave to go to the:
   a. musical?   b. historical play?   c. drama?

MAKING DECISIONS

6. The students are studying about the local history of their city. Where should they decide to go?

7. The students want to leave school as early as possible and return as close to 3:00 P.M. as possible. Where should they decide to go?

8. **What if** the students want to spend as much time outdoors as possible? Where should they go? What problems might they have?

9. **Write a list** of some other things the students should think about when deciding where to go for their class trip.

10. Which trip would you choose? Why?

Curriculum Connection

Math and Science

Scientists know that sound travels at about 4,900 feet per second in sea water. They use this knowledge and a technique called *echo location,* or sonar, to measure ocean depth. The word *sonar* is an acronym, a word made by combining the first letter or letters of several words (SOund NAvigation Ranging). Sonar can measure the time it takes for sound waves to travel from a ship to the ocean floor and back.

What if a sonar sounding takes 8 seconds? What depth would this indicate in feet?

Think: The total elapsed time is 8 seconds. This means it took 4 seconds to reach the ocean floor and 4 seconds to return.

Multiply the time it takes the sound wave to travel one way—down to the ocean floor—by the rate of speed (about 4,900 ft per second in sea water).

4 [×] 4900 [=] | 19600. |

A sonar sounding of 8 seconds indicates a depth of 19,600 ft.

ACTIVITIES

1. Use an encyclopedia or almanac to find the greatest known ocean depth. Then determine about how many seconds its sonar sounding would take.

2. Find out about other methods for measuring ocean depth. Write a short report for the class.

Calculator: Finding Factors

Baseball fever is spreading throughout the solar system. On Mars, 635 of the best players are trying out for the new Martian League. League officials want the number of teams to be a multiple of 8. If a team will consist of 24 players, what is the greatest number of teams that can be formed from this pool of players?

You can use a calculator and the Guess, Test, and Revise strategy to find out. First write a number sentence with a missing factor.

■ × 24 is less than 635

Then guess what multiple of 8 can replace ■.

| 16 × 24 = 384 | 24 × 24 = 576 | 32 × 24 = 768 |
| OK, but try 24. | OK, but try 32. | Too much. |

The greatest multiple of 8 that works is 24. Form 24 teams.

USING THE CALCULATOR

One of the factors is missing in each number sentence below. Use a calculator to find the greatest multiple of 8 that will replace ■ to make each statement true.

1. ■ × 6 is less than 235.
2. ■ × 5 is less than 308.
3. ■ × 8 is less than 488.
4. ■ × 12 is less than 638.
5. ■ × 16 is less than 795.
6. ■ × 43 is less than 2,000.

Use a calculator to find the smallest multiple of 8 that will make each sentence true.

7. ■ × 5 is greater than 425.
8. ■ × 7 is greater than 555.
9. ■ × 14 is greater than 1,202.
10. ■ × 32 is greater than 1,876.

11. A football frenzy is also sweeping the solar system. On Jupiter, 18,524 players are trying out for the new league that will have 35-player teams. If the number of teams has to be a multiple of 100, what is the greatest number of teams that can be formed?

Multiplying Whole Numbers   **169**

EXTRA PRACTICE

Meaning of Multiplication, page 139
Multiply.

1. 8 × 4
2. 3 × 6
3. 2 × 7
4. 6 × 5
5. 9 × 3
6. 4 × 0
7. 9 × 7

Complete. What property did you use?

8. 3 × 5 = 5 × ■
9. ■ × 131 = 0
10. 54 × 1 = ■
11. 3 × (4 + 7) = (■ × 4) + (3 × 7)

Basic Facts and Strategies, page 141
Find the product.

1. 4 × 7
2. 8 × 8
3. 6 × 7
4. 9 × 6
5. 3 × 8
6. 7 × 7

Compare. Use >, <, or =.

7. 5 × 7 ● 15 + 17
8. 36 + 16 ● 9 × 8
9. 7 × 8 ● 24 + 32

Mental Math: Using Multiplication Patterns, page 143
Find the product mentally.

1. 4 × 100
2. 8 × 10
3. 6 × 1,000
4. 3 × 200
5. 5 × 400
6. 7 × 3,000
7. 40 × 20
8. 60 × 500
9. 300 × 600
10. 20 × 40 × 30
11. 4 × 5 × 6
12. 20 × 50 × 30

Compare. Use >, <, or =.

13. 70 × 4,000 ● 400 × 700
14. 40 × 20 × 40 ● 600 × 30

Estimating Products, page 145
Estimate. Use front-end estimation.

1. 5 × 123
2. 19 × 43
3. 39 × 564
4. 54 × 897

Estimate. Use rounding.

5. 4 × 783
6. 36 × 46
7. 5 × $56.50
8. 54 × 678
9. 7 × $82.30
10. 72 × 37
11. 46 × 634
12. 29 × 62

Extra Practice

Multiplying by 1-Digit Numbers, page 147

Multiply.

1. 32 × 7
2. 43 × 5
3. 652 × 8
4. 764 × 6
5. 563 × 9
6. 804 × 2
7. 672 × 5
8. $3.09 × 7
9. 4,072 × 6
10. 6,789 × 3
11. 3 × 4,987
12. 8 × 765
13. 6 × 8,904
14. 4 × 93,054
15. 6 × 54
16. 8 × $32.09
17. 6 × 4,122
18. 5 × 79,320

Multiplying by 2-Digit Numbers, page 149

Multiply.

1. 87 × 20
2. 43 × 60
3. 52 × 15
4. 74 × 63
5. 23 × 13
6. 37 × 70
7. 68 × 75
8. $39 × 32
9. 66 × 78
10. 49 × 84
11. 43 × 65
12. 76 × 89
13. 86 × 32
14. 45 × 67
15. 27 × 74
16. 17 × 48
17. 54 × 89
18. 34 × 53

Problem Solving: Finding Needed Information, page 151

If there is not enough information to solve the problem, list what additional information is needed. If there is enough information, solve the problem.

1. The football team rides to away games in vans. Each van holds 6 players. How many vans are needed?

2. Students are selling tickets to a game. Each student has 4 tickets to sell. There are 466 students. How many tickets are there?

3. The first jetliner service between London and Johannesburg took 23 hours. How much faster is jetliner service between the two cities today?

4. The first woman to fly in a balloon did it in 1784 in France. The first woman to get a pilot's license got it in 1910 in France. How many years are there between the two?

Multiplying Whole Numbers

Extra Practice

Multiplying by 2-Digit Numbers, page 155

Multiply.

1. 347 × 43
2. 765 × 18
3. 456 × 24
4. $675 × 38
5. 436 × 32
6. 307 × 76
7. 786 × 25
8. 348 × 45
9. $67.78 × 54
10. 5,347 × 34
11. 8,065 × 78
12. 4,608 × 44

13. 36 × 437
14. 65 × 3,079
15. 49 × 4,327

Multiplying Large Numbers, page 157

Multiply. Use a calculator or paper and pencil.

1. 300 × 200
2. 564 × 312
3. 340 × 900
4. 896 × 509
5. 430 × 217
6. $5.08 × 458
7. 478 × 610
8. 865 × 332
9. 4,768 × 232
10. 3,980 × 459
11. 6,087 × 608
12. $45.78 × 557

13. 654 × 876
14. 302 × 789
15. 780 × 4,902

Problem-Solving Strategy: Solving a Multistep Problem, page 159

Solve the problem. Show the steps you used.

1. Mrs. Canon buys 28 packages of red Halloween masks with 4 masks in each package. She also buys 16 packages of black Halloween masks with 3 masks in each package. How many masks does she buy in all?

2. A group of 6 students decorated the gym with streamers. They worked for 60 minutes, hanging one streamer every 5 minutes. Each student received 3 coupons for the carnival for each streamer hung. How many coupons were distributed?

EXTRA PRACTICE

Perimeter, page 161
Find the perimeter.

1. 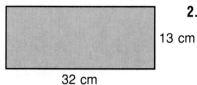 32 cm, 13 cm

2. 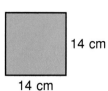 14 cm, 14 cm

3. 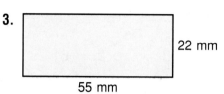 55 mm, 22 mm

Find the missing length.

4. 9 cm, ?

   Perimeter = 36 cm

5. 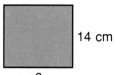 ?, 14 cm

   Perimeter = 60 cm

6. 5 cm, ?

   Perimeter = 26 cm

Area, page 163
Find the area.

1. 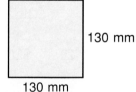 130 mm, 130 mm

2. 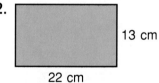 22 cm, 13 cm

3. 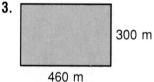 460 m, 300 m

4. a rectangle with length of 14 m, width of 12 m

5. a square with sides of 17 cm

Circumference, page 165
Find the approximate circumference.

1. 4 m

2. 63 mm

3. 112 cm

4. 49 cm

5. 12 m

6. 8 m

Multiplying Whole Numbers 173

Practice PLUS

KEY SKILL: Multiplying by a 1-Digit Number (Use after page 147.)

Level A
Multiply.

1. 38 × 3
2. 61 × 5
3. 85 × 7
4. 63 × 2
5. 49 × 8

6. $3.07 × 5
7. 107 × 8
8. 93 × 5
9. 603 × 7
10. 505 × 8

11. 150 × 3
12. 62 × 4
13. $1.25 × 5
14. 402 × 6
15. 39 × 7

16. 2 × 49
17. 5 × 43
18. 7 × 120
19. 8 × 243

Level B
Multiply.

20. 72 × 8
21. 56 × 7
22. 438 × 5
23. 519 × 4
24. 683 × 7

25. $5.07 × 8
26. 325 × 6
27. 5,123 × 3
28. 6,487 × 5
29. $20.32 × 3

30. 3 × 1,284
31. 5 × $29.41
32. 7 × 3,104
33. 9 × $11.13

Level C
Multiply.

34. 807 × 5
35. 632 × 7
36. 3,075 × 6
37. 4,307 × 3
38. 8,605 × 8

39. 7,115 × 4
40. $253.04 × 6
41. 30,875 × 4
42. 68,954 × 5
43. 83,418 × 9

44. 6 × 43,187
45. 9 × 80,168
46. 7 × 43,263
47. 5 × 61,183
48. 8 × $179.30
49. 9 × 4,076
50. 6 × 8,040
51. 5 × 13,827

Practice PLUS

KEY SKILL: Multiplying by a 2-Digit Number (Use after page 155.)

Level A

Multiply.

1. 123 × 33
2. 314 × 20
3. 250 × 13
4. 139 × 21
5. 201 × 34

6. 330 × 42
7. 411 × 26
8. 502 × 32
9. 405 × 25
10. 380 × 32

11. $1.10 × 24
12. 933 × 12
13. 770 × 23
14. $425 × 58
15. 821 × 37

16. 32 × 191
17. 82 × 220
18. 35 × 213

Level B

Multiply.

19. $985 × 86
20. 850 × 62
21. $6.04 × 68
22. 743 × 52
23. 550 × 47

24. 4,872 × 45
25. 6,130 × 29
26. $87.32 × 47
27. 5,209 × 59
28. 7,103 × 74

29. 76 × 329
30. 32 × 4,320
31. 56 × 8,903
32. 64 × 640

Level C

Multiply.

33. 4,375 × 37
34. 5,117 × 44
35. 6,431 × 68
36. 2,783 × 54
37. $89.73 × 21

38. 42 × 8,007
39. 24 × $13.45
40. 53 × 4,073

Compare. Use >, <, or =.

41. 67 × 83 ● 55 × 96
42. 84 × 42 ● 36 × 98
43. 23 × 648 ● 8,369 + 6,665
44. 46 × 719 ● 19,628 + 13,446

Multiplying Whole Numbers 175

Chapter Review

LANGUAGE AND MATHEMATICS

Match the examples in Column A with the correct property in Column B. *(page 138)*

Column A	Column B
1. 38 × 1 = 38	A. commutative property
2. 4 × 3 = 3 × 4	B. distributive property
3. 0 × 149 = 0	C. associative property
4. 3 × (4 + 5) = (3 × 4) + (3 × 5)	D. identity property
	E. property of zero

5. **Write a definition** or give an example of the property you did not match in Column B.

CONCEPTS AND SKILLS

Multiply. *(pages 140–143, 146–149, 154–157)*

6. 3 × 9
7. 6 × 7
8. 40 × 20
9. 60 × 30
10. 385 × 7

11. 3,073 × 5
12. $263.45 × 9
13. 29 × 50
14. $36 × 24
15. 18 × 27

16. 65 × 32
17. 149 × 25
18. 287 × 75
19. 8,079 × 39
20. 4,007 × 22

21. 900 × 300
22. 819 × 238
23. $9.00 × 553
24. $29.89 × 426
25. 3,847 × 809

Estimate. Use front-end estimation. *(page 144)*

26. 813 × 5
27. 4,817 × 4
28. $19.38 × 6
29. 86 × 21
30. 38 × 43

Estimate. Use rounding. *(page 144)*

31. 432 × 6
32. 8,163 × 3
33. $4.69 × 7
34. 27 × 45
35. $27.89 × 5

176 Chapter 4

Find the perimeter. *(page 160)*

36.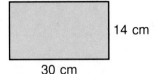
37.
38.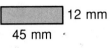

Find the area. *(page 162)*

39.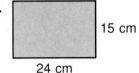
40.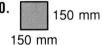
41.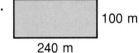

Use your formula to find the approximate circumference. *(page 164)*

42.
43.
44.

CRITICAL THINKING

45. How can you use the distributive property to find 5 × 104 mentally?

MIXED APPLICATIONS

46. Daniel sold 3,000 tickets to the opening night of the concert. If each ticket cost $9, how much money did the theatre take in that night? *(page 142)*

47. One section of the auditorium has 12 rows with 23 seats in a row. Another section has 15 rows with 31 seats in a row. How many seats are in the two sections? *(page 158)*

48. Joe earned $26.95 mowing lawns. How much did he charge for each lawn? What information is needed to solve this problem? *(page 150)*

49. Diana is planning a birthday party. She orders 2 pieces of pizza for each guest. How much pizza should she buy? What information is needed to solve this problem? *(page 150)*

Multiplying Whole Numbers 177

Chapter Test

Multiply.

1. 8 × 5
2. 27 × 6
3. 807 × 8
4. $12.53 × 4

5. 6,217 × 7
6. 10 × 3
7. 20 × 9
8. 50 × 40

9. 8,000 × 40
10. 900 × 500
11. 425 × 63
12. $9.05 × 39

13. 207 × 20
14. 653 × 84
15. 230 × 900
16. $48.93 × 638

Estimate by rounding to the nearest whole number or using front-end estimation.

17. $19.50 × 7
18. 415 × 3
19. $43.80 × 9
20. 389 × 38

Complete.

21.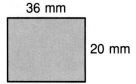
36 mm, 20 mm

Perimeter: ■

22.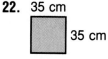
35 cm, 35 cm

Area: ■

23.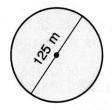
125 m

Circumference: ■

Solve.

24. The scouts sold greeting cards for a fund-raising project. They sold the cards for $6.89 a box. What information do you need if you want to find out how much money was earned the first two days of the fund raiser?

25. Ryan is packing boxes of pet food. One box contains 6 rows of cat food with 12 cans in each row. Another box contains 8 rows of dog food with 13 cans in each row. How many cans of pet food are there in all? What operations did you use?

ENRICHMENT FOR ALL

ENLARGING FIGURES ON A COORDINATE GRAPH

Nicky wants to enlarge triangle ABC to get a triangle with sides twice as long.

Here is what he started to do.

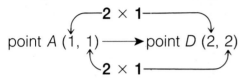

point B (3, 1) ⟶ point E (6, 2)

point C (4, 4) ⟶ point F (■, ■)

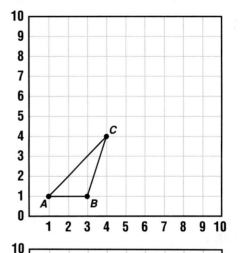

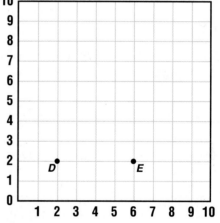

1. What did Nicky do to double the length of each side?

2. What would be the coordinates for point F?

3. Copy the graph and complete the figure at the right to draw triangle DEF with sides twice as long as triangle ABC.

4. How can you check that the sides of triangle DEF are twice as long as the sides of triangle ABC?

5. How would you enlarge triangle ABC to get a triangle with sides three times as long?

Enlarge the figure on a coordinate graph. Check your enlargement by measuring.

6. A quadrilateral has these coordinates: W (3, 1), X (3, 2), Y (5, 2), and Z (5, 1). Draw an enlargement of quadrilateral WXYZ so that each side is twice as long. Label the enlargement RSTU.

7. Draw a quadrilateral on a coordinate graph. Trade papers with another student and enlarge each other's quadrilateral.

8. **What if** you had to draw a quadrilateral with sides that are one-fourth as long as the sides of quadrilateral RSTU in Problem 6? How would you do it?

Multiplying Whole Numbers 179

Cumulative Review

Choose the letter of the correct answer.

1. 6.36
   − 3.8
   a. 2.06
   b. 2.46
   c. 2.56
   d. not given

2. Compare: 369,407 ● 369,047
   a. <
   b. >
   c. =
   d. not given

3. Estimate. Use front-end estimation.
   $33.80
   × 8
   a. $230
   b. $210
   c. $220
   d. not given

4. Which unit would you use to measure the height of a ladder?
   a. millimeter
   b. centimeter
   c. meter
   d. not given

5. Round 14.753 to the nearest tenth.
   a. 14.8
   b. 10
   c. 14.75
   d. not given

6. Round 48,167 to the nearest 10,000.
   a. 40,000
   b. 50,000
   c. 48,000
   d. not given

7. Compare: 5 × 7 ● 14 + 21
   a. <
   b. >
   c. =
   d. not given

8. 9 × 8,000
   a. 720
   b. 7,200
   c. 72,000
   d. not given

9. $47.89
   38.52
   + 13.63
   a. $100.04
   b. $100.40
   c. $101.04
   d. not given

10. 9,155
    × 7
    a. 64,805
    b. 64,085
    c. 64.087
    d. not given

11. 38
    × 75
    a. 2,850
    b. 2,800
    c. 2,100
    d. not given

12. 2,385
    × 307
    a. 730,095
    b. 731,195
    c. 732,195
    d. not given

13. How much fencing is needed to enclose a rectangular garden 13 m long and 5 m wide?
    a. 18 m
    b. 36 m
    c. 65 m
    d. not given

14. Janet made an afghan that is a 100-cm by 21-cm rectangle. What is the area of the afghan?
    a. 2,100 cm²
    b. 242 cm²
    c. 640 cm²
    d. not given

Dividing Whole Numbers— 1-Digit Divisors

CHAPTER 5

MATH CONNECTIONS: ALGEBRA • GRAPHING • STATISTICS • PROBLEM SOLVING

Juice Jamboree
small cup (6 oz.) 75¢
large cup (8 oz.) 90¢

1. What is happening in this picture?
2. What information do you see in the picture?
3. How can this information be useful to you?
4. Write a problem using the information.

EXPLORING A CONCEPT
Meaning of Division

Amanda has 26 stickers. She wants to display them in rows of 4.
How many rows will she make?
How many stickers will be left over?

WORKING TOGETHER

1. Use place-value models to represent the 26 stickers. Separate the models to make equal sets of 4. How many sets did you make? How many were left over?

2. Record your work on a chart like this.

Number	Number in Each Set	Number of Sets	Number Left Over
26	4		

3. Repeat the activity for 32, 49, and 63 stickers.

By separating the models into equal sets you are dividing.

$$\underset{\text{dividend}}{26} \div \underset{\text{divisor}}{4} = \underset{\text{quotient}}{6} \text{ R2} \leftarrow \text{remainder}$$

$$\text{quotient} \rightarrow 6 \text{ R2} \leftarrow \text{remainder}$$
$$\underset{\text{divisor}}{4)\overline{26}} \leftarrow \text{dividend}$$

4. Add another column to your chart. Write the division sentence for each modeling you recorded.

What if Amanda decides to divide the 26 stickers equally among 4 friends? How many stickers will each friend get? How many stickers will be left over?

5. Use place-value models to solve this problem. How many did you put in each set? How many were left over?

6. Record your work on a chart like this.

Number	Number of Sets	Number in Each Set	Number Left Over

7. Repeat the activity for 32, 49, and 63 stickers.

8. When you are making equal sets, you are also dividing. Add another column to your chart. Write the division sentence for each modeling you recorded.

SHARING IDEAS

Compare the division sentences in each chart.

9. Are they the same or different? What two questions does a division sentence answer?

10. What happens to the quotient when the dividend gets larger but the divisor remains the same?

11. What do you think happens if the divisor increases but the dividend remains the same? Why?

ON YOUR OWN

Use place-value models to solve. Write the division sentence that shows your work. Which question does the division answer?

12. Elliot has 43 sports magazines and 4 racks. He puts the same number of magazines on each rack. How many magazines does he put on each rack? How many does he have left over?

13. Melissa has 38 snapshots to put into frames. She can fit 5 snapshots in each frame. How many frames can she fill? How many snapshots will she have left?

14. Andrew has 86 baseball cards. He puts them in piles of 9. How many piles does he have? How many cards are left over?

15. Three sisters want to share 72 beads equally. How many beads should each sister get?

UNDERSTANDING A CONCEPT

Informal Algebra: Basic Division Facts

A. Ellen and Sonya are making bracelets, and they use 9 beads for each one. They have 36 beads to use. How many bracelets can the girls make?

You can solve by dividing.

Think: $36 \div 9 = \blacksquare$
$36 \div 9 = 4$
$\blacksquare = 4$

You can solve by finding a missing factor.

Think: $9 \times \blacksquare = 36$
$9 \times 4 = 36$
$\blacksquare = 4$

The girls can make 4 bracelets.

1. What do you notice about the numbers in the multiplication and division facts? What does this tell you about multiplication and division?

You can write four related facts using 4, 9, and 36. The facts form a **fact family** that can help you multiply and divide.

$9 \times 4 = 36$ $4 \times 9 = 36$ $36 \div 9 = 4$ $36 \div 4 = 9$

2. Write a related multiplication fact that can help you divide.
   a. $21 \div 3 = \blacksquare$ b. $32 \div 8 = \blacksquare$ c. $48 \div 6 = \blacksquare$

B. Study these number sentences to see how 0 and 1 are special in division.

$7 \div 1 = 7$ $8 \div 8 = 1$ $0 \div 9 = 0$
$4 \div 1 = 4$ $6 \div 6 = 1$ $0 \div 5 = 0$
$2 \div 1 = 2$ $4 \div 4 = 1$ $0 \div 3 = 0$

3. What happens when you divide a number by 1?
4. What happens when you divide a number by itself?
5. What happens when you divide 0 by a number other than 0?
6. How do the following multiplication sentences show that you cannot divide by 0?
   a. $\blacksquare \times 0 = 5$ b. $\blacksquare \times 0 = 0$

TRY OUT

Write a fact family for the group of numbers.

7. 3, 9, 27

8. 5, 7, 35

Find the quotient. Think of a related fact.

9. 8)40 **10.** 9)54 **11.** 7)63 **12.** 0 ÷ 4 **13.** 6 ÷ 1 **14.** 7 ÷ 7

PRACTICE

Divide.

15. 7)21 **16.** 6)12 **17.** 7)14 **18.** 5)15 **19.** 5)40 **20.** 2)12

21. 6)0 **22.** 1)5 **23.** 9)9 **24.** 8)0 **25.** 1)9 **26.** 3)3

27. 7)56 **28.** 8)72 **29.** 6)24 **30.** 5)30 **31.** 9)63 **32.** 1)3

33. 63 ÷ 7 **34.** 45 ÷ 9 **35.** 16 ÷ 8 **36.** 20 ÷ 5 **37.** 16 ÷ 4

38. 27 ÷ 9 **39.** 42 ÷ 7 **40.** 38 ÷ 1 **41.** 0 ÷ 75 **42.** 19 ÷ 19

Find the answer.

43. The dividend is 35 and the divisor is 5. What is the quotient?

44. The dividend is 20 and the quotient is 5. What is the divisor?

Write a fact family for the group of numbers.

45. 5, 8, 40 **46.** 4, 7, 28 **47.** 6, 7, 42 **48.** 4, 8, 32

Complete.

49. Rule: ÷ 3

Input	Output
24	
18	
12	
27	

50. Rule: ÷ 6

Input	Output
36	
18	
30	
54	

51. Rule: ■

Input	Output
24	6
12	3
36	9
28	7

52. Rule: ■

Input	Output
28	4
35	5
63	9
56	8

Mixed Applications

53. Tim has 4 sets of color markers. Each set has 12 markers. How many markers does he have?

54. **Write a problem** that can be solved using the division sentence 18 ÷ 6 = ■.

DEVELOPING A CONCEPT
Mental Math: Using Division Patterns

You can use division facts to help you divide large numbers.
Use a calculator to find the quotients in each set of examples.

Set A	Set B
60 ÷ 3 = ■ 20	420 ÷ 7 = ■
600 ÷ 3 = ■	4,200 ÷ 7 = ■
6,000 ÷ 3 = ■	42,000 ÷ 7 = ■
60,000 ÷ 3 = ■	

SHARING IDEAS

1. What division fact appears in Set A? Set B?

2. Look at the quotients you found. Compare the number of zeros in each quotient with the number of zeros in the dividend. What pattern do you see?

3. Does the pattern work for 2,000 ÷ 5? Why or why not?

4. Find the following quotients: 56,000 ÷ 7 and 30,000 ÷ 6.

TRY OUT   Write the letter of the correct answer. Divide mentally.

5. 800 ÷ 2   a. 4   b. 40   c. 160   d. 400

6. 3,000 ÷ 5   a. 6   b. 60   c. 600   d. 6,000

7. 3)24,000   a. 8,000   b. 800   c. 80   d. 8

PRACTICE

Divide mentally.

8. 2)40
9. 4)80
10. 2)60
11. 3)60
12. 2)80
13. 3)120
14. 5)250
15. 7)630
16. 4)240
17. 8)400
18. 6)1,800
19. 7)5,600
20. 9)8,100
21. 8)6,400
22. 4)2,000
23. 4)32,000
24. 9)45,000
25. 3)27,000
26. 5)10,000
27. 6)54,000
28. 160 ÷ 4
29. 210 ÷ 7
30. 3,500 ÷ 5
31. 42,000 ÷ 6
32. 3,000 ÷ 6
33. 90 ÷ 3
34. 6,300 ÷ 7
35. 240 ÷ 6
36. 3,600 ÷ 4
37. 60,000 ÷ 2
38. 810 ÷ 9
39. 2,700 ÷ 3

Find the missing number mentally.

40. ■ × 3 = 150
41. ■ × 2 = 4,000
42. ■ × 5 = 20,000
43. ■ × 4 = 2,800
44. ■ × 6 = 36,000
45. ■ × 7 = 49,000

Mixed Applications

Solve. Which method did you use?

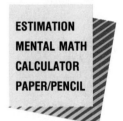

ESTIMATION
MENTAL MATH
CALCULATOR
PAPER/PENCIL

46. Workers at the Topper Company packed 3,000 shirts in one week. They packed 6 shirts in each box. How many boxes did they fill?

47. Pete packs 450 sweatshirts in 9 cartons. He puts the same number of sweatshirts in each carton. How many sweatshirts does he put in each carton?

48. The Topper Company sold 43,850 T-shirts in June. This was 725 more T-shirts than it sold in May. How many T-shirts did the company sell in May?

Mixed Review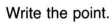

Write the ordered pair that names the point.

49. A
50. D
51. B

Write the point.

52. (3, 2)
53. (6, 1)
54. (2, 6)

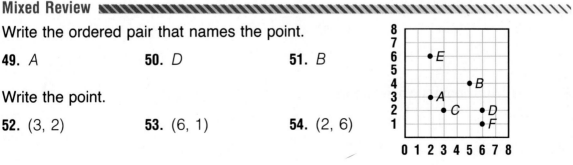

EXTRA Practice, page 210

UNDERSTANDING A CONCEPT
Estimating Quotients

The Jacksons plan to take a 5-day biking trip that is 362 km. The Jacksons plan to bike about the same distance each day. About how many kilometers will they bike each day?

Estimate: 362 ÷ 5

Here are two methods you can use.

Rounding
Round the dividend to its greatest place.

362 ÷ 5
↓   ↓
Think: 400 ÷ 5 = 80

They will bike about 80 km each day.

Compatible Numbers
Use a basic fact.

362 ÷ 5
↓   ↓
Think: 350 ÷ 5 = 70

They will bike about 70 km each day.

1. Which estimate is greater than the exact answer? Which estimate is less? How do you know?

2. Estimate 2,210 ÷ 6 by using compatible numbers. Then estimate by rounding. Which estimate is more helpful? Why?

TRY OUT   Write the letter of the correct answer.

Estimate by rounding.

3. 2)̅4̅3̅   **a.** 2   **b.** 20   **c.** 80   **d.** 200
4. 3,973 ÷ 8   **a.** 50   **b.** 400   **c.** 500   **d.** 5,000

Estimate. Use compatible numbers.

5. 7,439 ÷ 4   **a.** 2,000   **b.** 800   **c.** 200   **d.** 20
6. 6)̅2̅3̅,̅1̅0̅2̅   **a.** 40   **b.** 400   **c.** 4,000   **d.** 7,000

PRACTICE

Estimate by rounding.

7. 3)59
8. 5)228
9. 8)437
10. 7)6,805
11. 6)27,876

12. 2)9,601
13. 8)3,876
14. 5)23,085
15. 5)32,975
16. 8)44,128

Estimate by using compatible numbers.

17. 4)312
18. 7)546
19. 9)1,735
20. 5)3,803
21. 6)15,041

22. 5)489
23. 6)425
24. 9)572
25. 6)311
26. 4)825

27. 7)4,948
28. 5)1,491
29. 8)3,605
30. 3)5,237
31. 4)7,580

Estimate the quotient.

32. 63 ÷ 2
33. 623 ÷ 6
34. 15,214 ÷ 9
35. 2,678 ÷ 5

36. 4,498 ÷ 8
37. 14,205 ÷ 3
38. 853 ÷ 9
39. 512 ÷ 7

40. 5,729 ÷ 6
41. 3,626 ÷ 4
42. 16,901 ÷ 3
43. 27,021 ÷ 6

44. 752 ÷ 8
45. 43,311 ÷ 5
46. 1,279 ÷ 2
47. 8,037 ÷ 9

Use estimation to predict which division gives the greater quotient. Check your prediction on a calculator.

48. 4)296 or 5)280
49. 3)1,974 or 2)1,974

50. 8)24,032 or 6)32,700
51. 2)9,998 or 9)10,008

52. 7)50,843 or 5)47,193
53. 4)29,012 or 8)65,321

Mixed Applications Find the answer. Did you estimate or find an exact answer?

54. All 3 performances of the Top Trio at the Summer Festival were sold out. The festival theater has 1,860 seats. About how many people in all bought tickets to see the Top Trio?

55. A ticket to Will's Wild Animal Park costs $7. One day Will collected $5,712 in admission fees. About how many people came to the park that day?

56. Lodging for the Jacksons' first night costs $154 for the 4 of them. About how much is the cost for each person?

57. The Jacksons see 8 tour buses in the parking lot of a restaurant. Each bus carries 38 people. How many tourists are on the buses?

UNDERSTANDING A CONCEPT
Dividing 2- and 3-Digit Numbers

Louise has 284 tulip bulbs to pack in boxes. She can fit 8 bulbs in a box. How many boxes can she fill? How many bulbs will she have left over?

Divide: 284 ÷ 8

Step 1
Decide where to place the first digit of the quotient.

8)284   ***Think:*** 8 > 2 Not enough hundreds
         8 < 28 There are enough tens to divide.

Step 2
Divide the tens.

```
    3
8)284     Think: 8)28
- 24          Multiply. 3 × 8 = 24
  ---         Subtract. 28 − 24 = 4
   4          Compare. 4 < 8
```

Step 3
**Bring down the ones.
Divide the ones.
Write the remainder.**

```
   35 R4
8)284
- 24↓            5
  ---     Think: 8)44
   44           Multiply. 5 × 8 = 40
 - 40           Subtract. 44 − 40 = 4
   ---          Compare. 4 < 8
    4
```

**Check.
Multiply, then add.**

```
    35
  ×  8
  ----
   280
  +  4
  ----
   284
```

Louise can fill 35 boxes. She will have 4 tulip bulbs left over.

1. ***What if*** you got a remainder of 9 in the division above? How would you know that the quotient was wrong?

2. Where would you place the first digit of the quotient?

   a. 4)51   **b.** 3)29   **c.** 4)327   **d.** 5)569

TRY OUT
Divide.

3. 8)75   4. 2)45   5. 4)90   6. 5)627   7. 7)368

PRACTICE
Divide.

8. 5)29   9. 8)78   10. 4)73   11. 6)94   12. 2)87

13. 2)712   14. 7)824   15. 9)563   16. 4)207   17. 3)401

18. 5)279   19. 4)39   20. 8)649   21. 2)322   22. 6)707

23. 9)55   24. 3)68   25. 7)125   26. 6)350   27. 4)861

28. 5)412   29. 2)57   30. 8)928   31. 6)309   32. 4)96

33. 67 ÷ 7   34. 215 ÷ 6   35. 82 ÷ 3   36. 190 ÷ 4   37. 733 ÷ 9

38. 73 ÷ 9   39. 578 ÷ 3   40. 307 ÷ 5   41. 90 ÷ 2   42. 608 ÷ 8

Find the answer.

43. The dividend is 97. The divisor is 3. What is the quotient and the remainder?

44. The divisor is 4. The quotient is 89, and the remainder is 1. What is the dividend?

Mixed Applications

45. Some volunteers need to raise $2,000 to landscape a park. They have raised $855 already. How much more money do they need?

46. Hiroo spent 3 hours each morning working in the garden. If he worked 51 hours all together, how many mornings did he work?

47. The volunteers have 250 rosebushes. They plant 9 in each section of the park. How many sections does the park have? How many bushes are left over?

48. There are 15 volunteers working in each section of the park. There are 6 sections. How many volunteers are there?

Mixed Review
Find the answer. Which method did you use?

49.   24,000
    + 5,387

50.   2,000
    ×   300

51.   5.008
    − 3.954

52.   312
    ×   258

MENTAL MATH
CALCULATOR
PAPER/PENCIL

EXTRA Practice, page 211; Practice PLUS, page 214   Dividing Whole Numbers: 1-Digit Divisors

UNDERSTANDING A CONCEPT
Dividing 4- and 5-Digit Numbers

There are 1,368 balcony seats in the Music Hall. There are 4 balconies, and each has the same number of seats. How many seats are in each balcony?

Divide: 1,368 ÷ 4

Step 1
Decide where to place the first digit of the quotient.

4)1,368   **Think:** 4 > 1   Not enough thousands
          4 < 13 There are enough hundreds to divide.

Step 2
Divide the hundreds.

```
    3
4)1,368     Think: 4)13
 - 12       Multiply. 3 × 4 = 12
    1       Subtract. 13 - 12 = 1
            Compare. 1 < 4
```

Step 3
**Bring down the tens.
Divide the tens.**

```
   34
4)1,368
 - 12↓         Think: 4)16
   16          Multiply. 4 × 4 = 16
 - 16          Subtract. 16 - 16 = 0
    0          Compare. 0 < 4
```

Step 4
**Bring down the ones.
Divide the ones.
Write the remainder.**

```
   342
4)1,368
 - 12↓
   16
 - 16↓                Think: 4)8
    08               Multiply. 2 × 4 = 8
 -   8               Subtract. 8 - 8 = 0
     0               Compare. 0 < 4
```

There are 342 seats in each balcony.

1. Where would you place the first digit of the quotient?

   a. 3)2,165   **b.** 2)5,044   **c.** 6)11,218   **d.** 4)53,790

Divide money as you would divide whole numbers. Then write the dollar sign and decimal point in the quotient.

2. Place the dollar sign and decimal point in the quotient.

   a. 4)$148.48 (37 12)   **b.** 6)$25.86 (4 31)   **c.** 5)$225.55 (45 11)   **d.** 8)$38.48 (4 81)

Lesson 5-6

Try Out
Write the letter of the correct answer.

3. 3)4,857 a. 160 b. 1,600 c. 1,610 R2 d. 1,619
4. 6)15,098 a. 251 R3 b. 2,500 c. 2,516 d. 2,516 R2
5. 7)$115.71 a. $16.53 b. $165.30 c. 16.53 d. 1,653
6. 5)$46.25 a. 925 b. $9.25 c. 9.25 d. 9,250

Practice
Divide.

7. 4)8,492
8. 9)6,230
9. 9)7,304
10. 8)21,733
11. 3)38,207
12. 6)$511.38
13. 2)$51.92
14. 5)$317.05
15. 7)$82.53
16. 8)$421.12
17. 3)1,877
18. 2)$572
19. 6)5,709
20. 4)17,999
21. 7)$36.47
22. 9)3,002
23. 4)5,672
24. 3)$125.79
25. 8)60,092
26. 6)1,378
27. $8.46 ÷ 3
28. 1,069 ÷ 5
29. 42,875 ÷ 6
30. 3,961 ÷ 8
31. 17,325 ÷ 5
32. $7,136 ÷ 4
33. 8,019 ÷ 7
34. 11,234 ÷ 9

Critical Thinking
Make up a word problem for the division. Then solve. Tell why some quotients and some remainders have a dollar sign.

35. 4)$1,368
36. 4)$1,369
37. $4)$1,368
38. $4)$1,369

Mixed Applications
39. A stagehand uses 1,285 ft of rope to hold up 5 props. He uses the same amount of rope for each prop. How much rope does he use for each one?

40. Maria spends $19.14 on 6 yd of fabric for costumes. What is the cost for each yard of fabric?

41. Don kept the tally at the right to show the number of tickets he sold for the concert. How many tickets did he sell over five days?

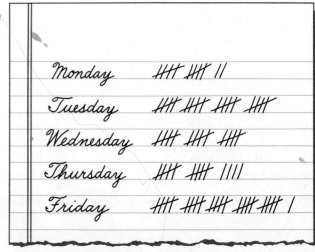

Monday											
Tuesday																	
Wednesday													
Thursday													
Friday	|||| |||| |||| |||| ||||	

Sum-Thing Strange Is

Investigating Patterns

Here is an interesting math trick for you to try.

Step 1
Start with a square like the one at the right. Pick any six numbers; for example, 3, 7, 10, 14, 20, 25. Place the numbers along the outside of the square as shown.

+			
14			
20			
25			

+	3	7	10
14			
20			
25			

Step 2
Find the sum of each pair of numbers just as you would in a regular addition table.

+	3	7	10
14	17	21	24
20	23	27	30
25	28	32	35

Now you are ready to do some "mathemagic."

Step 3
Circle any one of the numbers in the inside part of the square; for example, suppose you circle 17. Then cross out all the other numbers in the same row and column as 17.

+	3	7	10
14	(17)	~~21~~	~~24~~
20	~~23~~	27	30
25	~~28~~	32	35

Step 4
Circle one of the remaining numbers; for example, suppose you circle 30. Then cross out all the other numbers in the same row and column as 30. Circle the one number that is not crossed out.

+	3	7	10
14	(17)	~~21~~	~~24~~
20	~~23~~	~~27~~	(30)
25	~~28~~	(32)	~~35~~

Lesson 5-7

HAPPENING

Thinking MATHEMATICALLY

1. What is the sum of the circled numbers?

2. Go back to Step 3. Instead of 17, start the crossing out with a different number. What is the sum of the three circled numbers this time?

3. What is the sum of the six numbers you put around the outside of the square to start with?

4. Try to explain why this trick works. Remember how each number in the inner square is formed.

5. Build another square using six different numbers. Predict what the magic sum will be, then follow the steps and see if you are right.

6. Build a larger square using the same method. Pick eight numbers to put around the outside. Circle numbers and cross out three times, until four circled numbers remain. Does the trick work?

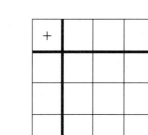

7. Try the trick on a friend. Use the square you made for Problem 6. Have your friend circle and cross out the numbers. Then you can surprise your friend by guessing the sum of the circled numbers.

UNDERSTANDING A CONCEPT
Zeros in the Quotient

The Northside Park Volunteers have 1,528 letters to address. If 5 people share the job equally, how many envelopes will each person address? How many envelopes will they have left over?

Divide: 1,528 ÷ 5

Step 1
Decide where to place the first digit of the quotient.

5)1,528   *Think:* 5 > 1   Not enough thousands
         5 < 15   There are enough hundreds to divide.

Step 2
Divide the hundreds.

```
    3
5)1,528      Think: 5)15
 - 1 5       Multiply. 3 × 5 = 15
   ---       Subtract. 15 − 15 = 0
    0        Compare. 0 < 5
```

Step 3
Bring down the tens. Divide the tens.

```
    30
5)1,528
 - 1 5↓
   ----
     02      Think: 2 < 5   There are not enough
                            tens to divide.
                            Write 0 in the
                            quotient.
```

Step 4
Bring down the ones. Divide the ones. Write the remainder.

```
    305 R3
5)1,528
 - 1 5↓↓
   ----          5
    028       Think: 5)28
   - 25       Multiply. 5 × 5 = 25
     ---      Subtract. 28 − 25 = 3
      3       Compare. 3 < 5
```

Each person addresses 305 envelopes. Three are left over.

1. **What if** you forgot to write the 0 in the quotient? What would your answer be? How would you know that the answer was not reasonable?

TRY OUT Divide.

2. 4)439   3. 5)3,009   4. $72.48 ÷ 8   5. 27,769 ÷ 9

PRACTICE

Divide.

6. 4)43
7. 3)62
8. 6)645
9. 7)495
10. 2)$4.18

11. 5)5,306
12. 9)2,712
13. 8)3,205
14. 3)$150.90
15. 6)12,039

16. 7)352
17. 2)941
18. 4)2,430
19. 6)$12.54
20. 9)4,507

21. 3)21,285
22. 5)10,300
23. 6)1,855
24. $7)$2,849
25. 9)3,624

26. 92 ÷ 3
27. $3.21 ÷ 3
28. 4,123 ÷ 4
29. 12,015 ÷ 6

30. 9)5,450
31. 6,305 ÷ 7
32. $36.72 ÷ 9
33. 23,041 ÷ 5

Mixed Applications

Solve. You may need to use the Databank on page 518.

34. The Volunteers held a festival for the opening of a garden. On Saturday 345 people attended the festival. On Sunday 739 people attended. How many more people attended on Sunday?

35. **Write a problem** about postal rates. Solve the problem. Then ask others to solve it.

MENTAL MATH

You do not need to write all the steps when dividing by a 1-digit number. You can do some of the steps mentally. This is called **short division**.

Divide: 5,248 ÷ 6

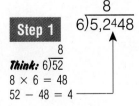

Step 1
       8
6)5,2⁴48

Think: 6)52
8 × 6 = 48
52 − 48 = 4

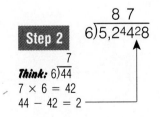

Step 2
       8 7
6)5,2⁴4²8

Think: 6)44
7 × 6 = 42
44 − 42 = 2

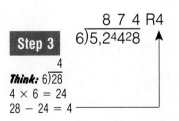

Step 3
       8 7 4 R4
6)5,2⁴4²8

Think: 6)28
4 × 6 = 24
28 − 24 = 4

Use short division to find the quotient.

1. 9)325
2. 7)7,610
3. 5)1,297
4. 8)28,869

Problem Solving

Strategy: Using Estimation

This year the Nortown Girl Scouts are selling muffins. Their goal is to sell at least 1,500 packages of muffins. If each of the 24 scouts in the troup sells 85 packages, will they meet their goal?

You can solve the problem by estimating the product of the number of scouts multiplied by the number of packages each scout sells.

$$\begin{array}{r} 85 \\ \times\,24 \\ \end{array} \rightarrow \begin{array}{r} 80 \\ \times\,20 \\ \hline 1{,}600 \end{array}$$

1. Why would you underestimate in this situation?

2. Will the scouts meet their goal? How can you tell?

The troop members can spend no more than $170 for flour. If they buy 78 bags of flour for $1.92 a bag, will they have enough money to pay for the flour?

To solve the problem you can estimate the product of the number of bags of flour multiplied by the cost of one package.

3. Would you overestimate or underestimate in this situation? Why?

4. Will they have enough money to pay for the flour? How can you tell?

PRACTICE

Use estimation to solve the problem. Did you overestimate or underestimate? Why?

5. The members of a scout squad want to buy a tent for $119.95. They plan to pay for it by earning $3.25 an hour doing chores. Last month they worked for 45 hours. Did they earn enough to pay for the tent?

6. To earn the swimming badge, a scout must swim three laps in 260 seconds. Tina swam the first lap in 78 seconds, the second lap in 86 seconds, and the third lap in 89 seconds. Did she earn the badge?

7. Each scout must deliver 150 packages of muffins in one week. If a scout delivers 21 packages a day for 7 days, will she deliver enough packages?

8. The scout troop members want to buy official sweaters for $19.39 each. They can spend up to $500. Do they have enough money to buy 24 sweaters?

Strategies and Skills Review

Solve. Use mental math, estimation, a calculator, or paper and pencil.

9. If a scout does volunteer work for 300 hours a year, she wins a special award. Raphaella did volunteer work for 32 hours a month for a year. Did she earn a special award? How can you use estimation to tell?

10. At 9 A.M. the scouts went on a three-hour nature hike. They walked 2.1 kilometers the first hour, 2.2 kilometers the second hour, and 1.8 kilometers the third hour. What distance did they walk?

11. Maggie is storing packages of muffins on a shelf. She has 8 cartons with 12 packages each and 15 cartons with 9 packages each. How many packages does she have in all?

12. One of the scouts delivers newspapers. In February she delivered 88 papers a day on each of the 28 days of the month. How many newspapers did she deliver?

13. Gwen has a list of 125 houses to visit. She has visited 97 so far. How many more houses does she have left to visit?

14. **Write a problem** that can be solved by using estimation. Solve the problem. Ask others to solve your problem.

DEVELOPING A CONCEPT

Using Pictographs

You can make a **pictograph** to display the data at the right. A pictograph uses symbols to represent numbers.

TEACHERS IN ELEMENTARY, JUNIOR HIGH, AND HIGH SCHOOLS

State	Number of Teachers
Alaska	6,814
Montana	9,705
Nevada	7,751
Vermont	6,397
Wyoming	7,162

WORKING TOGETHER

Use the data in the table and these steps to make a pictograph.

Step 1 List each item in the table.

Step 2 Choose a symbol. Round the data to a convenient place to determine the value of the symbol.

Step 3 Draw the symbols for each item.

Step 4 Write a title above the graph.

Step 5 Write the value of the symbol below the graph.

1. To which place did you round the data?
2. What is the value of the symbol you used?
3. Which state has the fewest teachers? How many symbols did you use for that state? Why?

Marsha made this pictograph for the data.

TEACHERS IN ELEMENTARY, JUNIOR HIGH, AND HIGH SCHOOL

Alaska	👤👤👤👤👤👤👤
Montana	👤👤👤👤👤👤👤👤👤👤
Nevada	👤👤👤👤👤👤👤👤
Vermont	👤👤👤👤👤
Wyoming	👤👤👤👤👤👤👤

👤 = 1,000 Teachers

200 Lesson 5-10

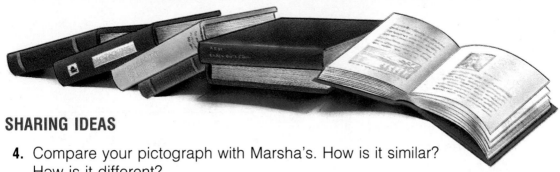

SHARING IDEAS

4. Compare your pictograph with Marsha's. How is it similar? How is it different?

5. What symbol could Marsha use to represent 500 teachers?

6. According to Marsha's pictograph, which states have the same number of teachers?

7. Do those states actually have the same number of teachers?

8. What are some advantages and disadvantages of displaying data in a pictograph?

PRACTICE

Use the pictograph at the right to answer Problems 9–12.

9. What does ▢ represent?

10. What does ▯ represent?

11. About how many art books are there?

12. About how many more fiction books than science books are there?

BOOKS IN PUBLIC LIBRARY

Art	▢ ▯
Biography	▢ ▢
Fiction	▢ ▢ ▢ ▢ ▯
History	▢ ▢ ▯
Medicine	▯
Science	▢ ▢ ▢

▢ = 100 books

Use the data in the table at the right to make a pictograph.

13. To which place did you round the data?

14. What is the value of the symbol you used?

15. Use the Databank on page 517 to make a pictograph of the amount of money spent by Americans on tickets to various events in the United States. Write a sentence explaining what the pictograph shows.

STUDENT POPULATION AT ALBANY ELEMENTARY SCHOOL

Grade	Number of Students
K	132
1	124
2	115
3	130
4	112

16. Find out how many students are in each grade of your school. Make a pictograph for the data. Then write a sentence explaining what the graph shows.

DEVELOPING A CONCEPT
Mean, Median, Mode, and Range

Edith and Fred are studying the habits of birds at the Northwoods Nature Camp. One week they kept a daily count of the blackbirds that came to the bird feeder.

Number of Blackbirds at Birdfeeder Between 7:00 and 7:30 a.m.

Sunday	17	Thursday	8
Monday	13	Friday	13
Tuesday	18	Saturday	17
Wednesday	12		

WORKING TOGETHER
Build stacks with centimeter cubes to represent the data for each day.

1. Compare the tallest and the shortest stacks. The **range** of the set is the difference between the tallest and the shortest stacks. How would you calculate the range? What is the range of the set?

2. Arrange the stacks of cubes in order from shortest to tallest. The number of cubes in the middle stack is the **median**. What is the median of the set?

3. Do any of the stacks have the same number of cubes? The number that occurs most often is the **mode**. There may be one mode, more than one, or none at all. What are the modes of the set?

4. Make 7 even stacks using all the centimeter cubes you used to represent the data for each day. How many did you put in each stack? The number you found is the **average,** or **mean,** of the set.

Here is another way to find the mean.

Step 1
First add to find the total.

17 + 13 + 18 + 12 + 8 + 13 + 17 = 98

The mean is 14.

Step 2
Then divide the total by the number of addends.

98 ÷ 7 = 14

SHARING IDEAS

5. **What if** Edith and Fred had seen 16 instead of 17 blackbirds on Sunday and 14 instead of 13 on Monday? What would have been the mean, median, mode, and range?

PRACTICE

Find the mean, median, mode, and range.

6. 129, 156, 138
7. 237, 231, 240
8. 48, 63, 21, 17, 36
9. 54, 97, 54, 101, 54
10. $56, $35, $28, $19, $47
11. $97, $123, $110, $122, $123
12. 425, 350, 425, 275, 350
13. 14, 10, 10, 36, 24, 24, 36
14. 100, 50, 85, 75, 25, 150, 75
15. 20, 114, 73, 100, 80, 48, 97

16. **CAMPERS IN SWIMMING GROUPS**

Tadpoles	Minnows	Dolphins	Barracudas	Sharks
25	37	36	32	25

17. **CAMPERS IN HORSEBACK RIDING CLASS**

Week 1	Week 2	Week 3	Week 4	Week 5	Week 6	Week 7
29	22	22	19	28	22	26

Mixed Applications

18. Campers saw these numbers of robins: 18, 13, 21, 18, 15. What are the mean, median, mode, and range of the set of numbers?

19. During the first 7 weeks of camp, 847 campers enrolled. On the average how many campers were enrolled for each of the 7 weeks?

20. The campers slept in cabins, with 8 campers in each cabin. One week there were 120 campers. How many cabins were filled?

21. The camp costs $225 per week. How much would it cost for Edith to spend 3 weeks at the camp?

Mixed Review

Find the answer. Which method did you use?

22. 2,500 × 30
23. 2,005 − 765
24. 3.50 + 5.25
25. 7)4,900

MENTAL MATH
CALCULATOR
PAPER/PENCIL

PROBLEM SOLVING

UNDERSTAND
✓ PLAN
✓ TRY
✓ CHECK
✓ EXTEND

Strategy: Guess, Test, and Revise

Zena buys young dogwood trees and silver maple trees to plant in her backyard. She buys 12 trees in all and pays $88 for them. The dogwoods cost $12 each and the silver maples cost $5 each. How many of each kind does she buy?

One way to solve the problem is to start with a guess, test it, and make more guesses until you find the correct answer.

Suppose that Zena buys 6 of each kind of tree.

Guess: 6 for $12 and 6 for $5

1. How much would she pay for the 12 trees? Is 6 of each kind of tree the correct answer? Why or why not?

   Test: 6 × $12 = $72
   6 × $5 = $30
   Total = $■

You need to revise the guess.

2. Should you increase or decrease the number of dogwood? Why?

Suppose that Zena buys 5 dogwood trees and 7 maple trees.

Guess: 5 for $12 and 7 for $5

3. How much would she pay for the 12 trees? Is this the correct answer? Why or why not?

   Test: 5 × $12 = $60
   7 × $5 = $35
   Total = $■

Continue to guess and test until you find the correct answer.

4. How many trees of each kind did she buy? How can you tell?

5. **What if** dogwood trees were on sale for $9 each? How many of each kind of tree would she buy if she still paid $88 for 12 trees?

204 Lesson 5-12

PRACTICE

Use the Guess, Test, and Revise strategy to solve the problem.

6. Carol buys seeds at the garden center. She receives 80¢ in change, all nickels and dimes. There are 10 coins. How many nickels and how many dimes does she get?

7. Danny is planting 192 bulbs in 14 rows. Some rows have 15 bulbs each, and the rest have 12 bulbs each. How many rows have 15 bulbs each?

8. Grace's garden book is open. The product of the page numbers on the two facing pages is 210. What are the page numbers?

9. Marvin pays $24 for a flower box and a flat of marigolds. The flower box costs $5 more than the marigolds. How much does the flower box cost?

Strategies and Skills Review

Solve. Use mental math, estimation, a calculator, or paper and pencil.

10. A bag of annual grass seeds costs $14.95, and a bag of perennial grass seeds costs $26.25. How much more does the bag of perennial seed cost than the bag of annual seed?

11. Zena buys daisies and petunias—15 plants in all. Daisies cost 20¢ each, and petunias cost 35¢ each. She pays $3.90 for the plants. How many of each kind does she buy?

12. Danny has planted 192 bulbs. He wants to plant more bulbs, so he buys 4 packs of daffodil bulbs with 15 in each pack and 3 packs of crocus bulbs with 25 in each pack. When he finishes the job, how many bulbs will he have planted?

13. Mr. Rosen adds 2.3 kilograms of fertilizer to 13.5 kilograms of soil. How much does the soil weigh now?

14. The outdoor garden light is set to go on at 8:05 P.M. What other information do you need to find how long the light will stay on?

15. **Write a problem** that can be solved by the Guess, Test, and Revise strategy. Ask other students to solve your problem.

EXTRA Practice, page 213

DECISION MAKING

Problem Solving: Determining Prices

SITUATION

At Beechwood Elementary School the fifth-grade students are planning a Juice Jamboree. They will be selling fruit juice to raise $50 at the school's 25th-anniversary open house.

PROBLEM

How much should they charge for the juice?

DATA

Grocery store will let us return any unopened can.

Napkins: 200 for $1.99

Straws: 70 for $1.40

Grape Juice — Add water. Makes 48 oz of juice. $2.40

Apple Juice — Add water. Makes 48 oz of juice. $1.92

Orange Juice — $2.88 — Add water. Makes 48 oz of juice.

Paper Cups — 100 cups for $2. Perfect size for a 6-oz. drink

Paper Cups — Perfect size for an 8-oz. drink. 100 cups for $3.

Lesson 5-1

USING THE DATA

How many servings of juice are in each can?

1. 6-oz servings
2. 8-oz servings

How much does 6 oz of each juice cost the fifth graders?

3. orange juice
4. grape juice
5. apple juice

How much does 8 oz of each juice cost the fifth graders?

6. orange juice
7. grape juice
8. apple juice

What is the total cost of each small serving of juice? Total cost includes the cost of napkin, straw and cup.

9. orange juice
10. grape juice
11. apple juice

What is the total cost of each large serving of juice?

12. orange juice
13. grape juice
14. apple juice

15. **What if** the students charge 50¢ for a small juice drink? How much profit will they earn if they sell 30 small servings of orange juice? of grape juice? of apple juice?

16. **What if** the students charge 65¢ for a large juice drink? How much will they earn if they sell 30 large servings of orange juice? of grape? of apple?

MAKING DECISIONS

17. How can the students predict how many drinks they might sell?

18. **What if** they estimate that they can sell only 150 drinks? How much profit will they need to make per drink in order to earn the $50?

19. **Write a list** of the things the students should think about before determining the price of a drink.

20. What price would you charge for the juice? Why?

21. **Write a list** of other things the students need to think about when planning the Juice Jamboree.

CURRICULUM CONNECTION

Math and Literature

In mathematics you have seen patterns in numbers and geometry. Poetry, too, has patterns.

Many poems have a beat and a pattern that please the ear. The rhythm in some poems comes from the regular, repeated pattern of accented and unaccented syllables in the words. Read this first line of a poem. Hear the accents in your head.

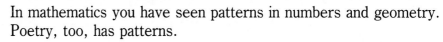

OH how I LOVE to go UP in a SWING

Can you "hear" the *ONE, two, three, ONE, two, three* beat?

Here are some common rhythm patterns in poems:

A. ◡ / ◡ / ◡ / ◡ /   **B.** / ◡ / ◡ / ◡ / ◡

C. / ◡ ◡ / ◡ ◡ / ◡ ◡ / ◡ ◡   **D.** ◡ ◡ / ◡ ◡ / ◡ ◡ / ◡ ◡ /

Read and "listen to" the first four lines of the nursery rhyme "Pussy-Cat and Queen."

> "Pussy-cat, pussy-cat,
>   Where have you been?"
> "I've been to London
>   To look at the Queen."

What if you wanted to note the accented syllables in this nursery rhyme? What is the pattern?

Think: As you read, you can hear these syllables accented:
(line 1) Pus-, Pus-
(line 2) Where, been
(line 3) been, Lon-
(line 4) look, Queen

ACTIVITIES

1. Find a poem whose rhythm pleases you. Read it aloud. Then try writing a poem using that pattern. Share your poem with the rest of your class.

2. Write a familiar nursery rhyme, or your favorite poem, on your own paper and mark the accents.

TECHNOLOGY

Calculator: Order of Operations

Enter these numbers and operation signs from left to right in your calculator. $12 + 6 \div 3 - 4$

What answer do you get?

If you do the operations in the order in which they appear, your calculator will display 2.

If you use the correct *order of operations:*

- do all operations within parentheses first
- multiply or divide
- add or subtract

$12 + 6 \div 3 - 4 = 10$, because first you would divide 6 by 3.

You *can* do a series of operations on a calculator and obtain the correct answer according to the order of operations. One way is to use the calculator's memory keys. Follow these steps:

Adds the number in the display to memory.

Subtracts the number in the display from memory.

Enter: 12 (M+) 6 (÷) 3 (=) (M+) 4 (M−) (MRC)

Recalls the numbers in the memory to display.

The calculator will display 10.

USING THE CALCULATOR

Find the answer. Use the calculator's memory keys.

1. $3 + 8 \div 2 + 5$
2. $15 - 3 \div 3 + 2$
3. $24 + 18 \div 9 + 6$
4. $0 + 7 \times 7 - 8 \times 4$
5. $10 + (3 + 14 \times 3)$
6. $27 - 32 \div 4 + 11$
7. $18 - 2 + (6 \times 5)$
8. $24 + 6 - 3 \times 6$

9. How could you obtain the correct answer to $32 + 18 + 24 \div 8 - 7$ *without* using the memory keys? What is the answer?

EXTRA PRACTICE

Basic Division Facts, page 185

Divide.

1. 3)$\overline{18}$
2. 7)$\overline{14}$
3. 5)$\overline{20}$
4. 4)$\overline{16}$
5. 2)$\overline{18}$
6. 8)$\overline{64}$
7. 6)$\overline{36}$
8. 7)$\overline{42}$
9. 1)$\overline{8}$
10. 5)$\overline{5}$
11. 5)$\overline{30}$
12. 6)$\overline{24}$
13. 9)$\overline{81}$
14. 7)$\overline{56}$
15. 6)$\overline{48}$
16. 49 ÷ 7
17. 28 ÷ 7
18. 17 ÷ 17
19. 0 ÷ 55

Complete.

20. Rule: ÷ 4

Input	Output
16	■
24	■
36	■
12	■

21. Rule: ÷ 5

Input	Output
15	■
30	■
45	■
10	■

22. Rule: ÷ 8

Input	Output
24	■
32	■
40	■
16	■

Mental Math: Using Division Patterns, page 187

Divide.

1. 2)$\overline{20}$
2. 3)$\overline{60}$
3. 4)$\overline{80}$
4. 6)$\overline{60}$
5. 2)$\overline{40}$
6. 3)$\overline{240}$
7. 5)$\overline{200}$
8. 7)$\overline{420}$
9. 4)$\overline{320}$
10. 8)$\overline{640}$
11. 3)$\overline{1,800}$
12. 6)$\overline{2,400}$
13. 4)$\overline{3,600}$
14. 5)$\overline{2,000}$
15. 9)$\overline{81,000}$
16. 360 ÷ 6
17. 2,800 ÷ 7
18. 30 ÷ 3
19. 64,000 ÷ 8

Estimating Quotients, page 189

Estimate by rounding.

1. 5)$\overline{53}$
2. 7)$\overline{72}$
3. 4)$\overline{82}$
4. 3)$\overline{85}$
5. 2)$\overline{84}$
6. 4)$\overline{385}$
7. 5)$\overline{495}$
8. 5)$\overline{365}$
9. 4)$\overline{759}$
10. 3)$\overline{645}$

Estimate by using compatible numbers.

11. 5)$\overline{4,563}$
12. 3)$\overline{2,754}$
13. 4)$\overline{7,807}$
14. 8)$\overline{3,975}$
15. 7)$\overline{4,225}$
16. 42,681 ÷ 5
17. 825 ÷ 8
18. 4,073 ÷ 2
19. 27,607 ÷ 7

EXTRA PRACTICE

Dividing 2- and 3-Digit Numbers, page 191
Divide.

1. 4)18
2. 8)34
3. 6)83
4. 2)67
5. 3)74

6. 3)746
7. 6)739
8. 9)438
9. 4)307
10. 5)608

11. 8)43
12. 3)67
13. 8)124
14. 8)249
15. 7)807

16. 6)438
17. 2)79
18. 7)137
19. 4)847
20. 5)317

21. 8)915
22. 2)77
23. 5)413
24. 6)310
25. 4)97

26. 53 ÷ 4
27. 417 ÷ 6
28. 950 ÷ 8
29. 734 ÷ 9

Dividing 4- and 5-Digit Numbers, page 193
Divide.

1. 5)5,625
2. 8)4,225
3. 9)6,509
4. 8)22,784

5. 6)$143.64
6. 6)$7,452
7. 3)4,863
8. 8)42,336

9. 3)1,875
10. 2)659
11. 7)6,747
12. 4)18,999

13. 8)2,053
14. 4)$14,312
15. 5)41,070
16. 6)1,437

17. 3)1,752
18. 4)2,976
19. 7)39,699
20. 9)68,239

21. $9.84 ÷ 4
22. 1,066 ÷ 5
23. $12,585 ÷ 3
24. 43,623 ÷ 8

Zeros in the Quotient, page 197
Divide.

1. 5)52
2. 3)61
3. 2)$8.18
4. 7)354
5. 7)422

6. 7)284
7. 6)647
8. 4)$4.32
9. 8)1,650
10. 7)1,756

11. 7)493
12. 5)5,308
13. 6)1,857
14. 5)4,504
15. 5)20,300

16. 7)7,720
17. 3)9,914
18. 4)8,295
19. 5)4,538

20. 82 ÷ 4
21. 4,125 ÷ 2
22. 36,080 ÷ 9
23. 6,303 ÷ 7

Dividing Whole Numbers: 1-Digit Divisors

Extra Practice

Problem-Solving Strategy: Using Estimation, page 199

Use estimation to solve the problem. Did you overestimate or underestimate? Why?

1. A troop has $112.69 for art supplies. If it buys 47 tubes of paint at $1.85 a tube, will it have enough money to pay for the supplies?

2. To earn a nature badge, a Scout must collect 85 items found outside. If Laurie collected 34 maple leaves, 22 pine cones, and 41 sunflowers, did she earn her badge?

3. Each Scout must deliver 126 books a week to hospital patients. If a Scout delivers 11 books a day for 7 days, will she deliver enough books?

4. The Scout troop wants to buy sleeping bags for $68.00 each. It plans to pay for them by selling cookies at $4.28 each box. If each Scout sells 24 boxes, will the troop have enough money to buy the bags?

Using Pictographs, page 201

Use the pictograph at the right to answer Questions 1–5.

1. What does ■ represent?

2. What does ▌ represent?

3. About how many news magazines were sold?

4. About how many photography magazines were sold?

5. About how many more sports magazines than puzzle magazines were sold?

MAGAZINES SOLD AT THE NEWSSTAND

Sports	■ ■ ■ ■ ■
Photography	■ ■ ■
News	■ ■ ■ ▌
Puzzles	▌
Science Fiction	■ ■

Each ■ = 100 magazines

Use the data in the table at the right to make a pictograph.

6. To which place did you round the data?

7. What is the value of the symbol you used?

ATTENDANCE AT THE BOOK FAIR

Day	Number of Students
Monday	121
Tuesday	140
Wednesday	129
Thursday	135
Friday	115

EXTRA PRACTICE

Mean, Median, Mode, and Range, page 203

Find the mean, median, mode, and range.

1. 130, 140, 135
2. 237, 221, 250
3. 47, 61, 19, 23, 40
4. 34, 85, 34, 103, 34
5. $67, $25, $18, $17, $143
6. $82, $145, $111, $145, $127
7. 65, 110, 100, 25, 65, 85, 40
8. 30, 143, 85, 100, 70, 38, 87

9. **MATH TEST SCORES**

Test 1	Test 2	Test 3	Test 4	Test 5
85	93	87	74	81

10. **FLOWERS IN A GARDEN**

Petunias	Roses	Daisies	Marigolds	Geraniums
24	23	23	18	27

Problem-Solving Strategy: Guess, Test, and Revise, page 205

Use the Guess, Test, and Revise strategy to solve the problem.

1. Ted's flower catalog is open. The product of the page numbers on the two facing pages is 306. What are the page numbers?

2. Lilian bought geraniums and hanging flowerpots for $52. The flowerpots cost $7 more than the geraniums. How much do the flowerpots cost?

3. Betty is planting 180 heads of lettuce in 18 rows. Some rows have 12 heads each and the rest have 9 heads each. How many rows have 12 heads of lettuce?

4. Jack bought beans from the farmer. He received $1.95 in change, all in quarters and dimes. There are 12 coins. How many quarters and how many dimes did he get?

Practice PLUS

KEY SKILL: Dividing 2- and 3-Digit Numbers (Use after page 191.)

Level A
Divide.

1. 5)18
2. 7)22
3. 6)10
4. 2)17
5. 4)19
6. 3)16
7. 4)25
8. 3)11
9. 3)22
10. 3)26
11. 5)79
12. 7)81
13. 5)66
14. 6)95
15. 2)59
16. 87 ÷ 5
17. 87 ÷ 2
18. 697 ÷ 3
19. 34 ÷ 4
20. 484 ÷ 2

Level B
Divide.

21. 6)43
22. 8)63
23. 4)372
24. 7)46
25. 3)233
26. 6)29
27. 7)638
28. 5)692
29. 9)729
30. 5)671
31. 191 ÷ 4
32. 774 ÷ 6
33. 114 ÷ 5
34. 779 ÷ 7
35. 359 ÷ 5
36. 325 ÷ 6
37. 794 ÷ 5
38. 857 ÷ 7
39. 649 ÷ 9
40. 498 ÷ 8

41. The dividend is 87. The divisor is 4. What are the quotient and the remainder?

42. The divisor is 4. The quotient is 98 and the remainder is 2. What is the dividend?

Level C
Divide.

43. 9)105
44. 3)746
45. 7)804
46. 8)497
47. 4)327
48. 4)916
49. 9)427
50. 6)676
51. 6)853
52. 5)916
53. 587 ÷ 3
54. 630 ÷ 4
55. 734 ÷ 8
56. 949 ÷ 3
57. 401 ÷ 7
58. 835 ÷ 7
59. 285 ÷ 9
60. 574 ÷ 6
61. 581 ÷ 8
62. 733 ÷ 9

63. The dividend is 398. The divisor is 7. What is the quotient and the remainder?

64. The divisor is 6. The quotient is 113 and the remainder is 3. What is the dividend?

Practice PLUS

KEY SKILL: Zeros in the Quotient (Use after page 197.)

Level A

Divide.

1. 6)65
2. 4)81
3. 9)96
4. 3)32
5. 2)81
6. 5)$5.50
7. 5)52
8. 3)62
9. 3)$3.15
10. 2)611
11. 6)$6.60
12. 3)3,075
13. 7)707
14. 5)1,031
15. 2)$4.08
16. 91 ÷ 3
17. 1040 ÷ 4
18. 351 ÷ 7
19. 651 ÷ 6
20. 801 ÷ 4

Level B

Divide.

21. 5)511
22. 9)544
23. 2)500
24. 4)807
25. 7)$2.10
26. 4)4,203
27. 8)4,814
28. 3)325
29. 5)2,050
30. 9)1,829
31. 6)3,610
32. 7)7,089
33. 9)6,381
34. 8)7,253
35. 2)8,414
36. 4)8,436
37. 7)9,456
38. 6)$96.30
39. 8)$56.08
40. 9)8,190
41. 4,322 ÷ 4
42. 6,349 ÷ 7
43. $45.54 ÷ 9
44. 24,017 ÷ 5

Level C

Divide.

45. 5)$155.40
46. 4)12,034
47. 9)3,649
48. 6)54,626
49. 8)48,028
50. 7)$14.42
51. 9)7,266
52. 9)15,350
53. 6)$984.06
54. 5)$88,510
55. 9)78,750
56. 7)91,624
57. 9)81,220
58. 7)85,429
59. 8)16,666
60. 6)29,762
61. 8)79,847
62. 7)17,570
63. 26,517 ÷ 5
64. 14,016 ÷ 7
65. $816.30 ÷ 9
66. 34,402 ÷ 8

Dividing Whole Numbers: 1–Digit Divisors **215**

Chapter Review

LANGUAGE AND MATHEMATICS

Match the definition in Column A with the correct term in Column B. *(page 182, 202)*

Column A
1. the difference between the greatest and the least numbers in a set of data
2. the number that appears most often in a set of data
3. the middle number of an ordered set of data
4. the average of a set of data

Column B
A. dividend
B. quotient
C. mean
D. mode
E. range
F. median

5. **Write a definition** or give an example of the words you did not use in Column B.

CONCEPTS AND SKILLS

Divide. *(pages 182–187, 190–193 196)*

6. $4\overline{)36}$
7. $7\overline{)42}$
8. $2\overline{)18}$
9. $5\overline{)25}$

10. $8\overline{)56}$
11. $4\overline{)32}$
12. $9\overline{)81}$
13. $7\overline{)49}$

14. $8\overline{)64}$
15. $3\overline{)9}$
16. $3\overline{)30}$
17. $4\overline{)120}$

18. $7\overline{)560}$
19. $9\overline{)810}$
20. $6\overline{)420}$
21. $3\overline{)92}$

22. $6\overline{)46}$
23. $9\overline{)912}$
24. $7\overline{)652}$
25. $5\overline{)627}$

26. $7\overline{)5,604}$
27. $9\overline{)8,642}$
28. $8\overline{)\$47,408}$
29. $6\overline{)\$59,724}$

30. $3\overline{)37}$
31. $6\overline{)565}$
32. $8\overline{)2,749}$
33. $4\overline{)\$1,980}$

34. $4\overline{)15,249}$
35. $7\overline{)12,431}$
36. $9\overline{)11,111}$
37. $2\overline{)5,999}$

38. $6\overline{)32,169}$
39. $8\overline{)29,989}$
40. $9\overline{)56,194}$
41. $7\overline{)19,898}$

42. $3\overline{)97,111}$
43. $4\overline{)39,827}$
44. $5\overline{)39,697}$
45. $8\overline{)94,878}$

Estimate the quotient. *(page 188)*

46. 6)62 **47.** 7)489 **48.** 4)325 **49.** 7)427

50. 6)312 **51.** 8)649 **52.** 8)72,305 **53.** 6)24,017

54. 5)4,095 **55.** 8)56,280 **56.** 7)63,436 **57.** 6)36,531

Find the mean, median, mode, and range. *(page 202)*

58. 22, 41, 53, 65, 19
59. 37, 56, 81, 73, 23
60. 105, 127, 113, 141, 139
61. 101, 103, 105, 107, 109
62. 37, 107, 93, 123, 116, 52, 60
63. $12, $13, $14, $12, $18, $17, $19
64. 115, 128, 145, 171, 126
65. 96, 72, 141, 84, 72
66. $36, $67, $15, $53, $21, $59, $15
67. 139, 188, 164, 150, 144

CRITICAL THINKING

68. How can you use mental math to find the mode of data such as 8, 12, 13, 13, 6, 21, 17, and 8?

MIXED APPLICATIONS

69. Barb is 7 years older than Ernie. In 5 years she will be twice as old as Ernie. How old are Barb and Ernie now? *(page 204)*

70. Sam works as a hospital volunteer. His goal is to work 70 hours per month. This month he has worked 52 hours. If he works 23 more hours, will he meet his goal? Did you overestimate or underestimate? *(page 198)*

71. The sum of three consecutive numbers is 252. What are the 3 numbers? *(page 204)*

72. Donald collected $6.50 for recycling cans. He has twice as many dimes as nickels. He has 8 more quarters than dimes. How many of each coin does he have? *(page 204)*

73. Bonnie spends $50 on sweaters, $29 on socks, and $18 on a belt. Is $100 enough to pay for these items? *(page 198)*

Chapter Test

Divide.

1. 3)60
2. 5)1,000
3. 8)64,000
4. 4)160

5. 6)98
6. 7)81
7. 2)417
8. 9)135

9. 5)219
10. 3)5,730
11. 4)8,025
12. 8)1,931

Estimate the quotient.

13. 8)85
14. 6)426
15. 7)4,937
16. 6)27,899

Use the pictograph at the right to answer Questions 17–20.

17. What does ■ represent?
18. What does ❙ represent?
19. About how many people said baseball was their favorite sport?
20. About how many more people liked basketball than bowling?

Favorite Sports

Football	■ ■ ■ ■ ■
Baseball	■ ■ ■ ❙
Soccer	■
Golf	❙
Bowling	■ ■ ■
Basketball	■ ■ ■ ■

■ = 100 people

Find the mean, median, mode, and range.

21. 32, 54, 78, 89, 67

22. 16, 24, 31, 16, 53

23. 100, 57, 89, 112, 60, 89, 102

Solve.

24. Walt bought 6 tulips and 6 roses for $32. The roses cost $6 more than the tulips. How much did the 6 tulips cost?

25. Althea needs $325 for a new bike. Will she save enough money if she saves $7 a week for a year? Did you overestimate or underestimate?

ENRICHMENT FOR ALL

DIVISIBILITY

When a whole number is **divisible** by another whole number, there is no remainder.

Divisible by 2	Divisible by 5	Divisible by 10
12, 14, 16, 18, 20	5, 10, 15, 20, 25	10, 20, 30, 40, 50

1. What digits appear in the ones place of the numbers that are divisible by 2? by 5? by 10?
2. Write a rule that can help you decide if a number is divisible by 2.
3. Write rules for divisibility by 5 and by 10.
4. If a number is divisible by 10, is it also divisible by 5? How do you know?

Here are divisibility rules for 3, 6, and 9.

Rule A number is divisible:	Example
by 3 if the sum of its digits is divisible by 3.	483 is divisible by 3. 4 + 8 + 3 = 15, and 15 is divisible by 3.
by 6 if it is divisible by both 2 and 3.	48 is divisible by 6. 48 is divisible by both 2 and 3.
by 9 if the sum of its digits is divisible by 9.	882 is divisible by 9. 8 + 8 + 2 = 18, and 18 is divisible by 9.

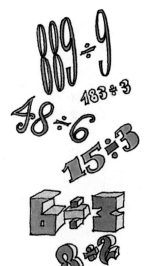

5. If a number is divisible by 9, by what other number is it also divisible? How do you know?

Is the number divisible by 2? by 3? by 5? by 6? by 9? by 10?

6. 33	**7.** 65	**8.** 94	**9.** 132	**10.** 516
11. 693	**12.** 1,096	**13.** 1,825	**14.** 5,470	**15.** 8,277
16. 9,010	**17.** 3,526	**18.** 7,401	**19.** 8,554	**20.** 6,480
21. 10,371	**22.** 16,080	**23.** 41,295	**24.** 24,608	**25.** 50,001

Cumulative Review

Choose the letter of the correct answer.

1. Which is the decimal for nine and thirty-eight thousandths?
   a. 0.938
   b. 9.38
   c. 9.038
   d. not given

2. 9 h 21 min
   − 4 h 26 min
   a. 5 h 55 min
   b. 4 h 55 min
   c. 4 h 47 min
   d. not given

3. 3,007
   − 1,213
   a. 1,694
   b. 1,894
   c. 1,794
   d. not given

4. 88
   × 70
   a. 6,160
   b. 660
   c. 61,160
   d. not given

5. 90 × 7,000
   a. 6,300,000
   b. 63,000
   c. 630,000
   d. not given

6. Compare: 19.07 ● 18.39
   a. <
   b. >
   c. =
   d. not given

7. The dividend is 42 and the divisor is 6. What is the quotient?
   a. 6
   b. 5
   c. 7
   d. not given

8. Find the mean: 35, 62, 40, 46, 52
   a. 47
   b. 40
   c. 46
   d. not given

9. 8)683
   a. 85 R3
   b. 85
   c. 85 R1
   d. not given

10. 5)$36,255
    a. $6,253
    b. $7,251.10
    c. $7,251
    d. not given

11. 9)2,656
    a. 295 R1
    b. 295 R3
    c. 259 R1
    d. not given

12. 5)5,200
    a. 1,400
    b. 1,040
    c. 1,004
    d. not given

13. 2,700 ÷ 3
    a. 90
    b. 900
    c. 9
    d. not given

14. The workers at Sudsy Shampoo Company packed 6,000 bottles of shampoo in one week. They packed 5 bottles of shampoo in each box. How many boxes did they fill?
    a. 1,000
    b. 1,200
    c. 1,400
    d. not given

Dividing Whole Numbers— 2-Digit Divisors

CHAPTER 6

MATH CONNECTIONS: TIME • VOLUME • ALGEBRA • PROBLEM SOLVING

1. What is happening in this picture?
2. What information do you see in the picture?
3. How can this information be useful to you?
4. Write a problem using the information.

UNDERSTANDING A CONCEPT

Mental Math: Using Division Patterns

Every day, Monday through Friday, the *Greenville Daily News* prints 150,000 copies. The newspapers are tied in bundles of 50. How many bundles are made each day?

Divide: 150,000 ÷ 50

Use the following pattern to help you find the answer.

```
    150 ÷ 50 = 3
  1,500 ÷ 50 = 30
 15,000 ÷ 50 = 300
150,000 ÷ 50 = 3,000
```

So 3,000 bundles are made each day.

1. Count the number of zeros in the dividend, the divisor, and the quotient. What is the pattern?

2. Does the pattern work for 40,000 ÷ 80? Why or why not?

3. Use the pattern to find the following quotients: 270 ÷ 90 and 49,000 ÷ 70.

TRY OUT Write the letter of the correct answer.
Find the quotient mentally.

4. 630 ÷ 70   a. 9   b. 90   c. 900   d. 9,000

5. 2,500 ÷ 50   a. 5   b. 50   c. 500   d. 5,000

6. 3,000 ÷ 60   a. 5   b. 50   c. 500   d. 5,000

7. 24,000 ÷ 30   a. 8   b. 80   c. 800   d. 8,000

PRACTICE

Find the quotient mentally.

8. 150 ÷ 30
9. 600 ÷ 30
10. 350 ÷ 70
11. 810 ÷ 90
12. 1,400 ÷ 20
13. 3,200 ÷ 40
14. 2,800 ÷ 70
15. 5,600 ÷ 80
16. 32,000 ÷ 80
17. 48,000 ÷ 60
18. 20,000 ÷ 50
19. 36,000 ÷ 90
20. 450 ÷ 50
21. 5,400 ÷ 60
22. 12,000 ÷ 20
23. 720 ÷ 90
24. 18,000 ÷ 30
25. 6,000 ÷ 30
26. 28,000 ÷ 40
27. 360 ÷ 40
28. 72,000 ÷ 80
29. 4,500 ÷ 90
30. 540 ÷ 90
31. 56,000 ÷ 70

Which two divisions have the same quotient?

32. **a.** 28,000 ÷ 40   **b.** 630,000 ÷ 90   **c.** 4,200 ÷ 60   **d.** 49,000 ÷ 70

33. **a.** 320 ÷ 40   **b.** 56,000 ÷ 70   **c.** 40,000 ÷ 50   **d.** 7,200 ÷ 90

Mixed Applications

Solve. Which method did you use?

ESTIMATION
MENTAL MATH
CALCULATOR
PAPER/PENCIL

34. Each week 20,000 copies of a town newspaper are printed. If the newspapers are tied together in batches of 50, how many batches are there?

35. Harper's Hardware Store runs an ad in the newspaper every Friday. The ad measures 18 cm by 12 cm. How many square cm of space does the ad take up?

36. Mrs. Coppola has a copy of the town newspaper for the day on which she was born. The page measures 37 cm by 27 cm. How many centimeters of wood are needed to make a frame for the page?

37. Make a pictograph for the data below.

DAILY NEWSPAPERS DELIVERED

Name	Number of Newspapers
Angie	75
Charlie	60
Willie	105

UNDERSTANDING A CONCEPT
Estimating Quotients

This year 1,012 students want to audition for the All-County Band. If 48 students can audition each day, about how many days will it take to audition all the students? You can estimate the quotient by using compatible numbers.

Estimate: 1,012 ÷ 48

Round the divisor to the nearest ten. Then use a basic fact.

1,012 ÷ 48
↓   ↓
Think: 1,000 ÷ 50 = 20

It will take about 20 days to audition all of the students.

1. Which division fact made the estimation easy?

2. Estimate 1,852 ÷ 38 using compatible numbers. Why is there more than one possible estimate?

3. Would rounding the dividend to the nearest thousand be a useful way to estimate 81)6,017? Why or why not?

TRY OUT Write the letter of the correct answer. Estimate the quotient.

4. 786 ÷ 19   a. 4   b. 40   c. 400   d. 4,000
5. 29,708 ÷ 62   a. 5   b. 50   c. 500   d. 5,000
6. 4,217 ÷ 71   a. 6   b. 60   c. 600   d. 6,000
7. 22,134 ÷ 68   a. 30   b. 300   c. 3,000   d. 30,000

PRACTICE

Estimate the quotient.

8. 225 ÷ 12	9. 948 ÷ 32	10. 6,231 ÷ 19	11. 29,140 ÷ 63
12. 575 ÷ 87	13. 1,776 ÷ 29	14. 3,488 ÷ 43	15. 55,879 ÷ 69
16. 645 ÷ 91	17. 38,401 ÷ 74	18. 4,873 ÷ 67	19. 847 ÷ 92
20. 28,521 ÷ 33	21. 1,689 ÷ 39	22. 50,432 ÷ 75	23. 37,299 ÷ 68
24. 4,015 ÷ 81	25. 66,427 ÷ 82	26. 2,609 ÷ 85	27. 295 ÷ 37
28. 25,765 ÷ 48	29. 17,421 ÷ 77	30. 846 ÷ 23	31. 9,879 ÷ 49
32. 39,524 ÷ 83	33. 979 ÷ 91	34. 47,905 ÷ 59	35. 6,363 ÷ 78
36. 80,746 ÷ 93	37. 17,821 ÷ 58	38. 615 ÷ 33	39. 7,109 ÷ 78

Mixed Applications

Solve. Which method did you use?

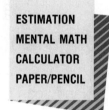

ESTIMATION
MENTAL MATH
CALCULATOR
PAPER/PENCIL

40. The music director spent $14.81 on sheet music, $29.35 on a record set, and $2.19 on a music notebook. What was his change from a $50 bill?

41. The music department owns 296 cassette tapes. If 48 tapes can fit in each box, about how many boxes does Paula need to pack all the tapes?

42. Of the students who auditioned for Mr. Lyon, 75 played trumpet, 115 played flute, and 59 played French horn. How many students auditioned for Mr. Lyon?

43. Mr. Lyon read that in one year the sale of compact discs in the United States rose from 48 million units to 97 million units. By how many units did sales increase?

Mixed Review

Find the answer. Which method did you use?

44. 562 × 7	45. 9,050 − 3,981	46. 0.689 + 7.051	47. 28 × 39	48. 9)521
49. 4,729 + 855	50. 5,400 ÷ 60	51. 8.042 − 0.93	52. 3,500 + 1,500	
53. 40.1 − 8.93	54. 100 × 250	55. 16 × 37	56. 3.87 + 5.9	

EXTRA Practice, page 252

Dividing Whole Numbers: 2-Digit Divisors

DEVELOPING A CONCEPT

Division with 2-Digit Divisors

Nancy is decorating posters for a social studies fair. She has 97 stickers and wants to decorate each poster with 23 stickers. How many posters can she decorate? How many stickers will she have left?

Nancy uses place-value models to solve the problem.

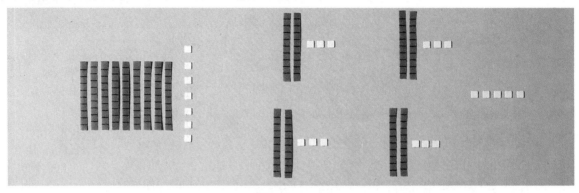

Nancy can decorate 4 posters. She will have 5 stickers left.

1. What does Nancy have to regroup in order to make equal sets of 23?

Use your place-value models to solve the problems below.

What if Nancy has 175 stickers and she wants to decorate each poster with 23 stickers?

2. How many posters can she decorate?
3. How many stickers will she have left?
4. What will she have to regroup?

What if Nancy has 250 stickers and she wants to decorate each poster with 18 stickers?

5. How many posters can she decorate?
6. How many stickers will she have left?
7. What will she have to regroup?

SHARING IDEAS

8. How is dividing by 2-digit numbers similar to dividing by 1-digit numbers?

PRACTICE

Use your place-value models to divide.

9. 22)67
10. 41)82
11. 13)156
12. 15)107
13. 12)139
14. 84 ÷ 31
15. 125 ÷ 15
16. 136 ÷ 12
17. 63 ÷ 28
18. 118 ÷ 17
19. 89 ÷ 21
20. 107 ÷ 15
21. 96 ÷ 43
22. 139 ÷ 12
23. 135 ÷ 11

Solve.

24. Althea and Nick are making models of city buildings for a social studies display. It takes 12 cubes to make one apartment building, and Althea has 67 cubes. How many apartment buildings can she make? How many cubes will she have left over?

25. It takes 18 cubes to build a skyscraper, and Nick has 82 cubes. How many skyscrapers can he build? How many cubes will he have left over?

26. Mrs. Murray has 129 poster boards to distribute equally among her 32 students. How many poster boards should each student get? How many poster boards will she have left over?

27. Jacob has 156 pushpins. He allots 12 pushpins for each bulletin board. For how many bulletin boards does he have enough pushpins? How many pushpins will he have left?

Mixed Review

Find the perimeter and area of the square or rectangle.

28.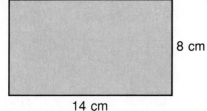
14 cm, 8 cm

29.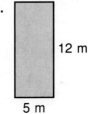
5 m, 12 m

30.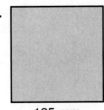
125 mm

UNDERSTANDING A CONCEPT
Dividing with 2-Digit Divisors

A. The fifth graders made 75 sandwiches. They divided the sandwiches equally among 32 students. How many sandwiches did they give each student? How many sandwiches did they have left over?

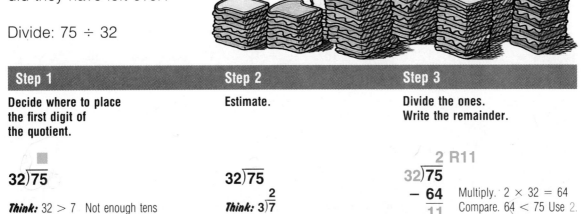

Divide: 75 ÷ 32

Step 1	Step 2	Step 3
Decide where to place the first digit of the quotient.	Estimate.	Divide the ones. Write the remainder.
$32\overline{)75}$ ■	$32\overline{)75}$ Think: $3\overline{)7}$ Try 2.	$\begin{array}{r} 2\text{ R}11 \\ 32\overline{)75} \\ -64 \\ \hline 11 \end{array}$ Multiply. $2 \times 32 = 64$ Compare. $64 < 75$ Use 2. Subtract. $75 - 64 = 11$
Think: $32 > 7$ Not enough tens. $32 < 75$ Divide the ones.		

The fifth graders gave each student 2 sandwiches. They had 11 sandwiches left over.

1. Is your answer reasonable? Why?

B. Sometimes you get a quotient that has more than 1 digit when you divide.

Divide: 459 ÷ 38

Step 1	Step 2	Step 3
Decide where to place the first digit of the quotient.	Divide the tens. Estimate.	Divide the ones. Estimate.
$38\overline{)459}$ ■	$\begin{array}{r} 1 \\ 38\overline{)459} \\ -38 \\ \hline 7 \end{array}$ Think: $3\overline{)4}$ Try 1.	$\begin{array}{r} 12\text{ R}3 \\ 38\overline{)459} \\ -38\downarrow \\ \hline 79 \\ -76 \\ \hline 3 \end{array}$ Think: $3\overline{)7}$ Try 2.
Think: $38 > 4$ Not enough hundreds. $38 < 45$ Divide the tens.		

2. Where would you place the first digit of the quotient?

    a. $48\overline{)125}$ **b.** $26\overline{)317}$ **c.** $32\overline{)1{,}426}$ **d.** $21\overline{)2{,}578}$

TRY OUT

3. 36)85 **4.** 57)349 **5.** 42)893 **6.** 21)6,826 **7.** 35)1,087

PRACTICE

Divide.

8. 23)94 **9.** 31)77 **10.** 24)59 **11.** 52)321 **12.** 42)265

13. 34)853 **14.** 27)569 **15.** 52)1,375 **16.** 45)6,391 **17.** 23)3,014

18. 42)97 **19.** 63)456 **20.** 31)660 **21.** 27)5,697 **22.** 76)5,478

23. 44)397 **24.** 62)942 **25.** 74)4,663 **26.** 28)5,939 **27.** 33)82

28. 22)89 **29.** 46)739 **30.** 76)325 **31.** 31)1,178 **32.** 69)9,109

33. 75 ÷ 23 **34.** 608 ÷ 43 **35.** 6,040 ÷ 41 **36.** 897 ÷ 39

37. 1,619 ÷ 73 **38.** 258 ÷ 32 **39.** 496 ÷ 44 **40.** 97 ÷ 21

Critical Thinking

41. What is the greatest number of digits you can have in a quotient if you divide a 3-digit number by a 2-digit number? Give an example to support your answer.

Mixed Applications

42. The fifth graders have 108 containers of juice. They have 36 containers of each kind of juice. How many kinds of juice do they have?

43. The fifth graders pack 21 containers of juice in each cooler. How many coolers do they fill if they have 108 containers? How many containers are left over?

44. Each container of juice costs 25¢. How much do 12 containers cost?

45. *Write a problem* using the following information: Your teacher has 425 crackers to divide equally among the students in the class. Use the number of students in your class. Ask others to solve your problem.

UNDERSTANDING A CONCEPT

Changing Estimates

A. Mrs. Ramos baked 195 oatmeal cookies. She put 24 cookies on each plate. How many plates did she fill? How many cookies did she have left over?

Sometimes you have to change an estimate when you divide.

Divide: 195 ÷ 24

Step 1
Decide where to place the first digit of the quotient.

24)‾195‾

Think: 24 > 1 Not enough hundreds
24 > 19 Not enough tens
24 < 195 Divide the ones.

Step 2
Estimate. Change the estimate if necessary.

24)‾195‾   **Think:** 2)‾19‾ with 9 on top
Try 9.
Multiply. 9 × 24 = 216
Compare. 216 > 195   Too much

Try 8.
Multiply. 8 × 24 = 192
Compare. 192 < 195   Use 8.

Step 3
Divide the ones.
Write the remainder.

```
      8 R3
24)195
  -192
     3    Subtract. 195 - 192 = 3
```

Check.
```
    24
  ×  8
   192
  +  3
   195
```

Mrs. Ramos filled 8 plates. She had 3 cookies left over.

B. Kristin did this division to find 627 ÷ 45.

```
     13 R42
45)627
  - 45↓
    177
  - 135
     42
```

1. In which place did Kristin have to change her estimate? Why?

TRY OUT Write the letter of the correct answer.

2. 26)68   a. 2   b. 2 R16   c. 3 R10   d. 3 R16
3. 35)230   a. 6 R15   b. 6 R20   c. 7 R5   d. 7 R15
4. 38)901   a. 23 R27   b. 24   c. 24 R27   d. 25
5. 59)8,120   a. 137 R37   b. 138 R37   c. 147 R37   d. 148 R37

PRACTICE

Divide.

6. 12)69   7. 26)83   8. 47)251   9. 38)507
10. 55)6,201   11. 28)81   12. 17)56   13. 48)321
14. 25)188   15. 39)1,829   16. 34)612   17. 29)97
18. 14)72   19. 17)207   20. 46)3,161   21. 55)406
22. 37)129   23. 39)256   24. 16)235   25. 24)3,278
26. 45)243   27. 56)451   28. 27)856   29. 23)87
30. 34)1,952   31. 56)670   32. 18)59   33. 27)729
34. 49)208   35. 35)4,062   36. 27)82   37. 65)839
38. 297 ÷ 16   39. 171 ÷ 39   40. 2,580 ÷ 57   41. 505 ÷ 39
42. 85 ÷ 13   43. 176 ÷ 47   44. 933 ÷ 18   45. 3,154 ÷ 25

Mixed Applications

46. A committee of 28 students spent a total of 952 hours organizing the bake sale. On the average, how many hours did each student spend on the sale?

47. Bob has to pack 180 bran muffins in boxes. He puts 16 muffins in each box. How many boxes does he fill? How many bran muffins does he have left?

48. Each of 18 students worked 22 hours the first week. How many hours did they work all together?

49. **Write a problem** that can be solved using 425 ÷ 75 = ■. Ask others to solve it.

EXTRA Practice, page 253; Practice PLUS, page 256

UNDERSTANDING A CONCEPT

Zeros in the Quotient

A toy manufacturer has 5,000 baseballs that have to be packed in boxes. Each box can hold 48 baseballs. How many boxes can the manufacturer fill? How many baseballs will the manufacturer have left over?

Divide: 5,000 ÷ 48

Step 1
Decide where to place the first digit of the quotient.

$$48\overline{)5{,}000}$$

Think: 48 > 5 Not enough thousands
48 < 50 Divide the hundreds.

Step 2
Divide the hundreds. Estimate. Change the estimate if necessary.

```
      1
48)5,000
  - 4 8
      2
```

Think: $4\overline{)5}$
Try 1.
Multiply. 1 × 48 = 48
Compare. 48 < 50 Use 1.
Subtract. 50 − 48 = 2

Step 3
Divide the tens. Estimate. Change the estimate if necessary.

```
     10
48)5,000
  - 4 8↓
      20
```

Think: 20 < 48
There are not enough tens to divide. Write 0 in the quotient.

Step 4
Divide the ones. Estimate. Change the estimate if necessary.

```
     104 R8
48)5,000
  - 4 8↓↓
      200
    - 192
        8
```

Think: $4\overline{)20}$
Try 5.
Multiply. 5 × 48 = 240
Compare. 240 > 200 Too much
Try 4.
Multiply. 4 × 48 = 192
Compare. 192 < 200 Use 4.
Subtract. 200 − 192 = 8

The manufacturer can fill 104 boxes with 8 baseballs left over.

1. ***What if*** you had written the answer as 14 R8? What error would you have made? How would an estimate have helped you discover the error?

2. Divide. Which examples have a zero in the ones place of the quotient? in the tens place? in the ones and tens places?

   a. $23\overline{)926}$ b. $42\overline{)4{,}239}$ c. $19\overline{)9{,}633}$ d. $35\overline{)7{,}014}$

TRY OUT   Divide.

3. 37)748   4. 54)3,245   5. 26)7,814   6. 43)$45.15   7. 35)7,251

PRACTICE

Divide.

8. 22)665   9. 57)596   10. 43)875   11. 34)711   12. 13)401

13. 38)1,912   14. 24)4,820   15. 65)6,552   16. 29)8,713   17. 42)$3,360

18. 27)8,156   19. 38)7,719   20. 69)$34.50   21. 22)9,258   22. 41)4,428

23. 32)964   24. 28)1,130   25. 35)3,783   26. 12)4,892   27. 43)8,628

28. 562 ÷ 55   29. 2,177 ÷ 27   30. 7,209 ÷ 18   31. 850 ÷ 21

32. $63.86 ÷ 31   33. 4,983 ÷ 47   34. 725 ÷ 72   35. 9,640 ÷ 31

36. 493 ÷ 16   37. 1,157 ÷ 23   38. 8,569 ÷ 79   39. 5,532 ÷ 27

Mixed Applications   Solve. You may need to use the Databank on page 519.

40. The heights of five stacks of toy boxes are 127 cm, 158 cm, 123 cm, 127 cm, and 140 cm. Find the mean and median heights of the boxes.

41. By how much has the average weekly earnings for workers in the United States increased from 1955 to 1988?

MENTAL MATH

You can divide by 5 and 50 mentally.

Divide: 300 ÷ 5

Think: Dividing by 5 is the same as multiplying by 2 and dividing by 10.

300 × 2 = 600   600 ÷ 10 = 60
So 300 ÷ 5 = 60.

Similarly, dividing by 50 is the same as multiplying by 2 and then dividing by 100.

Divide mentally.

1. 5)800   2. 5)1,000   3. 50)1,500   4. 50)1,200

UNDERSTANDING A CONCEPT
Dividing Larger Numbers

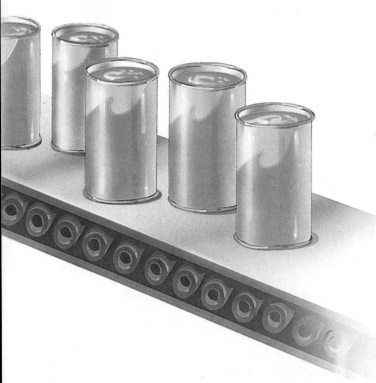

The Valley Juice Company processed 12,848 oz of orange juice. How many 32-oz cans will the company be able to fill with this amount of juice?

Gwen used a calculator.

12848 ⊕ 32 ⊜ | 401.5 |

Tony used paper and pencil to divide.

```
       401 R16
   32)12,848
     −12 8
         48
        −32
         16
```

1. Compare Tony's and Gwen's answers. How are they the same? How are they different?

2. Gwen says that she never bothers to estimate when she uses a calculator because calculators never make mistakes. Do you agree or disagree with Gwen about not needing to estimate? Why?

3. **What if** Tony and Gwen also had to find how many ounces of juice the company would have left? Whose answer would be more useful? Why?

4. **What if** Tony and Gwen had to find how many cans the company needs to hold all the juice? Whose method do you think would be easier to use? Why?

TRY OUT Write the letter of the correct answer.

5. 46)38,201   a. 83 R21   b. 803 R21   c. 830 R21   d. 8,300 R21

6. 23)16,819   a. 731 R6   b. 831 R6   c. 7,310 R6   d. 8,310 R6

7. 34)45,370   a. 133 R14   b. 1,334 R14 c. 1,335 R14 d. 13,340 R1

PRACTICE

Divide. Use mental math, a calculator, or paper and pencil.

8. 72)3,780   9. 80)56,000   10. 27)15,295   11. 34)41,888   12. 48)$14,736

13. 60)4,800   14. 73)29,523   15. 38)62,004   16. 52)31,726   17. 45)16,000

18. 19)1,905   19. 56)42,081   20. 30)18,030   21. 22)56,399   22. 33)25,111

23. 28)5,040   24. 10)51,727   25. 25)$10,000   26. 79)32,678   27. 41)12,500

28. 32)2,307   29. 70)63,000   30. 49)37,546   31. 30)90,900   32. 50)25,000

33. 7,192 ÷ 47   34. 18,119 ÷ 62   35. 27,090 ÷ 90   36. 45,164 ÷ 39

37. $12,000 ÷ 24   38. 8,013 ÷ 64   39. 11,669 ÷ 37   40. 5,454 ÷ 90

Mixed Applications

41. The workers at Richfield Orchards packed 36,000 shipping crates of oranges in only 48 days. They packed the same number of crates each day. How many crates of oranges did the workers pack each day?

42. Grains such as wheat, rye, and barley make up a part of the American diet. In a given year the average person ate 128.0 lb of wheat, 13.4 lb of rye, and 17.1 lb of barley. How much grain did the average person eat that year?

43. In a given year the average person ate about 200 lb of fresh fruits and vegetables. About how many pounds of fresh fruits and vegetables did the person eat each week?

44. Use the Databank on page 519 to make a line graph showing the average weekly earnings to the nearest $10 from 1955 to 1988 for workers in the United States. What can you conclude from the graph?

Mixed Review

Solve. Which method did you use?

MENTAL MATH
CALCULATOR
PAPER/PENCIL

45. 10,509
    − 7,468

46.   4,200
    ×   20

47. 897,316
    + 65,943

48.   4,600
    ×   11

49. 18.79
    − 5.068

50. 426.1
    +   7.93

51.   597
    ×   7

52.   308
    × 209

53. 8)4,132   54. 3)27,000   55. 5)602   56. 32)5,731

Extra Practice, page 254

PROBLEM SOLVING

✓ UNDERSTAND
✓ PLAN
✓ TRY
✓ CHECK
✓ EXTEND

Strategy: Choosing the Operation

A carton of film has 20 rolls of film in it. The camera store is selling cartons for a special price of $86.00. At this price, how much does one roll cost?

Study how the five-step process is used to solve this problem.

UNDERSTAND What do I know?	I know that a carton of film has 20 rolls of film and that it costs $86.00.
What do I need to find out?	I need to find the cost of 1 roll of film.
PLAN What can I do?	I can divide the cost of 1 carton by the number of rolls in each carton.
TRY Let me try my plan.	The carton costs $86.00. There are 20 rolls in each carton. $$86.00 \div 20 = 4.30$$ So each roll of film costs $4.30.
CHECK Have I answered the question?	Yes. I know that at $86.00, each roll of film in a carton costs $4.30.
EXTEND What have I learned?	I learned that when I know the cost of a group of items, I can divide to find the cost of one of the items.

Lynn is a professional photographer who buys cartons of film. Each roll of film has 36 pictures. How many pictures will Lynn be able to take with a carton of film?

1. Which operation did you use to solve the problem?

PRACTICE

Write which operation you will use. Then solve the problem.

2. Joan has 288 photos to put into an album. Each page of the album holds 12 pictures. How many pages will she fill?

3. It costs 85¢ to make an enlargement of a photograph. Herb had 7 pictures enlarged. How much did he pay?

4. Alice wants to buy film for her camera. At Camera World, she can buy 4 rolls of film for $16.48. What does 1 roll of film cost at Camera World?

5. Admission to the photography exhibit is $1.75. Attendance at the exhibit was 2,217 people on the weekend. How much money was collected?

Strategies and Skills Review

Solve. Use mental math, estimation, a calculator, or paper and pencil.

6. It costs $24.50 to buy a year's subscription to a photography magazine. A single issue costs $3 at a newsstand. How much will you save by buying a year's subscription over buying 12 single issues?

7. Connie buys color film for $3.99 a roll and black-and-white film for $2.99 a roll. She pays $21.94 for the film. How many rolls of each kind did she buy?

8. Hugo wants to buy a camera for $129.95 and a camera bag for $17.30. Will $150 be enough to pay for these items? How can you use estimation to tell?

9. There are 19 booths at the photography exhibit, with 20 photographs on display in each booth. How many photographs are being shown at the exhibit?

10. A roll of film costs $3.99. What other information do you need in order to find the cost of 1,000 pictures?

11. **Write a problem** that can be solved by using multiplication or division. Solve the problem. Ask others to solve your problem.

Division Thinkies

Using Number Concepts

A. Andy's favorite relative is his Uncle Sandy. Andy's uncle always brings gifts that he turns into puzzles for Andy to solve. Uncle Sandy says, "Today I brought you a box of baseball cards. If you guess how many cards are in the box, you can have them all. Listen carefully.

"If I put all the cards in 2 equal piles, I have 1 left over.
If I put all the cards in 3 equal piles, I have 1 left over.
If I put all the cards in 5 equal piles, I have 1 left over.

"What is the least number of cards I could have in the box?"

1. Help Andy get his gift. Solve the problem.

2. How did you find the answer?

Uncle Sandy and Andy solved these puzzles together.

3. Suppose there is 1 cap left over when some bottle caps are put into 2 equal piles, 1 cap left over when they are put into 5 equal piles, and 1 cap left over when they are put into 7 equal piles.

What is the least number of bottle caps there could be?

4. Suppose some peanuts are put into 6 equal piles, then 7 equal piles, then 8 equal piles. Each time there is 1 peanut left over.

What is the least number of peanuts there could be?

5. Suppose some raisins are divided into 2, 3, 4, 5, and 6 equal piles. Each time there is 1 raisin left over.

   What is the least number of raisins there could be?

   If the answer did not have to be the least number, what are some other possibilities?

B. Uncle Sandy knows lots of number tricks. Here is one he showed Andy.

	Example
• Pick a 3-digit number.	345
• Write your number twice to form a 6-digit number.	345,345
• Divide your number by 7. It should divide evenly.	$345{,}345 \div 7 = 49{,}335$
• Divide the quotient you got by 11. Again, there should not be a remainder.	$49{,}335 \div 11 = 4{,}485$
• Once more, divide the last quotient by 13. What number do you get?	$4{,}485 \div 13 = 345$

6. Try the trick with different 3-digit numbers. What happens each time?

7. Why does this trick work?
   (*Hint:* Dividing by 7, then 11, then 13 is the same as dividing by what single number? How do you know?)

8. What is the product of the number you just found and any 3-digit number ABC?
   Do the multiplication. What is the result?

UNDERSTANDING A CONCEPT

Changing Measures of Time

A. Kate recorded 120 minutes of music on a cassette tape. How many hours of music did she record?

| 60 seconds (s) = 1 minute |
| 60 minutes (min) = 1 hour |
| 24 hours (h) = 1 day |
| 7 days (d) = 1 week |
| 52 weeks (wk) = 1 year (y) |
| 12 months (mo) = 1 year |
| 365 days = 1 year |
| 366 days = 1 leap year |

You can divide to change a smaller unit to a larger unit.

120 min = ■ h

Think: 60 min = 1 h
 120 ÷ 60 = 2
 120 min = 2 h

Kate recorded 2 hours of music.

B. Jordan wanted to find how many seconds of music Kate recorded.

You can multiply to change a larger unit to a smaller unit.

120 min = ■ s

Think: 60 s = 1 min
 120 × 60 = 7,200
 120 min = 7,200 s

Kate recorded 7,200 seconds of music.

1. Would you multiply or divide to change from months to years? Why?

2. Would you multiply or divide to change from weeks to days? Why?

TRY OUT Write the letter of the correct answer.

3. 360 s = ■ min **a.** 6 **b.** 15 **c.** 30 **d.** 60
4. 84 d = ■ wk **a.** 7 **b.** 12 **c.** 14 **d.** 15
5. 4 h = ■ min **a.** 15 **b.** 96 **c.** 240 **d.** 360
6. 6 d = ■ h **a.** 72 **b.** 108 **c.** 144 **d.** 240

PRACTICE

Write *s, min, h, d, wk, mo,* or *y* to complete.

7. The basketball game lasted 2 ■.
8. Baseball season lasts about 7 ■.
9. The movie was 90 ■ long.
10. Eva's cold lasted 1 ■.
11. The baby napped for 45 ■.
12. It took Ed 1 ■ to ring the doorbell.
13. The lightning flashed for about 2 ■.
14. The thundershower lasted 10 ■.
15. Ron spent 1 ■ scraping and painting his room.
16. Jane trained for 1 ■ before entering the swimming contest.

Complete.

17. 120 s = ■ min
18. 72 h = ■ d
19. 24 mo = ■ y
20. 366 d = 1■
21. 3 h = ■ min
22. 4 wk = ■ d
23. 2 y = ■ wk
24. 3 y = ■ mo
25. 96 h = ■ d
26. 2 y = ■ d
27. 8 min = ■ s
28. 36 mo = ■ y
29. 156 wk = ■ y
30. 63 d = ■ wk
31. 7 d = ■ h
32. 15 min = ■ s
33. 4 y = ■ mo
34. 360 s = ■ min
35. 380 d = ■ y ■ d
36. 150 s = ■ min ■ sec
37. 59 d = ■ wk ■ d
38. 90 h = ■ d ■ h
39. 65 mo = ■ y ■ mo
40. 200 wk = ■ y ■ wk
41. 3 h 12 min = ■ min
42. 4 y 8 mo = ■ mo
43. 50 min 20 s = ■ s

Mixed Applications

44. Pablo recorded two pieces of music. The first piece was 5 min 13 s long. The second piece was 1 min 18 s shorter than the first piece. How long was the second piece?

45. Tom has a 90-minute blank cassette tape. Does he have enough tape to record a 2-hour music special from television?

46. Arlene spends $5.75 on a record and $7.95 on a tape. How much change does she get if she pays with a $20 bill?

PROBLEM SOLVING

Strategy: Finding a Pattern

Dr. Manuel is an archaeologist. He found an ancient picture made up of circles and triangles. The pattern in the picture shows 1 circle with 2 triangles, 2 circles with 4 triangles, 3 circles with 6 triangles, and so on. How many triangles will there be with 6 circles?

One way to solve the problem is to find a pattern by making a table.

Circles	1	2	3	4	5	6
Triangles	2	4	6	■	■	■

1. What is the rule for the pattern?

2. How many triangles will there be with 4 circles? with 5 circles? with 6 circles?

3. **What if** the table had looked like this? What is the rule for the pattern in this table?

Circles	1	2	3	4	5	6
Triangles	1	4	9	■	■	■

4. How many triangles would there be with 4 circles? with 5 circles? with 6 circles?

PRACTICE

Look for a pattern. Then solve the problem.

5. Dr. Manuel found a model of a temple. It was made of layers of stones. The bottom layer had 96 stones, the next layer had 88 stones, and the third layer had 80 stones. How many stones did the sixth layer have?

6. Another time he dug out a staircase. The first step was 9 inches above the ground, the second step was 18 inches above the ground, and the third step was 27 inches above the ground. How high off the ground was the eighth step?

7. Inside a temple he found a set of clay bowls that fit inside one another. The distance across the smallest bowl was 2 inches, across the next larger bowl was 6 inches, and across the third bowl was 18 inches. How many inches across was the fifth largest bowl?

8. He also discovered a pile of coins and stored them in boxes by size. He stored the smallest coins 96 to a box, the next size 48 to a box, and the next size 24 to a box. How many of the next largest coins did he store to a box?

Strategies and Skills Review

Solve. Use mental math, estimation, a calculator, or paper and pencil.

9. The archaeologists found a set of poles standing in a row. They were probably used as a primitive measuring instrument. The shortest pole was 2 inches high, the next one was 8 inches high, and the next was 32 inches high. How high was the next larger pole?

10. Dr. Johnson found an obelisk that is 78.5 feet tall. Dr. Lewis found one that is 18.25 feet tall. About how many times taller is the one Dr. Johnson found?

11. The largest standing obelisk in the world weighs 502.3 tons. The largest single block from a pyramid weighs 319.2 tons. How much more does the obelisk weigh than the block?

12. Melissa earns $7 an hour. Last week she worked 27 hours. What other information do you need in order to know whether she has earned enough money to buy a set of archaeological tools?

13. Teri's archaeology handbook has 12 chapters, each with the same number of pages. If the book is 264 pages long, how many pages are in each chapter?

14. **Write a problem** that can be solved by finding a pattern. Solve the problem. Ask others to solve your problem.

DEVELOPING A CONCEPT

Volume

The number of cubes it takes to build a box is the **volume** of the box. Volume is often measured in cubic centimeters, cubic decimeters, or cubic meters.

A liter container holds 1,000 cubic centimeters. There are 1,000 cubic centimeters in a cubic decimeter. So a liter is equal to a cubic decimeter.

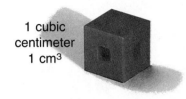

1 cubic centimeter
1 cm³

1 cubic decimeter
1 dm³

1. How many cubic centimeters are in a milliliter? How do you know?

WORKING TOGETHER

2. Fill a pencil box with various size cubes. How many did you use? Was this number the same for each size cube?

3. Why is it better to use a standard unit, such as a cubic centimeter, to measure volume?

Now use centimeter cubes to find the volume of small objects such as chalkboard erasers.

What is the volume of a chalkboard eraser in your classroom? First estimate by placing a centimeter cube next to the eraser. About how many cubes will it take to build the eraser shape? Check your estimate. Use centimeter cubes to build a box the same size and shape as the eraser.

Step 1 Make a row the length of the eraser.

Step 2 Make as many of these rows as needed to form a layer of the eraser.

Step 3 Make as many layers as needed to complete the eraser. What can you do if you do not have enough cubes?

4. How many cubes are in each row?
5. How many rows are in each layer?
6. How many layers make up the eraser?
7. What is the volume of the eraser?
8. Record your answers in a chart like this one.

Number of Cubes in Each Row	Number of Rows in Each Layer	Number of Layers	Volume

9. Choose two other small objects in the classroom that are shaped like boxes. First estimate the volume of each. Then build each shape using centimeter cubes. Fill in the chart for the shapes you build.

SHARING IDEAS

10. Look at your chart. How does the product of the numbers in the first three columns compare with the number in the last column?
11. Make up a rule for finding the volume of a box.
12. Do you think your rule is true for boxes that you would measure with cubic decimeters? Why or why not?

Practice

Use centimeter cubes to find the volume of the object.

13. math book
14. calculator
15. pencil box

Use your rule to find the volume.

16.

17.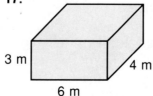

18.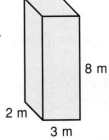

UNDERSTANDING A CONCEPT
Volume of a Rectangular Prism

A. A **rectangular prism** is a space figure with six rectangular faces.

When you find the volume of a rectangular prism, you find the number of cubic units that fill the prism.

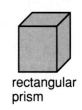

rectangular prism

Mr. Norton has a bin that is shaped like a rectangular prism. The bin is 5 m long, 4 m wide, and 2 m high. What is the volume of the bin?

You can use this formula to find the volume of the bin.

$V = \ell \times w \times h$ ℓ = length w = width h = height
$V = 5 \times 4 \times 2$
$V = 40$ The volume of the bin is 40 m³.

1. **What if** another bin is 4 m long, 2 m wide, and 5 m high? What would be the volume of this bin?

2. How does the volume of this bin compare to the volume of Mr. Norton's bin? Explain.

3. What formula would you write for finding the volume of a cube with a side whose length is s?

B. The volume of a box is 108 cm³. How long is the box?

You can divide to find the length.

$V = \ell \times w \times h$
$108 = \blacksquare \times 4 \times 3$
$108 = \blacksquare \times 12$ **Think:** 108 ÷ 12
$9 = \blacksquare$

The box is 9 cm long.

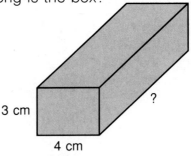

3 cm
4 cm
?

TRY OUT Find the missing number.

4.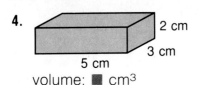
volume: ■ cm³

5. length: 4 m
   width: 2 m
   height: 4 m
   volume: ■ m³

6. length: ■ cm
   width: 3 cm
   height: 4 cm
   volume: 84 cm³

PRACTICE

Find the volume.

7.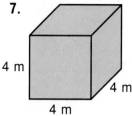

8.

9.

10.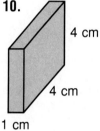

11.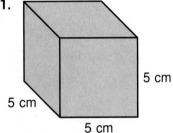

12.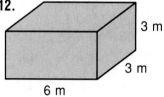

13. length: 5 m
    width: 2 m
    height: 6 m

14. length: 8 cm
    width: 4 cm
    height: 5 cm

15. length: 3 m
    width: 6 m
    height: 6 m

Complete the table.

Length	Width	Height	Volume
4 cm	2 cm	2 cm	16. ■ cm³
5 m	4 m	17. ■ m	160 m³
18. ■ m 5m	3 m	6 m	90 m³
7 cm	19. ■ cm	4 cm	56 cm³

Mixed Applications

20. Marsha owns these two jewelry boxes. Which one has the greater volume?

21. Gary has a sheet of wrapping paper that is 96 cm long and 50 cm wide. Does he have enough paper to wrap a present that requires 4,900 cm² of paper? If not, how much more paper does he need?

DECISION MAKING

Problem Solving: Buying Magazines

SITUATION

The fifth-grade students at Cotton Hill School would like to include magazine articles in their studies. They want to subscribe to several magazines that they think are interesting and that could be helpful as part of their lessons. Their teacher, Miss Latino, has suggested that they compare both magazine subscription and newsstand prices before making any decisions.

PROBLEM

Should the students buy single copies at a newsstand or a subscription to the magazines they select?

DATA

ALL-NEWS $.95
The weekly magazine of important news events written for students.
Subscription Price
28 weeks $19.60
52 weeks $28.60

SCIENCE MONTHLY $1.25
Once-a-month magazine of all science happenings.
Subscription Price
6 months $6.00
12 months $10.80

SPORTS NEWS $2.50
A look at the world of sports each month.
Subscription Price
6 months $12.00
12 months $21.00

NUMBER MAGIC $.50
A weekly magazine of math puzzles and games
Subscription Price
28 weeks $11.20
52 weeks $18.20

STORIES AND POEMS $3.50
A monthly magazine of new stories and poems.
Subscription Price
6 months $18.00
12 months $24.00

USING THE DATA

Using the shorter subscription, find the cost per issue of the magazine.

1. *All-News*
2. *Science Monthly*
3. *Sports News*
4. *Number Magic*
5. *Stories and Poems*

What is the cost per issue of a one-year subscription?

6. *All-News*
7. *Science Monthly*
8. *Sports News*
9. *Number Magic*
10. *Stories and Poems*

MAKING DECISIONS

11. Which way to buy magazines offers the lowest cost per issue?

12. **What if** the students have $31? Which magazines should they consider? List reasons for your answer.

13. **What if** Miss Latino wants to encourage her students to read for fun and recreation? Which magazines might she choose for the class? List reasons for your answer.

14. The contents of *Number Magic* and *Stories and Poems* will not become outdated in a few weeks or months. How could this fact affect your decision?

15. Which length of magazine subscription do you think would be best? Why?

16. **Write a list** of the things the students should think about when deciding which magazines to buy for the class.

17. **What if** you could choose any two of the magazines for your class? Which two would you choose? How would you buy them? List reasons for your decisions.

CURRICULUM CONNECTION

Math and Music

In music, a *scale* is a series of seven notes representing steps from one tone to the next tone. Many songs are based on a *major scale*. All major scales have the same pattern of steps.

Look at the picture of part of a piano keyboard. The distance between two keys right next to each other on the keyboard is a *half step*. The distance between two notes that have a key between them is a *whole step*.

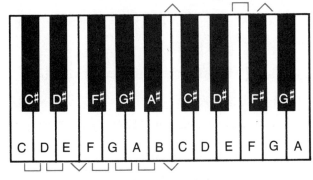

⊔ = whole step
∨ = half step

If you start on C and play every white key to the next C, you can see the pattern of whole and half steps that make a major scale. The steps follow this pattern:

    2 whole steps   1 half step   3 whole steps   1 half step

A major scale always follows this pattern of steps, no matter what note it starts on.

What if you wanted to know what keys to play for a major scale starting on G? Use the keyboard above and the major scale pattern.

Think: Start on G. Follow the pattern.

    2 whole steps: G to A
                           A to B
    1 half step: B to C
    3 whole steps: C to D
                           D to E
                           E to F♯
    1 half step: F♯ to G

The keys to play for a major scale starting on G are: G, A, B, C, D, E, F♯, G.

ACTIVITIES

1. Write the names of the notes of the major scale starting on D.

2. Do research to find the pattern of whole and half steps in a minor scale.

Computer Spreadsheet Informal Algebra: Operation Rules

You can think of the mathematical operations you use—addition, subtraction, multiplication, and division—as a machine. The machine has three important parts:

The machine uses its rule to change each number IN to exactly one other number OUT. For example, suppose the rule is "multiply by 4" and you begin with 7:

The number OUT is 28. What will the OUT number be if you put IN 20?

In the computer game RULES, you will try to discover what rule the machine is using to change each number you put IN.

AT THE COMPUTER

1. Enter a number in the IN column. The computer will display a number in the OUT column. It is probably too soon to guess the rule, but begin thinking about how the IN and OUT numbers are related.

2. Keep trying IN numbers until you know the rule the computer is using. Then enter the rule, for example:

$$\text{OUT} = \text{IN} * 4 \quad \text{* means multiply.}$$

The computer will let you know if your answer is correct.

3. **What if** the rule for the example at the top were "multiply by 4 and then subtract 5"? How would you show the rule?

4. Keep track of how many IN numbers you need to guess each rule. Why is it so hard to guess the rule with only one IN/OUT pair?

5. Play RULES with a friend. Take turns guessing the rules. The player who guesses five rules in the least number of tries wins.

EXTRA PRACTICE

Mental Math: Using Division Patterns, page 223
Find the quotient mentally.

1. 140 ÷ 20
2. 360 ÷ 60
3. 450 ÷ 90
4. 640 ÷ 80
5. 1,800 ÷ 20
6. 2,400 ÷ 80
7. 4,900 ÷ 70
8. 3,200 ÷ 40
9. 4,800 ÷ 80
10. 36,000 ÷ 40
11. 20,000 ÷ 40
12. 54,000 ÷ 90
13. 400 ÷ 80
14. 6,000 ÷ 20
15. 30,000 ÷ 50
16. 54,000 ÷ 60
17. 2,800 ÷ 70
18. 420 ÷ 70
19. 7,200 ÷ 90
20. 18,000 ÷ 30
21. 8,000 ÷ 40
22. 15,000 ÷ 30
23. 5,400 ÷ 60
24. 35,000 ÷ 50

Estimating Quotients, page 225
Estimate the quotient.

1. 312 ÷ 14
2. 827 ÷ 22
3. 3,125 ÷ 34
4. 28,139 ÷ 40
5. 627 ÷ 72
6. 468 ÷ 56
7. 1,275 ÷ 14
8. 46,110 ÷ 89
9. 812 ÷ 42
10. 1,849 ÷ 29
11. 1,114 ÷ 61
12. 84,217 ÷ 91
13. 924 ÷ 53
14. 2,758 ÷ 38
15. 624 ÷ 26
16. 67,921 ÷ 81
17. 3,724 ÷ 20
18. 2,227 ÷ 42
19. 52,319 ÷ 72
20. 1,619 ÷ 54

Division with 2-Digit Divisors, page 227
Use your place-value models to divide.

1. 34)79
2. 23)95
3. 56)78
4. 31)142
5. 16)108
6. 22)123
7. 11)120
8. 21)89
9. 43)96
10. 11)115
11. 36)82
12. 27)118
13. 85 ÷ 14
14. 98 ÷ 35
15. 135 ÷ 42
16. 121 ÷ 23
17. 94 ÷ 22
18. 396 ÷ 63
19. 149 ÷ 68
20. 89 ÷ 44

EXTRA PRACTICE

Dividing with 2-Digit Divisors, page 229
Divide.

1. 22)¯97
2. 41)¯335
3. 31)¯74
4. 27)¯579
5. 52)¯694
6. 21)¯886
7. 32)¯755
8. 36)¯869
9. 61)¯814
10. 46)¯509
11. 21)¯932
12. 32)¯96
13. 26)¯5,798
14. 89)¯3,859
15. 31)¯1,974
16. 73)¯987
17. 73)¯5,217
18. 62)¯9,383
19. 63)¯7,322
20. 41)¯2,981

Changing Estimates, page 231
Divide.

1. 14)¯86
2. 29)¯89
3. 57)¯359
4. 28)¯433
5. 34)¯98
6. 19)¯116
7. 56)¯359
8. 25)¯229
9. 36)¯197
10. 39)¯109
11. 25)¯868
12. 36)¯2,094
13. 38)¯3,668
14. 14)¯270
15. 68)¯4,378
16. 54)¯210
17. 629 ÷ 36
18. 91 ÷ 15
19. 804 ÷ 27
20. 2,564 ÷ 48

Zeros in the Quotient, page 233
Divide.

1. 26)¯525
2. 67)¯673
3. 43)¯869
4. 16)¯489
5. 39)¯1,957
6. 26)¯5,221
7. 45)¯4,532
8. 28)¯8,417
9. 33)¯6,766
10. 38)¯$22.80
11. 22)¯889
12. 13)¯7,897
13. 1,384 ÷ 46
14. 3,307 ÷ 32
15. $24.80 ÷ 62
16. 785 ÷ 13

Dividing Whole Numbers: 2-Digit Divisors

Extra Practice

Dividing Large Numbers, page 235

Divide.

1. 63)3,453
2. 90)81,000
3. 28)18,349
4. 52)$21,216
5. 35)26,629
6. 12)62,081
7. 25)$15,000
8. 50)30,000
9. 22)1,956
10. 80)24,160
11. 24)$18,000
12. 52)40,831
13. 4,069 ÷ 31
14. $25,420 ÷ 82
15. 60,600 ÷ 30
16. 35,675 ÷ 15

Problem-Solving Strategy: Choosing the Operation, page 237

Solve the problem. Which operation did you use?

1. Anthony bought 5 rolls of film. Each roll cost $5.98. How much did he pay for all the film?

2. Kenneth wants to buy a new camera. The camera he wants is on sale for $87.25. It usually costs $109.30. How much will Kenneth save if he buys the camera on sale?

3. There were 597 copies of *Photo World* magazine printed in May and 726 copies printed in June. How many magazines were printed in the two months?

4. Muriel has 255 slides to load on a carousel. Each carousel can hold 15 slides. How many carousels will she use?

Changing Measures of Time, page 241

Complete.

1. 180 s = ■ min
2. 48 h = ■ d
3. 365 d = ■ y
4. 144 h = ■ d
5. 5 y = ■ d
6. 6 min = ■ s
7. 35 d = ■ wk
8. 6 d = ■ h
9. 10 min = ■ s
10. 56 d = ■ wk
11. 300 s = ■ min
12. 3 wk = ■ d
13. 370 d = ■ y ■ d
14. 90 s = ■ min ■ s
15. 46 d = ■ wk ■ d
16. 4 h 13 min = ■ min
17. 3 y 6 mo = ■ mo
18. 45 min 13 s = ■ s
19. 35 mo = ■ y ■ mo
20. 300 wk = ■ y ■ wk
21. 80 h = ■ d ■ h
22. 2 h 25 min = ■ min
23. 5 y 3 mo = ■ mo
24. 30 min 17 s = ■ s
25. 27 mo = ■ y ■ mo
26. 215 wk = ■ y ■ wk
27. 62 h = ■ d ■ h

EXTRA PRACTICE

Problem-Solving Strategy: Finding a Pattern, page 243

Find a pattern. Then solve the problem.

1. At pottery class Julie made a set of clay bowls that fit one inside the other. The smallest bowl is 3 cm across. The next larger bowl is 5.5 cm across, and the third bowl is 8 cm across. How many centimeters across is the sixth largest bowl?

2. The pottery class built a pyramid out of clay squares. The bottom layer has 64 clay squares; the next layer, 49 clay squares; and the third layer, 36 clay squares. How many clay squares does the fifth layer have?

Volume, page 245

Use your rule to find the volume of the box.

1.
2.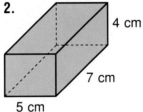
3.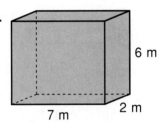

Volume of a Rectangular Prism, page 247

Find the volume.

1.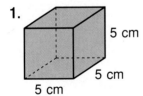
2.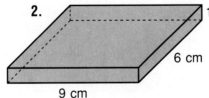
3.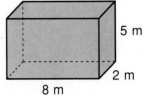

4. length = 6 m
   width = 3 m
   height = 7 m

5. length = 8 cm
   width = 3 cm
   height = 4 cm

6. length = 2 m
   width = 7 m
   height = 7 m

7. length = 9 m
   width = 5 m
   height = 6 m

8. length = 8 m
   width = 3 m
   height = 9 m

9. length = 7 m
   width = 6 m
   height = 8 m

10. length = 4 m
    width = 3 m
    height = 9 m

11. length = 5 m
    width = 6 m
    height = 8 m

12. length = 5 m
    width = 9 m
    height = 7 m

Dividing Whole Numbers: 2-Digit Divisors

Practice PLUS

KEY SKILL: Changing Estimates (Use after page 231.)

Level A

Divide.

1. 13)42
2. 11)81
3. 27)62
4. 19)76
5. 26)87
6. 28)88
7. 43)80
8. 24)91
9. 37)69
10. 25)82
11. 29)174
12. 33)211
13. 28)116
14. 35)185
15. 39)195
16. 32)275
17. 54)459
18. 42)250
19. 76)640
20. 68)553

Level B

Divide.

21. 37)321
22. 26)198
23. 58)203
24. 87)423
25. 15)126
26. 47)290
27. 77)449
28. 86)253
29. 29)158
30. 36)183
31. 26)988
32. 24)847
33. 19)442
34. 12)680
35. 62)801
36. 29)1,092
37. 46)2,844
38. 38)1,209
39. 27)1,108
40. 56)3,123

Level C

Divide.

41. 26)864
42. 36)941
43. 45)897
44. 28)620
45. 33)618
46. 82)6,427
47. 38)2,561
48. 41)3,264
49. 82)2,440
50. 75)5,003
51. 12)6,197
52. 28)4,509
53. 36)7,095
54. 19)4,060
55. 44)8,267
56. 73)14,087
57. 58)26,211
58. 67)50,192
59. 82)60,344
60. 91)33,303

Practice PLUS

KEY SKILL: Zeros in the Quotient (Use after page 233.)

Level A

Divide.

1. 17)346
2. 27)277
3. 62)1,249
4. 31)935

5. 48)2,432
6. 32)1,627
7. 53)1,064
8. 13)267

9. 23)692
10. 12)611
11. 73)2,201
12. 57)2,296

13. 16)486
14. 21)850
15. 46)921
16. 32)1,302

17. 24)1,223
18. 14)849
19. 22)1,992
20. 13)1,693

Level B

Divide.

21. 72)7,569
22. 43)8,917
23. 27)5,617
24. 36)3,859

25. 52)8,324
26. 62)9,325
27. 24)2,889
28. 66)6,735

29. 19)9,652
30. 23)10,350
31. 46)5,019
32. 56)28,566

33. 29)3,143
34. 31)9,427
35. 16)$17.44
36. $49.68 ÷ 46

37. 10,309 ÷ 51
38. 3,850 ÷ 11
39. 8,719 ÷ 29
40. 4,006 ÷ 39

Level C

Divide.

41. 89)9,430
42. 76)7,691
43. 96)9,893
44. 19)5,758

45. 47)9,463
46. 80)9,630
47. 71)9,238
48. 83)9,043

49. 54)7,020
50. 69)7,108
51. 35)4,201
52. 83)$920.47

53. 10,103 ÷ 33
54. 5,103 ÷ 49
55. 6,953 ÷ 17
56. 6,028 ÷ 15

57. $86.94 ÷ 42
58. 10,088 ÷ 21
59. $80.32 ÷ 16
60. 9,303 ÷ 31

Dividing Whole Numbers: 2–Digit Divisors

Chapter Review

LANGUAGE AND MATHEMATICS

Write the letter of the definition in Column A that matches the term in Column B.

Column A
1. unit used to measure volume *(page 244)*
2. 366 days *(page 240)*
3. $\ell \times w \times h$ *(page 244)*
4. space figure with six rectangular faces *(page 246)*
5. 60 minutes *(page 240)*

Column B
A. one leap year
B. volume
C. one minute
D. rectangular prism
E. one hour
F. cubic meter

CONCEPTS AND SKILLS

Divide. *(pages 222, 226–235)*

6. 140 ÷ 70
7. 400 ÷ 50
8. 360 ÷ 60
9. 270 ÷ 90
10. 16,000 ÷ 80
11. 7,200 ÷ 90
12. 5,400 ÷ 60
13. 64,000 ÷ 80
14. 39)97
15. 24)114
16. 56)156
17. 17)100
18. 23)304
19. 31)64
20. 62)256
21. 40)1,293
22. 56)6,548
23. 74)132
24. 27)2,342
25. 52)321
26. 17)2,444
27. 73)6,521
28. 64)4,370
29. 59)1,267
30. 62)732
31. 55)1,277
32. 39)6,241
33. 43)4,063
34. 21)632
35. 64)1,285
36. 72)2,164
37. 93)81,242
38. 27)56,491
39. 35)27,001
40. 52)18,030
41. 46)57,241
42. 85 ÷ 26
43. 145 ÷ 15
44. 139 ÷ 11
45. 78 ÷ 33
46. 113 ÷ 14
47. $83.43 ÷ 27
48. 8,635 ÷ 41
49. 5,062 ÷ 18
50. 4,984 ÷ 56
51. 46,118 ÷ 42

Estimate the quotient. *(page 224)*

52. 3,689 ÷ 41
53. 469 ÷ 24
54. 44,351 ÷ 74
55. 973 ÷ 12

Complete. *(page 240)*

56. 180 min = ■ h
57. 24 mo = ■ y
58. 499 s = ■ min ■ s
59. 67 d = ■ h
60. 222 d = ■ wk ■ d
61. 324 d = ■ wk ■ d

Find the volume. *(page 246)*

62. length: 2 m
    width: 7 m
    height 5 m:
63. length: 4 cm
    width: 3 cm
    height: 4 cm
64. length: 9 m
    width: 2 m
    height: 2 m

Find the volume. *(page 246)*

65.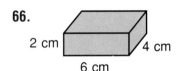
66.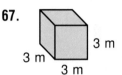
67.

CRITICAL THINKING

68. How would you divide 1,400 ÷ 50 mentally? What is the quotient?

69. How would you find the width of a box if you know its volume, length, and height?

MIXED APPLICATIONS

70. Dennis made $28 babysitting. Sue made $22. Janet made half as much as Dennis and Sue but spent half of her money on a compact disc. How much money does Janet have now? *(page 236)*

71. Arlene saved $6 the first week, $10 the second week, $14 the third week, and $18 the fourth week, and so on. How much did Arlene save the eighth week? *(page 242)*

72. The store manager stacks cans of cat food in a display. One can will be on the top row, 3 cans on the next row, 5 cans on the third row, and so on. How many cans will be on the twelfth row? *(page 242)*

73. James has to pack 160 dinner rolls into boxes. He puts 12 rolls into each box. How many boxes does he fill? How many dinner rolls does he have left? *(page 226)*

Chapter Test

Divide.

1. 800 ÷ 40
2. 5,400 ÷ 90
3. 450 ÷ 50
4. 81,000 ÷ 90
5. 36)942
6. 45)7,026
7. 75)5,862
8. 52)8,329

Estimate the quotient.

9. 927 ÷ 31
10. 1,785 ÷ 28
11. 54,322 ÷ 76
12. 2,619 ÷ 87

Complete.

13. 20 min = ■ s
14. 3 y = ■ mo
15. 300 s = ■ min
16. 156 wk = ■ y

Find the volume.

17.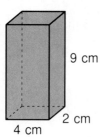

18.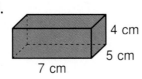

19.

20.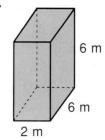

Solve.

21. Ray earns $15 for each lawn he mows. How many lawns must he mow to earn $270? Which operation did you use to solve?

22. A pair of running shoes costs $42.95 at the Foot Place. How much does the store make from selling 12 pairs? Which operation did you use to solve?

23. Frank has a job assembling the sections of the Sunday newspaper. He is a fast worker. It takes him 20 ■ to assemble one paper. Write s, min, h, d, wk, mo, or y to complete.

24. Mrs. Rogers owns a set of bowls that fit inside one another. The distance across the smallest bowl is 2 inches, across the next larger bowl is 4 inches, and across the third larger bowl is 6 inches. How many inches across is the fifth larger bowl?

25. Claudia has 12 diamonds, 24 rubies, and 48 emeralds. In what pattern can she arrange these to make a necklace?

ENRICHMENT FOR ALL

MAGIC SQUARES

Find the sum of the numbers in each row, column, and diagonal in the square at the right.

What do you notice about the sums?

The square is a 3-by-3 **magic square** for addition. The sum of the numbers in each row, column, and diagonal is the same. This sum is called a **magic sum.**

A

16	18	8
6	14	22
20	10	12

1. What is the magic sum for the square?

2. Copy the square at the right. Divide each number in the magic square above by 2 and then add 3 to complete the square. Is the square a magic square? If so, what is the magic sum?

B

		7
	10	

3. Copy the square at the right. Complete the square by adding 4 to each number in the magic square at the top of the page and then dividing by 2. Is this square a magic square? If so, what is the magic sum?

C

10		
	9	

You can predict the magic sum in a 3-by-3 magic square if you know the number in the center box. Study the three completed squares.

4. How does the magic sum of each square compare with the center number in each square?

5. Use what you learned to make a magic square with 12 as the center number.

Dividing Whole Numbers: 2-Digit Divisors **261**

Cumulative Review

Choose the letter of the correct answer.

1. 7)835
   a. 119 R4
   b. 119
   c. 119 R2
   d. not given

2. Estimate by rounding to the nearest whole number.
   3.07
   2.12
   6.53
   + 3.47
   a. 17
   b. 16
   c. 15
   d. 14

3. Round 23.807 to the nearest hundredth.
   a. 23.9
   b. 23.81
   c. 24.807
   d. not given

4. 9)3,681
   a. 400
   b. 409
   c. 490
   d. not given

5. 60 × 70 × 50
   a. 210,000
   b. 21,000
   c. 2,100,000
   d. not given

6. 719
   × 325
   a. 233,675
   b. 232,657
   c. 233,765
   d. not given

7. 1,400 ÷ 2
   a. 7
   b. 70
   c. 7,000
   d. not given

8. Estimate: 29,373 ÷ 32
   a. 900
   b. 1,000
   c. 980
   d. not given

9. 180 s = ■ min
   a. 2
   b. 30
   c. 3
   d. not given

10. 29)8,882
    a. 304 R8
    b. 300 R8
    c. 306 R8
    d. not given

11. 89)39,362
    a. 442 R24
    b. 424 R20
    c. 442 R2
    d. not given

12. 54)1,407
    a. 26 R3
    b. 27
    c. 27 R2
    d. not given

13. Find the volume of a rectangular prism with length = 6 m, width = 3 m, height = 8 m.
    a. 144 m
    b. 144 m²
    c. 144 m³
    d. not given

14. There are 126 cartons of juice. There are 42 cartons of each kind of juice. How many kinds of juice are there?
    a. 2
    b. 3
    c. 4
    d. not given

Multiplying and Dividing Decimals

CHAPTER 7

MATH CONNECTIONS: MEASUREMENT
• MASS AND CAPACITY • PROBLEM SOLVING

Sneakers	$10.98
Loafers	$14.89
Socks (Cotton) 3 pair for	$ 5.00
Sweat socks 4 pair for	$10.00

1. What information do you see in this picture?
2. How can this information be useful to you?
3. About how much will 2 pairs of sneakers cost?
4. Write a problem using the information.

UNDERSTANDING A CONCEPT

Mental Math: Multiply and Divide Decimals

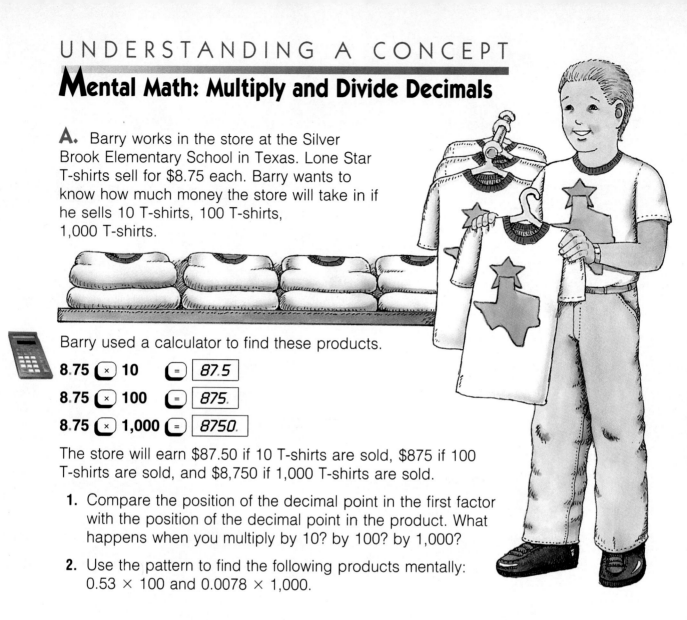

A. Barry works in the store at the Silver Brook Elementary School in Texas. Lone Star T-shirts sell for $8.75 each. Barry wants to know how much money the store will take in if he sells 10 T-shirts, 100 T-shirts, 1,000 T-shirts.

Barry used a calculator to find these products.

8.75 × 10 = 87.5
8.75 × 100 = 875.
8.75 × 1,000 = 8750.

The store will earn $87.50 if 10 T-shirts are sold, $875 if 100 T-shirts are sold, and $8,750 if 1,000 T-shirts are sold.

1. Compare the position of the decimal point in the first factor with the position of the decimal point in the product. What happens when you multiply by 10? by 100? by 1,000?

2. Use the pattern to find the following products mentally: 0.53 × 100 and 0.0078 × 1,000.

B. You can also use patterns to divide decimals mentally.

Tim used a calculator to find these quotients.

23.4 ÷ 10 = 2.34
23.4 ÷ 100 = 0.234
23.4 ÷ 1,000 = 0.0234

3. Compare the position of the decimal point in the dividend to the position of the decimal point in the quotient. What happens when you divide by 10? by 100? by 1,000?

4. What happens when there are not enough places to move the decimal point as far to the left as needed?

TRY OUT   Write the letter of the correct answer.

Multiply mentally.

5. 0.063 × 100   a. 0.63   b. 6.3   c. 63   d. 630
6. 78.03 × 1,000   a. 7.803   b. 780.30   c. 7,803   d. 78,030

Divide mentally.

7. 45.6 ÷ 1,000   a. 0.0456   b. 0.456   c. 456   d. 4,560
8. 0.038 ÷ 10   a. 0.0038   b. 0.38   c. 3.8   d. 38

PRACTICE

Multiply or divide mentally.

9. 3.0 × 10
10. 21.6 × 100
11. 0.73 × 10
12. $2.25 × 100
13. 0.24 × 100
14. $.13 × 10
15. 0.017 × 1,000
16. 33.9 × 1,000
17. 3.09 ÷ 10
18. 5.0 ÷ 100
19. 0.079 ÷ 10
20. 41.2 ÷ 100
21. 0.1008 ÷ 100
22. 54.9 ÷ 100
23. 0.6 ÷ 100
24. 0.97 ÷ 10
25. 35.8 ÷ 1,000
26. 0.00017 × 10
27. 0.64 ÷ 100
28. 1.009 × 1,000
29. $.03 × 1,000
30. $87.45 × 100
31. 0.05 ÷ 1,000
32. 7 ÷ 1,000

Compare. Use >, <, or =.

33. 0.008 × 1,000 ■ 0.08 × 100
34. 0.01 × 1,000 ■ 10 ÷ 100
35. 6.57 ÷ 100 ■ 65.7 ÷ 10
36. 8.3 × 100 ■ 0.83 × 1,000

Mixed Applications   Solve. You may need to use the Databank on page 519.

37. Barry and Tim live in a Texas town of 2,400 people. The town covers about 10 square miles. About how many people are there per square mile?

38. In Texas in 1980, about how many people lived in a 100 square mile area?

39. Pencils are 10 for $.50 in the school store. What is the cost of one pencil?

40. The school store needs a new floor covering that measures 12 square yards. At $10.00 a square yard, how much will a new floor cost?

UNDERSTANDING A CONCEPT
Estimating Products

Thalia needs 3.75 meters of fabric to make one soldier's costume for the play. About how many meters will she need for 6 costumes?

Estimate: 6 × 3.75

Here are two methods for estimating the product of a whole number and a decimal.

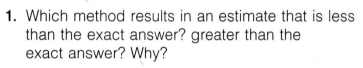

Front-End Estimation
Use the front digits.

6 × 3.75
↓   ↓
Think: 6 × 3 = 18

Rounding
Round to the greatest place.

6 × 3.75
↓   ↓
Think: 6 × 4 = 24

1. Which method results in an estimate that is less than the exact answer? greater than the exact answer? Why?

2. Which estimate should Thalia use to order enough fabric for the costumes? Why?

3. Estimate 14 × 6.38 by front-end estimation and rounding. How do the estimates compare? Why?

4. Estimate by rounding to the nearest whole number: 8 × 0.93. Is the exact answer greater than or less than the estimate? Why?

TRY OUT Write the letter of the correct answer.

Estimate. Use the front digits.

5. 3 × 5.186   a. 155.58   b. 15   c. 0.15   d. 8
6. 4 × 1.32   a. 4   b. 5.28   c. 40   d. 0.4

Estimate. Use rounding.

7. 6 × 7.607   a. 48   b. 4.8   c. 42   d. 4.2
8. 9 × 19.2   a. 90   b. 29   c. 9   d. 180

PRACTICE

Estimate. Use the front digits.

9. 5.22 × 3
10. 1.7 × 9
11. 3.8 × 7
12. 15.7 × 4
13. 91.5 × 8

14. 47.3 × 5
15. 5.799 × 8
16. 12.8 × 2
17. 62.8 × 9
18. 10.27 × 6

Estimate. Use rounding.

19. 15.43 × 4
20. 16.6 × 7
21. 4.83 × 15
22. 6.20 × 4
23. 8.42 × 9

24. 3.862 × 8
25. 7.886 × 5
26. 1.89 × 7
27. 8.270 × 12
28. 99.587 × 10

Estimate. Write the letter of the correct answer.

29. 8 × 0.6   a. less than 8   b. greater than 8
30. 9 × 3.41   a. less than 27   b. greater than 27
31. 12 × 2.97   a. less than 24   b. greater than 24
32. 10 × 0.54   a. less than 5   b. greater than 5

Between which two numbers will the product fall?

33. 3 × 4.7   12–15
34. 5 × 6.810
35. 8 × 3.59
36. 7 × 3.92

Mixed Applications

37. Thalia spends $2.89 per meter for the soldier costume fabric. About how much does she spend for 24 meters?

38. In addition to the 22.5 meters of fabric for the soldiers, Thalia needs 27.6 meters of fabric for the ballerinas. How much fabric does she need for both sets of costumes?

39. Thomas covers a 3-m by 4-m aluminum frame with canvas to make a backdrop for the play. What is the area of the backdrop?

40. Mrs. Economos is buying 10 tickets to the play for $27.50. How much does each ticket cost?

DEVELOPING A CONCEPT
Multiplying Whole Numbers and Decimals

Chris swam 0.14 mi each day for 4 days. How far did she swim all together?

Multiply: 4 × 0.14

Here is one way to solve the problem using graph paper and different-color markers.

Shade 0.14 of the 10-by-10 grid 4 times. Use a different-color marker each time.

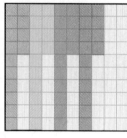

Chris swam 0.56 mi.

1. How does the model represent the problem?

2. How does the product of 4 × 0.14 compare with the product of the whole-number multiplication 4 × 14?

3. Find the whole-number products below. Use models to find the decimal products.
   a. 8 × 17 **b.** 5 × 19 **c.** 3 × 23
   8 × 0.17   5 × 1.9   3 × 0.23

4. Is your answer to Problem 2 true for the multiplications in Problem 3?

5. Look at the decimal multiplications in Problem 3. What do you notice about the number of decimal places in the factors and the products?

268 Lesson 7–3

SHARING IDEAS

6. How is multiplying a whole number and a decimal similar to multiplying two whole numbers? How is it different?

7. Write a rule that tells how to multiply a decimal by a whole number. Compare your rule to those of others.

PRACTICE

Use your rule to place the decimal point.

8. 0.34
    $\times2$
    $\overline{0{,}68}$

9. 0.9
    $\times5$
    $\overline{45}$

10. 1.28
    $\times3$
    $\overline{3{,}84}$

11. 3.07
    $\times4$
    $\overline{12{,}28}$

12. 2.5
    $\times1.1$
    $\overline{2{,}75}$

Use your rule to find the product. Use models to check your answers.

13. 0.36
    $\times2$

14. 0.22
    $\times3$

15. 0.15
    $\times6$

16. 0.21
    $\times5$

17. 0.8
    $\times4$

18. 1.3
    $\times6$

19. 3.7
    $\times4$

20. 1.06
    $\times5$

21. 7.85
    $\times2$

22. 1.01
    $\times9$

23. 0.9
    $\times15$

24. 0.34
    $\times21$

25. 2.9
    $\times11$

26. 1.8
    $\times12$

27. 4.06
    $\times14$

Solve.

28. Erica trains for the swim meet at a pool a total distance of 4.2 mi to and from her home. How far does she travel in 5 days?

29. Marty uses 0.65 yd of bandage to wrap each ankle when he is running. How much bandage does he use for both ankles?

30. Mrs. Wong drove for 3 hours at a speed of about 48.9 miles an hour. About how far did she drive?

31. Marty runs 3.7 mi each day. How far does he run in 6 days?

UNDERSTANDING A CONCEPT
Multiplying Decimals by Whole Numbers

A. Kevin has planted beans for a science project. He measures one plant over a 3-day period. He finds that on the average it grows 0.85 cm a day. How much has the plant grown after the 3 days?

Multiply: 3 × 0.85

Step 1	Step 2
Multiply as you would multiply whole numbers.	Write the decimal point in the product.

```
   0.85              0.85   ←   2 decimal places
 ×    3            ×    3   ← + 0 decimal places
   255              2.55   ←   2 decimal places
```
Think:

The plant has grown 2.55 cm.

1. Is 2.55 a reasonable answer? Why?

2. Multiply 6 × 3.412. What is the product? How many decimal places are in the product? How do you know?

B. Sometimes you need to write zeros in the product in order to place the decimal point.

Multiply: 3 × 0.014

Step 1	Step 2
Multiply as you would multiply whole numbers.	Add enough zeros to show the correct number of decimal places. Write the decimal point in the product.

```
   0.014             0.014   ←   3 decimal places
 ×     3           ×     3   ← + 0 decimal places
      42             0.042   ←   3 decimal places
```
Think:

3. Multiply 4 × 0.002. How many decimal places are in the product? How do you know?

Try Out

| 4. 0.7 × 4 | 5. 0.12 × 4 | 6. 0.013 × 3 | 7. 0.029 × 3 | 8. 1.376 × 5 |

Practice

Multiply.

9. 0.2 × 5	10. 0.6 × 6	11. 0.9 × 7	12. 0.8 × 8	13. 0.5 × 9
14. 0.15 × 3	15. $.27 × 4	16. 0.19 × 6	17. 0.54 × 8	18. 0.37 × 5.
19. 0.01 × 6	20. 0.012 × 5	21. 0.004 × 9	22. 0.019 × 2	23. 0.005 × 6
24. 3.27 × 3	25. 6.01 × 9	26. $3.95 × 4	27. 5.21 × 7	28. 8.09 × 6
29. 2.907 × 4	30. 6.342 × 8	31. 21.675 × 9	32. 36.896 × 4	33. 27.469 × 8

34. 3 × 2.17   35. 8 × 0.925   36. 5 × $12.06   37. 4 × 2.378
38. 5 × 0.016   39. 9 × 3.025   40. 9 × 0.019   41. 4 × 32.501

42. the product of six and twelve thousandths
43. thirty-seven hundredths multiplied by seven
44. eight times two and forty-one thousandths

Critical Thinking

45. Touraj said, "When I multiply a decimal less than 1 by a whole number, the product will always be greater than the whole number." Do you agree? Why or why not?

Mixed Applications

46. Edward's bean plants grow 3.36 cm each week. How tall will they be in 2 weeks?

47. In 10 weeks a plant grew 39.75 cm. On the average, how much did it grow each week?

UNDERSTANDING A CONCEPT
More Multiplying

Lauren helped her father put gasoline in their car. The fuel pump read 12 gallons, and the gasoline cost 94.9¢ per gallon or $.949 per gallon. To the nearest cent, how much did her father spend on gasoline?

Estimate first. Then find the answer.

Multiply: 12 × $.949

Calculator

12 ⊗ 0.949 ⊜ | 11.388 |

$11.388 → $11.39

Paper and Pencil

```
   $.949
 ×    12
   1 898
 + 9 490
  $11.388  →  $11.39
```

Lauren's father spent $11.39, to the nearest cent.

1. How does your estimate show that $11.39 is a reasonable answer?

2. ***What if*** you have to find the products below? Which method would you use: mental math, a calculator, or paper and pencil? Why?
   a. 65 × 0.10   **b.** 75 × 0.47   **c.** 25 × 0.2

3. ***What if*** you entered 28 ⊗ 1.729 on a calculator and saw | 4.8412 |? How would you know by estimating that you had made an error?

TRY OUT Multiply. Use mental math, a calculator, or paper and pencil.

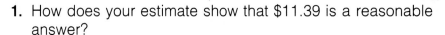

4. 6.4   5. 0.73   6. $2.56   7. 1.043   8. 0.007
  × 23   × 18   ×   32   ×   45   ×   80

PRACTICE

Multiply. Use mental math, a calculator, or paper and pencil.

9.	1.87 × 16	10.	0.946 × 31	11.	$6.89 × 25	12.	18.011 × 29	13.	0.009 × 11
14.	4.089 × 54	15.	26.29 × 10	16.	0.035 × 65	17.	0.918 × 78	18.	1.473 × 36
19.	6.95 × 4	20.	$18.03 × 21	21.	9.372 × 10	22.	0.109 × 36	23.	10.01 × 6
24.	0.009 × 88	25.	$56.37 × 64	26.	9.986 × 3	27.	0.568 × 85	28.	10.010 × 10

29. 7 × 8.099
30. 25 × 31.65
31. 10 × 9.981
32. 13 × 0.005
33. 26 × 45.36
34. 17 × 2.698
35. 20 × 0.004
36. 34 × 0.066

37. the product of thirty-four and six and forty-nine hundredths
38. eight and six hundredths multiplied by seventy

Mixed Applications Solve. Which method did you use?

39. Lauren's father bought 2 quarts of motor oil at $1.79 each. How much did he spend?

40. Dan's family spent $180 on food, $260 on lodging, $140 on sightseeing, and $48 on gas. How much were these expenses?

41. Lauren takes care of Dan's pets when he goes on a trip. She earns $3.75 each time she does this. How much will she earn if she takes care of Dan's pets 10 times?

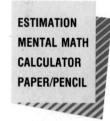

ESTIMATION
MENTAL MATH
CALCULATOR
PAPER/PENCIL

Mixed Review

Find the mean, median, mode, and range.

42. 36, 24, 33
43. 1, 8, 3, 5, 8, 5, 5

Find the starting or ending time.

44. 18 minutes before 12 noon
45. 2 hours 20 minutes after 3:18 P.M.

PROBLEM SOLVING

Strategy: Working Backward

Ian bought some shirts to give as presents. He gave 3 to his father and 4 to his uncle. He gave the remaining half to his brother. How many shirts did Ian buy?

One plan for solving this problem is to work backward. First try to find *half* of the number of shirts that Ian bought. His brother got half. So his father and uncle together must have gotten the other half.

Half the number bought = 3 + 4 = 7
                        ↑   ↑
                     father uncle

Next find the total number of shirts he bought.

Half = 7 shirts   Total must be twice this number: 2 × 7 = 14 shirts

Ian bought 14 shirts.

1. Was all the information in the problem used to find the answer?
2. Tell how you used the plan of working backward to solve this problem.
3. How can you check the answer?
4. Are there other ways to solve the problem? Tell how.

PRACTICE

Work backward to solve the problem.

5. Margot bought three sweaters. The wool sweater cost $8.04 more than the acrylic sweater. The acrylic sweater cost $3.97 more than the cotton sweater. The cotton sweater cost $17.98. How much did the wool sweater cost?

6. Margot withdrew $75 from her savings account two weeks ago to buy some summer clothes. She now has $207. Last week she deposited $50. How much was in her account before she withdrew the money?

7. Ian bought some socks for $3.25 and a pair of shoes. The shoes cost 5 times as much as the socks. He had $4.30 left. How much did Ian start out with?

8. Margot spent twice as much on lunch as Ian. Ian's sister spent $2.75, which was $.50 less than what Ian spent. How much did Margot spend?

Strategies and Skills Review

Solve. Use mental math, estimation, a calculator, or paper and pencil. You may need to use information from the table.

9. Carrie has $100. Can she buy a jacket, a shirt, a belt, and a pair of jeans?

10. Frank is going to shop at the Clothes Cave. It takes him $\frac{1}{2}$ hour to get to the store and back home again. If he allows himself one hour to shop, what time should he plan on going to the store in order to get home at 6 P.M.?

11. Maria has $50 to spend at the Clothes Cave. She bought a pair of jeans. How many belts can she buy with the money she has left?

THE CLOTHES CAVE SPECIAL SALE

Item	Price
Jacket	$49.99
Jeans	$24.99
Skirt	$21.99
Shirt	$19.99
Belt	$ 9.99
Socks	$ 1.99

12. Minh works at the Clothes Cave. At 1 P.M. Minh's sales totaled $520 more than her sales at noon. At noon her sales totaled $457 more than her sales at 11 A.M. Her total sales at 11 A.M. totaled $278. What were Minh's total sales at 1 P.M.?

13. Steve bought a pair of jeans and a shirt during the sale. He gave the clerk three bills and received $5.02 as change. What three bills did Steve give the clerk?

14. What is the greatest number of shirts a person could buy if he or she had $100 to spend during the sale?

15. *Write a problem* that can be solved by working backward. Solve the problem. Ask other students to solve your problem.

EXPLORING A CONCEPT
The Products of Two Decimal Factors

You can use a calculator to explore what happens when you multiply decimals.

WORKING TOGETHER

Use your calculator to find the following products.

  1. 1.8 × 34.2 **2.** 0.18 × 34.2 **3.** 1.8 × 3.42 **4.** 0.18 × 3.42

Record the multiplication sentences in a chart like this.

Factor	×	Factor	=	Product
1.8	×	34.2	=	61.56

5. How are the multiplication sentences similar?

6. How are they different?

7. How do the multiplications in the chart compare to the whole number multiplication 18 × 342?

Look at 1.8 × 34.2 = 61.56 in the chart.

8. How many decimal places are in the first factor?

9. How many decimal places are in the second factor?

10. How many decimal places are in the product?

11. What do you notice about the number of decimal places in the factors and the product?

12. Is what you found true for the other multiplication sentences?

SHARING IDEAS

13. Write a rule for finding the product of two decimals when using paper and pencil.

14. Using paper and pencil find the product of 0.4 and 3.9. How many decimal places are in the product? Check your answer on a calculator.

15. Using paper and pencil find the product of 0.36 and 0.25. Then use a calculator to find the same product. What do you notice? Why do you think this happens?

ON YOUR OWN

Use your rule to place the decimal point.

| 16. | 6.3
×0.4
———
2 5 2 | 17. | 7.6 8
× 1.9
———
1 4,5 9 2 | 18. | 0.4 1 9
× 0.3 8
———
0.1 5 9 2 2 | 19. | 5.7 2
×0.0 6
———
0 3 4 3 2 | 20. | 1 8.0 6
× 2.5
———
4 5 1 5 0 |

Find the product. Use your rule for multiplying decimals.

| 21. | 0.56
× 0.4 | 22. | 1.3
×2.7 | 23. | 4.26
× 1.3 | 24. | 2.57
×0.64 | 25. | 0.25
×0.39 |

| 26. | 7.09
× 0.6 | 27. | 0.258
× 0.7 | 28. | 4.7
×1.3 | 29. | 1.46
×0.29 | 30. | 0.364
× 0.23 |

31. 0.5 × 1.82   **32.** 7.2 × 3.4   **33.** 0.51 × 0.297   **34.** 1.9 × 24.6

I Remember Numbers

Applying Mathematics

A. Look at the following breakfast order:

5 orders of pancakes @ $1.19
3 orders of eggs @ $1.59
6 orders of hashbrowns @ $.69
8 orders of orange juice @ $.89

1. Suppose you had an old-fashioned adding machine with only ⊕ and ⊖ operation keys. How could you find the total cost of the order?

2. Now suppose you have a calculator with ⊕, ⊖, ⊗, and ⊘. You may also make notes on a piece of paper. What is a shorter way to find the total cost of the order? Try it. What is your answer?

There is another way to find the total, and you do not need to write anything. Follow these steps with your calculator:

Press: ⑤ ⊗ ① . ① ⑨ M+
③ ⊗ ① . ⑤ ⑨ M+
⑥ ⊗ . ⑥ ⑨ M+
⑧ ⊗ . ⑧ ⑨ M+ MR

What does the display show?

Each time you use the memory key, M+, it adds the displayed number to whatever is already in the calculator memory. Note that you do not have to clear the calculator after the M+ key is used. At the end of the computation, the total is displayed when you press the MR (memory recall) key.

Thinking MATHEMATICALLY

B. Practice using the (M+) and (MR) keys. Try these problems with your calculator. Find the total cost of each order.

3. 6 tubes of toothpaste @ $1.79
   4 toothbrushes @ $2.00
   3 boxes of tissues @ $.99
   2 bottles of shampoo @ $2.39
   5 rolls of mints @ $.45

4. 5 greeting cards @ $.95
   2 rolls of wrapping paper @ $2.59
   4 gift bows @ $1.09
   4 gift boxes @ $1.25
   3 pens @ $.79
   1 paperback book @ $4.99

5. Compare your answers to the above problems with a partner. If your answers differ, work the problems again together.

6. Write some problems of your own. Try them out with your partner.

MANIPULATIVES

EXPLORING A CONCEPT
Division of a Decimal by a Whole Number

Anita needs to make 3 bows to use as decoration for her crafts project. If she has 5.25 m of ribbon, how much ribbon can she use for each bow?

WORKING TOGETHER
Solve the problem using a model.

1. How does your model show 5.25?
2. How did you separate 5.25 into 3 equal groups?
3. How much ribbon can Anita use for each bow?

Here is a way to model the problem.

Step 1 Draw 10-by-10 grids on graph paper.

Step 2 Shade 5.25 of the grids.

Step 3 Divide into 3 equal groups by cutting the grids apart as necessary.

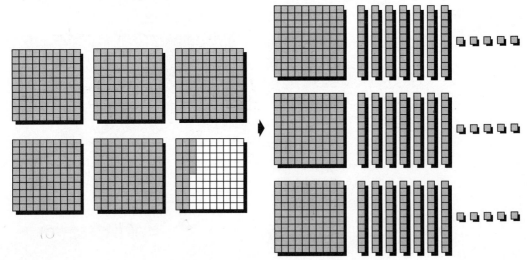

Anita can use 1.75 m of ribbon for each bow.

4. What regroupings did you make in order to divide?
5. **What if** Anita had 3.36 m of ribbon? How long could each of 4 bows be? Use models to find out.

280 Lesson 7-9

SHARING IDEAS

6. Did you use a graph-paper model to solve the problems? If not, why did you use a different model?

7. Compare your model and results to those of others. How are they the same? How are they different?

8. How does dividing a decimal compare to dividing a whole number?

ON YOUR OWN

Use models to solve.

9. Robert spends $4.50 on 6 airmail stamps. How much does each one cost?

10. Terry needs to cut a 2.85 m board into 3 equal pieces. How long will each piece be?

11. Marsha plants a garden in a strip of ground 3.75 m long. She divides the length into 5 equal parts. How long is each section?

12. Anita has 2.45 m of ribbon. If she cuts it into 7 equal pieces, how long will each one be?

13. 4)4.56   14. 6)2.46   15. 3)1.68   16. 2)0.64   17. 5)0.25

18. 7)3.01   19. 4)5.2   20. 3)0.27   21. 11)3.3   22. 12)4.8

23. 1 ÷ 5   24. 2.54 ÷ 2   25. 0.63 ÷ 3   26. 0.56 ÷ 8   27. 2.52 ÷ 12

DEVELOPING A CONCEPT
Dividing Decimals by Whole Numbers

A. Tony and his friends are making 4 kites. They have a strip of cloth 5.84 m long to make kite tails that are equal in length. How long can they make each one?

Divide: 5.84 ÷ 4

You can divide using a model.

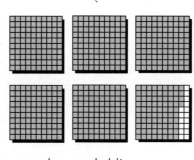

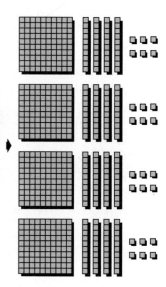

They can make each kite tail 1.46 m long.

B. Here is a way to divide without models. Divide the same way you divide whole numbers.

Divide: 5.84 ÷ 4

Step 1	Step 2	
Write the decimal point of the quotient above the decimal point of the dividend.	Divide as you would divide whole numbers.	Check.
$$4\overline{)5.84}$$	$$\begin{array}{r} 1.46 \\ 4\overline{)5.84} \\ -\underline{4} \\ 18 \\ -\underline{16} \\ 24 \\ -\underline{24} \\ 0 \end{array}$$	$$\begin{array}{r} 1.46 \\ \times4 \\ \hline 5.84 \end{array}$$

282   Lesson 7-10

SHARING IDEAS

1. Why is it important to write the decimal point in the quotient before you begin dividing?
2. Will the quotient of 40.15 ÷ 11 be greater than or less than 1? Why?
3. Will the quotient of 15.028 ÷ 26 be greater than or less than 1? Why?

PRACTICE

Divide.

4. 3)20.4
5. 4)28.8
6. 6)14.4
7. 8)24.8
8. 5)6.85
9. 3)$1.05
10. 6)4.14
11. 5)1.15
12. 7)3.99
13. 9)3.114
14. 21)10.92
15. 37)67.71
16. 54)$364.50
17. 62)11.78
18. 83)72.708
19. 5)$6.85
20. 32)29.952
21. 43)58.566
22. 8)48.736
23. 25)14.25
24. 7)22.757
25. 37)$346.32
26. 8)11.032
27. 22)12.342
28. 58)338.72
29. 49.12 ÷ 8
30. 5.925 ÷ 25
31. $112.50 ÷ 15
32. 8.739 ÷ 9

Mixed Applications

33. In 1919 a chain of 8 kites reached a record altitude of 9,739.88 m. The record altitude reached by a single kite was 8,534.4 m in 1967. How many meters higher did the kite chain go?

34. It took Tony 2 hours and 30 minutes to make his kite. He flew it for 1 hour and 45 minutes before it landed in a tree. How long did he spend making and flying the kite?

35. The 3 friends bought a ball of kite string for $2.85. They shared the cost equally. How much did each one pay?

36. Tony had 13.6 m of ribbon to make tails for his kites. He cut the ribbon into 16 strips of equal length. How long was each strip?

UNDERSTANDING A CONCEPT
Zeros in Division

A. While on vacation in Florida, Patty found 6 large shells to add to her collection. All together the 6 shells weighed 0.54 kg. On average how much did each shell weigh?

Divide: 0.54 ÷ 6

Step 1
Write the decimal point of the quotient.

```
      .
6)0.54
```

Step 2
Divide.

```
   0.09
6)0.54
 − 54
    0
```

Think: 6 > 5
Not enough tenths
Write 0 in the tenths place.
Divide the hundredths.

On the average each shell weighed 0.09 kg.

1. **What if** you left out the zero in the tenths place of the quotient? Would the answer make sense? Why or why not?

B. For some divisions you can write zeros in the dividend in order to keep on dividing until there is no remainder.

Divide: 46.4 ÷ 32

Step 1
Write the decimal point of the quotient. Divide until you have a remainder.

```
      1.4
32)46.4
  −32
   14 4
  −12 8
     1 6
```

Step 2
Write zeros in the dividend. Continue to divide until there is no remainder.

```
      1.45
32)46.40
  −32
   14 4
  −12 8
     1 60
    −1 60
        0
```

2. Why can you add zeros to a dividend without changing its value?

Try Out

3. 9)9.54   4. 8)$16.56   5. 22)67.98   6. 5)0.36   7. 12)1.02

Practice

Divide.

8. 2)2.16   9. 4)0.16   10. 21)1.47   11. 3)$12.09   12. 12)0.672

13. 6)1.89   14. 4)21.8   15. 15)21.3   16. 5)47.3   17. 28)3.5

18. 8)0.472   19. 37)$76.22   20. 8)0.84   21. 65)1.56   22. 5)0.2

23. 5)4.88   24. 18)$37.62   25. 6)31.5   26. 27)2.565   27. 30)90.3

28. 12.2 ÷ 4   29. $7.56 ÷ 7   30. 5.7 ÷ 38   31. 1.075 ÷ 25

Mixed Applications

Solve. You may need to use the Databank on page 520.

32. The largest shell ever found was a marine giant clam shell that was 43.3 inches long. The smallest shell was only 0.02 inches long. What is the difference in the sizes?

33. On the beach Mark found 5 large conch shells. He saw shells just like them in the Shell Shop for $.89 each. How much would his 5 shells have cost in the shop?

34. At the Shell Shop Patty bought 5 candy-striped snail shells for $.50. How much did one shell cost?

35. **Write a problem** about the annual rainfall in some southern cities in the United States. Ask others to solve the problem.

Mixed Review

Find the perimeter and area of the square or rectangle.

36.

33 cm

37.

58 mm

38.

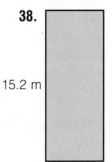

15.2 m

6.0 m

39.

4.28 cm

8.0 cm

UNDERSTANDING A CONCEPT
More Dividing Decimals

Refreshments for a class party cost $18.69. The 25 students in the class each contributed to the cost. To the nearest cent, how much did each one have to pay?

Divide: $18.69 ÷ 25

Kelly divided using a calculator.

18.69 ÷ **25** = **0.7476**

$.7476 → $.75

Sam divided using paper and pencil.

```
      0.7 4 7
 25)1 8.6 9 0
   -1 7 5↓
      1  1 9
     -1  0 0↓
         1 9 0
        -1 7 5
            1 5
```
$.747 → $.75

Kelly and Sam agreed that each student had to pay $.75.

1. How many decimal places did Sam divide to? Why?

2. How is Kelly's calculator display different from Sam's quotient? How is it the same?

3. **What if** Kelly's calculator displayed 0.07476 ? How would a correct estimate tell you that the calculator display is wrong?

4. **What if** you have to find each quotient below? Which method would you use: mental math, a calculator, or paper and pencil? What is the quotient?
   a. 6.72 ÷ 10   **b.** 75.25 ÷ 5   **c.** 34.56 ÷ 15

TRY OUT
Divide. Use mental math, a calculator, or paper and pencil.

5. 9)0.63   **6.** 5)1.2   **7.** 8)21.4   **8.** 25)12.25   **9.** 16)18.32

PRACTICE

Divide. Use mental math, a calculator, or paper and pencil.

10. 4)6.8   **11.** 5)18.5   **12.** 7)0.28   **13.** 6)1.056   **14.** 3)$15.09

15. 15)$12.75   **16.** 10)77.28   **17.** 32)32.64   **18.** 24)1.08   **19.** 9)35.73

20. 13)28.21   **21.** 81)2.268   **22.** 40)1.6   **23.** 19)$14.44   **24.** 15)1.89

25. 20)84.62   **26.** 6)0.018   **27.** 80)0.32   **28.** 43)$45.15   **29.** 75)4.95

30. $29.76 ÷ 12   **31.** $316.56 ÷ 8   **32.** 0.456 ÷ 19   **33.** 1.8 ÷ 20

34. $140.58 ÷ 66   **35.** 0.575 ÷ 23   **36.** 3.95 ÷ 79   **37.** 8.14 ÷ 4

Mixed Applications

Solve. Which method did you use?

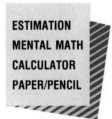

ESTIMATION
MENTAL MATH
CALCULATOR
PAPER/PENCIL

38. It cost $13.25 for muffins for the party. To the nearest cent, how much will each person have to pay if 10 people share the cost equally?

39. At the party the students played games for 30 minutes, had refreshments for 40 minutes, and sang songs for 20 minutes. How long did the party last?

40. The class members spent $18.69 on refreshments for the party, $7.68 on decorations, and $25.75 on a gift for their teacher. How much did the class members spend in all?

41. During vacation Gerry and his family will take a bicycle trip. They plan to cover 235 mi in 8 days. About how many miles a day is this?

EXTRA Practice, page 300

PROBLEM SOLVING

UNDERSTAND
✓ PLAN
✓ TRY
CHECK
✓ EXTEND

Strategy: Making a Table

Steve and Mary Ellen Henson are mixing blue and yellow paint to get exactly the shade of green they want. For every 2 quarts of blue, they use 3 quarts of yellow. How many quarts of each will they need to use to get 30 quarts of green?

To help them solve the problem, they started this table.

Color	Quarts of Paint					
Blue	2	4	6			
Yellow	3	6	9			
Green	5					

1. What pattern do you notice in the numbers in the row for blue? in the row for yellow?

2. How will you get the numbers in the row for green?

3. Complete the table.

4. How many quarts of blue and of yellow do they need to get 30 quarts of green?

5. **What if** the Hensons had needed 50 quarts of green? How many quarts of blue and of yellow would they have needed?

Make a table to solve the problem.

6. For doors and the wood trimming around windows, the Hensons will use an off-white mixture of 0.5 liter of gray to every 2.5 liters of white. How many liters of white will they need if they want a total of 9 liters of off-white?

7. Mary Ellen can paint 12 square feet in 5 minutes. Steve can paint 11 square feet in the same amount of time. To the nearest 5 minutes, about how long will it take them to paint 250 square feet if they work together?

8. At the house-warming, the Hensons served mixed nuts. They use 1 cup of cashews for every 2 cups of pecans and for every 2.5 cups of peanuts. How many cups of each did they use to get a total of 22 cups of mixed nuts?

9. For their house-warming party, the Hensons made punch. They used 0.75 liter of grape juice for every 2.25 liters of punch. How many liters of grape juice did they need to make 13.5 liters of punch?

Strategies and Skills Review

Solve. Use mental math, estimation, a calculator, or paper and pencil.

10. Mary Ellen figured it would cost 3 times as much to wallpaper her bedroom as to paint it. She estimated the cost of wallpapering it to be about $87.75. About how much did she estimate it would cost to paint the bedroom?

11. Steve estimated the cost of paintbrushes to be between $27 and $30. He has already spent $68.50 on paint and $3.50 on brush cleaners. He may need to put gas in his car on the way home. He has $100 in cash. Should he plan to pay by check or by cash at the paint store?

12. The Stones paid painters $900.20 to paint their apartment. The Hensons figured that the Stones paid 4 times what they did by doing it themselves. The Hensons, however, also spent $137.55 on a new carpet for a hallway. How much did the Hensons spend on decorating?

13. The Stones' monthly rent has increased each year as shown in the table. If the rent continues to increase at this rate, how much will they be paying each month two years from now?

Year	Monthly Rent
1	$395.50
2	$413.00
3	$430.50
4	$448.00

UNDERSTANDING A CONCEPT

Measuring Metric Capacity and Mass

A. Kareem is mixing plaster of paris for an art class. He needs to add 1 liter of water to the plaster.

The **liter (L)** and **milliliter (mL)** are metric units of capacity.

about 1 liter   about 1 milliliter

1. **What if** you had to measure the amount of water in a teaspoon? What metric unit of capacity would you use? Why?

B. The **kilogram (kg)**, **gram (g)**, and **milligram (mg)** are metric units of mass.

about 1 kilogram   about 1 gram   about 1 milligram

2. **What if** you had to measure the mass of a piece of chalk? of a leaf? Which metric unit of mass would you use for each? Why?

TRY OUT   Write the letter of the correct answer. Choose the appropriate metric unit of measure.

3. capacity of a thimble   **a.** mL   **b.** L   **c.** g   **d.** mg
4. capacity of a fish tank   **a.** mL   **b.** L   **c.** g   **d.** mg
5. mass of an apple   **a.** mg   **b.** g   **c.** kg   **d.** L
6. mass of a table   **a.** mg   **b.** g   **c.** kg   **d.** L

PRACTICE

What unit of capacity would you use to measure? Write *milliliter* or *liter*.

7. juice glass
8. paint can
9. baby bottle
10. test tube
11. bathtub
12. can of soup

Which unit of mass would you use to measure? Write *milligram*, *gram*, or *kilogram*.

13. bicycle
14. feather
15. bumblebee
16. box of cereal
17. car
18. pen

Which is the most reasonable estimate of capacity?

19. milk jug   a. 2 mL   b. 20 mL   c. 2 L   d. 20 L
20. bottle cap   a. 2mL   b. 2 L   c. 20 L   d. 200 L
21. bucket   a. 1 mL   b. 10 mL   c. 10 L   d. 100 L
22. soup bowl   a. 25 mL   b. 250 mL   c. 250 L   d. 25 L

Which is the most reasonable estimate of mass?

23. paper clip   a. 1 mg   b. 1 g   c. 1 kg   d. 10 kg
24. truck   a. 500 mg   b. 500 g   c. 5 kg   d. 500 kg
25. grain of salt   a. 1 mg   b. 1 g   c. 10 g   d. 1 kg
26. robin   a. 200 mg   b. 200 g   c. 20 kg   d. 200 kg

Mixed Applications

27. Kareem wants to add a small amount of food coloring to some play dough to color it. What should he use to measure the food coloring—a liter container or a dropper that holds 1 mL?

28. Nina mixes 1.8 L of red paint with 0.25 L of white paint to make pink paint. How many liters of pink paint does she make?

29. Nina mixes 1.5 L of plaster and uses it to fill 5 molds. How much plaster does each mold contain?

30. **Write a problem** comparing two units of mass. Use two of the units of mass from Exercises 23–26.

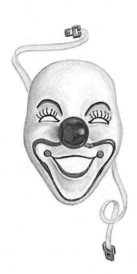

UNDERSTANDING A CONCEPT
Renaming Metric Measures

A. Masaru's calculator is 11 cm long. How can he rename the measure in millimeters?

1 cm = 10 mm
1 m = 100 cm
1 km = 1,000 m
1 L = 1,000 mL
1 g = 1,000 mg
1 kg = 1,000 g

To rename a larger metric unit as a smaller unit, you can multiply by 10, 100, or 1,000.

11 cm = ■ mm

Think: 1 cm = 10 mm
11 × 10 = 110
11 cm = 110 mm

1. A table measures 2 m in length. How many centimeters long is the table? What did you do to rename?

2. Complete. Tell how you renamed.

   a. 4.5 km = ■ m   **b.** 6 L = ■ mL

   c. 12.3 kg = ■ g   **d.** 8 g = ■ mg

3. Why do you multiply when you rename a larger unit as a smaller one?

B. To rename a smaller metric unit as a larger one, you can divide by a multiple of 10.

561 g = ■ kg   355 mL = ■ L   36.5 mm = ■ cm

Think: 1 kg = 1,000 g   *Think:* 1 L = 1,000 mL   *Think:* 1 cm = 10 mm
561 ÷ 1,000 = 0.561   355 ÷ 1,000 = 0.355   36.5 ÷ 10 = 3.65
561 g = 0.561 kg   355 mL = 0.355 L   36.5 mm = 3.65 cm

4. Why do you divide when you rename a smaller unit as a larger one?

TRY OUT Write the letter of the correct answer.

5. 6 cm = ■ mm   **a.** 0.6   **b.** 60   **c.** 600   **c.** 6,000
6. 5 kg = ■ g   **a.** 0.5   **b.** 50   **c.** 500   **d.** 5,000
7. 125 mL = ■ L   **a.** 0.125   **b.** 1.25   **c.** 12.5   **d.** 1,250
8. 33.5 cm = ■ m   **a.** 0.335   **b.** 3.35   **c.** 335   **d.** 3,350

PRACTICE

Complete.

9. 7 kg = ■ g
10. 14 L = ■ mL
11. 8 m = ■ cm
12. 12 cm = ■ mm
13. 7,360 mg = ■ g
14. 654 mL = ■ L
15. 298 g = ■ kg
16. 400 m = ■ km
17. 27.5 m = ■ cm
18. 832 mg = ■ g
19. 3.25 cm = ■ mm
20. 29 mL = ■ L
21. 19 g = ■ mg
22. 0.3 km = ■ m
23. 48 cm = ■ m
24. 560 mm = ■ cm
25. 58 cm = ■ mm
26. 612 g = ■ kg
27. 73 m = ■ km
28. 2.5 km = ■ m
29. 0.8 g = ■ mg
30. 506 cm = ■ m
31. 3.14 L = ■ mL
32. 720 mm = ■ cm
33. 8.2 kg = ■ g
34. 1.6 L = ■ mL
35. 0.27 km = ■ m
36. 839 cm = ■ m
37. 250 g = ■ kg
38. 1,350 mm = ■ cm
39. 6.8 g = ■ mg
40. 750 mL = ■ L
41. 8 kg = 8,000 ■
42. 4.7 km = 4,700 ■
43. 32 cm = 0.32 ■
44. 831.6 mm = 0.8316 ■

Mixed Applications

45. Victor measures the total mass of an apple, an orange, and a bunch of grapes, which is 1,257 g. How many kilograms is this?

46. One of Dana's pencils has a mass of 3.4 g. The pencils come in a package of 6. What is the total mass of the package of 6 pencils? How many milligrams is this?

47. At the beginning of the year, Meredith was 137.16 cm tall. Now her height is 1.52 m. How many centimeters did she grow?

48. **Write a problem** involving the length, capacity, or mass in metric units of two objects in your classroom. Ask others to solve your problem.

Mixed Review

Find the answer. Which method did you use?

MENTAL MATH
CALCULATOR
PAPER/PENCIL

49. 10.06 + 28.42
50. 163.9 − 89.27
51. 65 × 24
52. 1.29 × 5

53. 6)1,286
54. 50)250
55. 32)7,328
56. 48)11,060
57. 9)3.24

58. 189 + 27
59. 1,000 − 412
60. 30 × 60
61. 2,762 ÷ 3
62. 7 × 89

EXTRA Practice, page 301

DECISION MAKING

Problem Solving: Buying Shoes

SITUATION

The members of the Taylor family are going shopping for athletic shoes. Mr. and Mrs. Taylor need shoes for everyday wear and for aerobic walking. Janet and Joe need shoes for school and for playing basketball.

PROBLEM

Which shoes should they buy?

DATA

SPECIAL SALE on Basketball Shoes
Buy one pair at the sale price and get a second pair for **$5.00**

Girls
Hi-top shoe
~~$39.95~~
$25.00

Low-top shoe
~~$33.95~~
$22.50

Canvas Hi-top
~~$20.95~~
$12.50

Boys
Hi-top shoe
~~$39.95~~
$25.00

Low-top shoe
~~$33.95~~
$22.50

Canvas Hi-top
~~$19.95~~
$10.50

SPEEDY FEET SHOE STORE

All-Sport Shoes	Star Brand	Speedy's
Boys	$35.95	$19.95
Girls	$35.95	$19.95
Women	$42.95	$24.95
Men	$44.95	$25.95

Walking Shoes	Star Brand	Speedy's
Women Walking shoe	$51.95	$29.95
Men Walking shoe	$51.95	$29.95

USING THE DATA

How much will it cost to buy the shoes?

1. Star Brand all-sport shoes for Mr. and Mrs. Taylor
2. Speedy's all-sport shoes for Mr. and Mrs. Taylor
3. Star Brand all-sport shoes for Janet and Joe
4. Speedy's all-sport shoes for Janet and Joe
5. hi-top basketball shoes for Janet and Joe
6. Speedy's walking shoes for Mr. and Mrs. Taylor
7. How much more will a men's Star Brand walking shoe cost than a discontinued red walking shoe?

MAKING DECISIONS

8. Basketball shoes are on sale because it is the end of the season. Is it a good idea to buy now if you will not need them until next year? if you will wear them now?

9. **What if** many of the clearance shoes were soiled from being tried on in the store? Should Janet and Joe consider these shoes? Why or why not?

10. **What if** all shoes sold at clearance prices are marked Final Sale, meaning that they cannot be returned? Should the Taylors buy the sale shoes? Why or why not?

11. **What if** Mr. Taylor decides to buy two pairs of shoes, a Speedy's all-sport shoe and walking shoes from the table of discontinued styles? Why might he do this?

12. **Write a list** of other things the Taylor family should think about before buying a pair of shoes?

13. Which choice of type of shoe would you make? Why?

SALE on discontinued styles. Not all sizes and colors available.

Men's walking shoe, red only $42.95 $25.00

Women's walking shoe, yellow only $39.95 $20.00

Multiplying and Dividing Decimals

CURRICULUM CONNECTION

Math and Science

According to some experts, a lawn needs only one inch of water a week to stay healthy. Homeowners can use a simple method to find how much water they use when they water the lawn for one hour.

Place 6 same-size empty cans in a straight line from a sprinkler to the farthest point the water reaches. Water for 15 minutes. Measure the amount in each can. Add the depths together and divide by 6 to find the average amount of water. Since 15 minutes is $\frac{1}{4}$ hour, multiply by 4 to find the amount of water the lawn would receive in 1 hour, or 60 minutes.

What if you try the method on your lawn? These are the amounts in the cans: 0.125, 0.125, 0.5, 0.5, 0.625, and 0.5 inches. Would you be over or under the recommended one inch if you watered for an hour?

Think: Find the total in the 6 cans.

   0.125 + 0.125 + 0.5 + 0.5 + 0.625 + 0.5 = 2.375

Divide the total by 6 to find the average for 15 minutes.

   2.375 ÷ 6 = 0.3958

Multiply by 4 to find the amount of water the lawn would receive in an hour.

   4 × 0.3958 = 1.58

You would be over the recommended one inch.

ACTIVITIES

1. Measure about how much water you let run to get a cold drink of water. Think of ways to conserve or use that water. Share your findings and ideas with the class.

2. Find out what your community does to conserve water or other natural resources. Prepare an oral report for the class.

Calculator: Cooking With a Calculator

Raymond's dad is roasting a 3.8 pound chicken. How long will it take?

You can use a calculator and a roasting chart to find out.

ROASTING CHART (350°F)

Roast	Time per Pound
beef	18 minutes
chicken	20 minutes
pork	45 minutes
veal	25 minutes

Look at the chart. Chicken cooks at a rate of 20 minutes per pound.

Multiply 3.8 by 20 using a calculator.

20 ⊗ 3.8 ⊜ 76

It will take 76 minutes to roast the chicken.

USING THE CALCULATOR

How long will it take to roast each of the following? Give your answer to the nearest minute.

1. 4.6 pounds of roast beef
2. 6.3 pounds of veal
3. 5.9 pounds of pork
4. 8.5 pounds of chicken
5. 3.95 pounds of roast beef
6. 4.45 pounds of veal

Solve.

7. Mrs. Kidd cooked a 4.2-pound roast in 75.6 minutes. What was it she roasted?

8. It took just under 2 hours for Jack's 6.00-pound roast to cook completely. What did Jack cook?

9. Colleen cooked a veal roast for 2 hours and 42 minutes. How much did it weigh?

10. Which would you rather eat, a 2.5-pound chicken that has been roasting for half an hour, or a 3.1-pound roast beef that has been roasting for just under an hour? Why?

EXTRA PRACTICE

Mental Math: Multiply and Divide Decimals, page 265
Multiply or divide mentally.

1. 2.0 × 10
2. 31.5 × 100
3. 0.65 × 10
4. $3.15 × 100
5. 0.14 × 100
6. $.12 × 10
7. 0.016 × 1,000
8. 22.8 × 1,000
9. 4.08 ÷ 10
10. 5.0 ÷ 10
11. 0.069 ÷ 10
12. 32.3 ÷ 100
13. 0.1009 ÷ 100
14. 45.8 ÷ 100
15. 0.4 ÷ 100
16. 0.87 ÷ 10
17. 43.5 ÷ 1,000
18. 0.0014 × 10
19. 0.53 ÷ 100
20. 3.007 ÷ 1,000
21. $.70 × 1,000
22. $73.45 × 100
23. 0.07 ÷ 1,000
24. 9 ÷ 1,000

Estimating Products, page 267
Estimate. Use the front digits.

1. 6.23 × 2
2. 1.3 × 8
3. 4.8 × 6
4. 14.6 × 5
5. 82.6 × 7

Estimate. Use rounding.

6. 7.33 × 8
7. 17.5 × 8
8. 3.65 × 13
9. 6.10 × 5
10. 7.38 × 6

Multiplying Whole Numbers and Decimals, page 269
Use your rule to find the product.

1. 0.32 × 4
2. 0.15 × 6
3. 0.21 × 8
4. 0.16 × 5
5. 0.7 × 9
6. 1.29 × 4
7. 2.83 × 2
8. 1.07 × 8
9. 0.7 × 12
10. 3.09 × 17

Multiplying Decimals by Whole Numbers, page 271
Multiply.

1. 0.3 × 6
2. 0.16 × 7
3. $.67 × 8
4. 0.013 × 5
5. 0.017 × 3
6. $4.95 × 3
7. 3.21 × 8
8. 3.906 × 5
9. 31.765 × 9
10. 46.895 × 6

Chapter 7

EXTRA PRACTICE

More Multiplying, page 273
Multiply.

1. 2.73 × 14
2. 0.893 × 24
3. $5.87 × 35
4. 17.017 × 19
5. 0.007 × 12

6. $19.09 × 13
7. 8.374 × 10
8. 0.689 × 75
9. 10.017 × 12
10. $63.47 × 67

11. 8 × 7.098
12. 27 × 33.52
13. 14 × 0.009
14. 23 × 82.25
15. 16 × 36.74
16. 20 × 0.077

Problem-Solving Strategy: Working Backward, page 275
Work backward to solve the problem.

1. Lauren bought three books. The science book cost $4.50 more than the health book. The health book cost $6.76 more than the sports book. The sports book cost $9.79. How much did the science book cost?

2. Karen has $460 in her savings account this week. Last week she deposited $25. The week before that she withdrew $47. How much was in her account two weeks ago before she withdrew her money?

Dividing Decimals by Whole Numbers, page 283
Divide.

1. 3)13.5
2. 3)18.3
3. 2)16.2
4. 2)30.8

5. 5)$1.75
6. 7)1.54
7. 9)5.112
8. 3)16.8

9. 5)13.95
10. 63)431.55
11. 7)22.575
12. 30)29.85

13. 8)$256.32
14. 5)$7.85
15. 7)19.81
16. 34)25.84

17. 7)31.094
18. 6)$5.34
19. 20)26.080
20. 5)4.865

21. 9)$28.53
22. 6)94.932
23. 8)3.912
24. 7)73.633

Multiplying and Dividing Decimals

EXTRA PRACTICE

Zeros in Division, page 285
Divide

1. 3)3.18
2. 7)0.14
3. 6)0.42
4. 5)$15.05
5. 2)30.7
6. 5)$10.40
7. 65)1.495
8. 27)2.241
9. 85)4.42
10. 65)2.73
11. 63)$131.04
12. 30)61.5
13. 62)5.27
14. 95)2.185
15. 42)2.1
16. 55)47.3
17. 96)5.376
18. 54)3.726
19. 56)8.4
20. 32)2.72

More Dividing Decimals, page 287
Divide.

1. 9)54.45
2. 85)0.425
3. 63)5.04
4. 84)4.704
5. 23)1.242
6. 72)1.656
7. 4)$12.24
8. 60)3.6
9. 22)0.99
10. 5)7.65
11. 52)$160.68
12. 89)4.183
13. 5.616 ÷ 78
14. 98.8 ÷ 65
15. 0.3 ÷ 6
16. 14.31 ÷ 54
17. 34.16 ÷ 61
18. $107.10 ÷ 85
19. $98.16 ÷ 24
20. 5.886 ÷ 54
21. $94.38 ÷ 66
22. 96.672 ÷ 32
23. 3.968 ÷ 64
24. 16.932 ÷ 83

Problem-Solving Strategy: Making a Table, page 289
Make a table to solve the problem.

1. Gloria made some trail mix. She used 1 cup of raisins for every 3 cups of peanuts and for every 2 cups of almonds. How many cups of each did she use to get a total of 18 cups of trail mix?

2. Jennifer makes some vegetable stew. Her recipe will make 4 servings. She uses 3 cups of carrots for every 1.5 cups of peas. How many cups of carrots and peas will she need if she wants to make enough stew for 20 servings?

EXTRA PRACTICE

Measuring Metric Capacity and Mass, page 291

Which unit of capacity would you use to measure? Write *milliliter* or *liter*.

1. pitcher of iced tea
2. coffee mug
3. eye dropper
4. small can of juice
5. liquid bleach
6. bucket of water

Which unit of mass would you use to measure? Write *milligram*, *gram*, or *kilogram*.

7. truck
8. pencil
9. turtle
10. butterfly
11. box of cat food
12. motorcycle

Which is the most reasonable estimate of capacity or mass?

13. drinking glass   **a.** 25 mL   **b.** 250 mL   **c.** 250 L
14. bathtub   **a.** 50 mL   **b.** 50 mL   **c.** 500 L
15. rubber band   **a.** 1 mg   **b.** 1 g   **c.** 1 kg
16. person   **a.** 30 mg   **b.** 300 g   **c.** 30 kg

Renaming Metric Measures, page 293

Complete.

1. 6 kg = ■ g
2. 12 L = ■ mL
3. 6 m = ■ cm
4. 10 cm = ■ mm
5. 3,740 mg = ■ g
6. 398 mL = ■ L
7. 169 g = ■ kg
8. 300 m = ■ km
9. 32.5 m = ■ cm
10. 483 mg = ■ g
11. 2.35 cm = ■ mm
12. 19 mL = ■ L
13. 38 cm = ■ mm
14. 514 g = ■ kg
15. 7.13 L = ■ mL
16. 810 m = ■ cm
17. 3.5 kg = ■ g
18. 1.9 L = ■ mL
19. 0.54 km = ■ m
20. 789 cm = ■ m
21. 62.5 m = ■ cm
22. 479 mL = ■ L
23. 4.9 kg = ■ g
24. 69 cm = ■ mm

Multiplying and Dividing Decimals

Practice PLUS

KEY SKILL: Multiplying Decimals by Whole Numbers (Use after page 271.)

Level A
Multiply.

1. 0.3 × 3
2. 0.6 × 5
3. 0.4 × 2
4. 0.2 × 1
5. 0.5 × 4

6. 0.8 × 7
7. 0.7 × 6
8. 0.9 × 8
9. 0.14 × 3
10. 0.23 × 5

11. 0.64 × 9
12. 0.33 × 6
13. 0.41 × 4
14. 0.52 × 2
15. 0.56 × 6

16. 7 × 0.03
17. 4 × 0.014
18. 2 × 0.006
19. 8 × 0.007

Level B
Multiply.

20. 0.007 × 5
21. 0.13 × 4
22. 0.018 × 8
23. 0.011 × 9
24. 9.09 × 9

25. 1.706 × 3
26. 4.231 × 7
27. 9.565 × 6
28. 6.613 × 8
29. 32.316 × 4

30. 4 × $11.07
31. 5 × 4.364
32. 8 × 0.015
33. 6 × 5.044

Level C
Multiply.

34. 26.765 × 4
35. 6.915 × 7
36. 48.092 × 4
37. 36.234 × 8
38. $15.08 × 5

39. 1.264 × 4
40. 0.012 × 9
41. 32.04 × 5
42. 51.348 × 7
43. 72.633 × 6

44. 7 × 3.042
45. 8 × 3.57
46. 9 × 4.055
47. 6 × $19.95

48. Karen's maple tree grows 54.38 cm a month. How much will the tree grow in 3 months?

Practice PLUS

KEY SKILL: Zeros in Division (Use after page 285.)

Level A

Divide.

1. 9)0.54
2. 4)0.16
3. 2)2.16
4. 3)15.09
5. 5)2.6

6. 5)0.25
7. 8)24.56
8. 4)0.24
9. 8)8.56
10. 6)2.7

11. 4)100.2
12. 5)0.78
13. 5)8.4
14. 6)4.23
15. 5)8.73

16. 5)87.6
17. 3)9.27
18. 8)$96.80
19. 5)0.25
20. 6)0.372

21. 4.14 ÷ 2
22. 21.06 ÷ 3
23. $5.60 ÷ 4
24. 5.43 ÷ 5

Level B

Divide.

25. 9)0.63
26. 7)0.245
27. 8)2.56
28. 45)4.59
29. 6)54.06

30. 5)0.3
31. 8)7.16
32. 5)1.02
33. 85)3.57
34. 40)2.12

35. 8)0.6
36. 5)5.54
37. 12)0.708
38. 15)15.3
39. 25)0.575

40. 50)2.55
41. 8)3.608
42. 30)6.60
43. 7)0.518
44. 9)8.118

45. 20.4 ÷ 8
46. 5.22 ÷ 4
47. 8.1 ÷ 4
48. 6.099 ÷ 3

Level C

Divide.

49. 18)$73.08
50. 4)40.2
51. 60)3.36
52. 5)0.07
53. 53)0.159

54. 5)3.49
55. 6)4.842
56. 8)8.344
57. 36)2.52
58. 20)80.3

59. 36)9.9
60. 15)0.7830
61. 8)0.36
62. 7)$7.35
63. 45)18.09

64. 5.3 ÷ 5
65. 45.8 ÷ 8
66. 3.8 ÷ 50
67. 0.408 ÷ 8

68. The recycling club raised $450.75. The money they earned will be shared equally by 5 charities. How much money will each charity receive?

Multiplying and Dividing Decimals

Chapter Review

LANGUAGE AND MATHEMATICS
Complete the sentences. Use the words in the chart on the right.

1. A ■ is a metric unit of capacity. *(page 290)*

2. A ■ is a metric unit of mass. *(page 290)*

3. A milliliter is a metric unit of ■. *(page 290)*

4. A kilogram is a metric unit of ■. *(page 290)*

5. ***Write a definition*** or give an example of the words you did not use from the chart.

VOCABULARY
meter
liter
milligram
capacity
mass

CONCEPTS AND SKILLS

Multiply. *(pages 264, 268–273)*

6. 4.0 × 10
7. 0.85 × 100
8. 0.018 × 1,000
9. 1.007 × 1,000

10. 0.24 × 3
11. 0.73 × 4
12. 2.8 × 12
13. 5.07 × 15
14. $4.85 × 6

15. 0.03 × 7
16. 3.887 × 8
17. 0.016 × 4
18. 24.948 × 9
19. $62.43 × 57

20. 0.007 × 99
21. 0.634 × 73
22. 10.01 × 8
23. 8.623 × 10

24. 5 × 26.201
25. 16 × 3.006
26. 32 × 0.043
27. 30 × 0.002

Estimate. Use the front digits. *(page 266)*

28. 6.11 × 4
29. 4.7 × 3
30. 80.6 × 5
31. 18.3 × 2
32. 6.803 × 7

Estimate. Use rounding. *(page 266)*

33. 3.11 × 7
34. 1.33 × 7
35. 4.21 × 8
36. 5.97 × 4
37. 85.503 × 11

Divide. *(pages 282–287)*

38. 15.2 ÷ 4 **39.** 0.322 ÷ 14 **40.** 1.794 ÷ 69 **41.** $1.52 ÷ 4 **42.** 1.84 ÷ 23
43. 0.275 ÷ 5 **44.** $14.42 ÷ 7 **45.** $8.72 ÷ 8 **46.** 0.63 ÷ 70 **47.** 83.64 ÷ 41
48. 1.638 ÷ 63 **49.** 17.22 ÷ 6 **50.** 34.84 ÷ 13 **51.** 1.5 ÷ 40

Which is the most reasonable estimate of capacity or mass? *(page 290)*

		a.	b.	c.	d.
52.	kitten	600 mg	600 g	60 kg	600 kg
53.	button	1 mg	1 g	1 kg	10 kg
54.	bucket	1 mL	10 mL	10 L	100 L
55.	can of juice	25 mL	250 mL	10 L	25 L

Complete. *(page 292)*

56. 3,160 mg = ■ g **57.** 89 mL = ■ L **58.** 3.6 km = ■ m **59.** 0.38 m = ■ km
60. 68 cm = ■ mm **61.** 333 mg = ■ g **62.** 0.139 kg = ■ g **63.** 24 g = ■ mg

CRITICAL THINKING

64. Without calculating, determine which number sentence has the greatest product and which number sentence has the least product.

a. 4.0 × 5
b. 0.4 × 5
c. 0.04 × 5
d. 0.004 × 5

MIXED APPLICATIONS

65. Clarisa delivered fliers for three days. Each day she earned twice as much as the day before. If she earned $12 on the third day, how much did she earn for all three days? *(page 288)*

66. Gary withdrew money from the bank. He spent $8.95 on a baseball cap and twice that amount on a program book. He had $13.15 left. How much money did he start with? *(page 274)*

67. Jeffrey buys 4 packages of hot dogs. Each package sells for $3.60. How much does Jeffrey spend? *(page 270)*

68. At the souvenir shop, Debbie bought 5 postcards for $.75. How much did each one cost? *(page 282)*

Chapter Test

Multiply or divide mentally.

1. 32.7 × 10
2. 0.023 × 1,000
3. 42.3 ÷ 100
4. 0.62 ÷ 1,000

Estimate. Use the front digits.

5. 24.7
   ×   9

6. 6.877
   ×   6

Estimate. Use rounding.

7. 3.16
   ×   5

8. 7.876
   ×   5

Multiply.

9. 0.7
   × 5

10. 2.6
    × 7

11. 8.89
    × 13

12. 0.76
    × 27

Divide.

13. 4)10.4
14. 9)3.15
15. 34)$185.30
16. 74)19.906

Which is the most reasonable estimate?

17. staple   a. 1 mg   b. 1 g   c. 1 kg   d. 10 kg

18. juice carton   a. 2 mL   b. 20 mL   c. 2 L   d. 20 L

19. bowling ball   a. 5 mg   b. 5 g   c. 50 g   d. 5 kg

Complete.

20. 26 mL = ■ L
21. 0.7 cm = ■ mm
22. 45 m = ■ km
23. 1.8 kg = ■ g

Solve.

24. Linda bought three items at the bookstore. The hardcover book cost three times as much as the paperback. The paperback cost $5 less than the poster. The poster cost $8.95. How much did the paperback cost? the hardcover?

25. To make a fizzy fruit punch, Cassandra used 0.75 liter of club soda for every 2 liters of fruit punch. How many liters of club soda did she use for 10 liters of fruit punch?

Enrichment For All

ORDER OF OPERATIONS

David and Beth each tried to solve the same number sentence. They did not get the same answer.

1. Which operation did David do first, subtraction or division?
2. Which operation did Beth do first?

When there is more than one operation in a number sentence, the order in which you perform the operations is very important.

- Do the work inside the parentheses first.
- Raise to a power.
- Multiply and divide in order from left to right.
- Add and subtract in order from left to right.

3. Who solved the number sentence correctly?

Use the order of operations to solve.

4. $(5 + 3) \times 6$
5. $(81 \div 9^2) + 7$
6. $17 - (48 \div 6)$
7. $(27 - 18) \times 12$
8. $18 - (9 \times 2)$
9. $(15 \times 14) \div 5$
10. $52 - 6 \times 5$
11. $9 \times 4 + 8$
12. $14 + 12 \div 4$
13. $36 \div 6 - 3$
14. $21 - 2 \times 4$
15. $24 \div 3 + 8$
16. $1 + 9 - 8 + 2$
17. $2 \times 8 - 9 - 1$
18. $1 + 9 - (8 + 2)$
19. $(9 - 8) \times (2^2 - 1)$

Copy the exercise. Put in parentheses to make the sentence true.

20. $6 + 3 \times 7 = 63$
21. $14 \div 7 \times 9 = 18$
22. $9 \times 5 - 4 = 9$
23. $54 \div 9 \times 8 = 48$
24. $6 \times 11 - 4 = 42$
25. $45 \div 9 \times 3 = 15$

Multiplying and Dividing Decimals

Cumulative Review

Choose the letter of the correct answer.

1. 4)877
   a. 220
   b. 219
   c. 219 R1
   d. not given

2. 893
   × 9
   a. 8,073
   b. 8,027
   c. 8,037
   d. not given

3. 670
   × 47
   a. 31,490
   b. 30,490
   c. 31,040
   d. not given

4. 15 min = ■ s
   a. 300
   b. 600
   c. 900
   d. not given

5. 5)$7,595
   a. $1,519
   b. $1,015
   c. $1,115
   d. not given

6. 27)8,119
   a. 300 R9
   b. 300 R19
   c. 309 R1
   d. not given

7. 25.8 ÷ 1,000
   a. 0.258
   b. 0.025
   c. 0.0258
   d. not given

8. Estimate. Use the front digits.
   6.23
   × 4
   a. 10
   b. 24
   c. 28
   d. not given

9. 1.6 ÷ 20
   a. 0.8
   b. 0.08
   c. 8.0
   d. not given

10. $3.87
    × 5
    a. $18.05
    b. $19.25
    c. $19.35
    d. not given

11. 8.395
    × 42
    a. 352.59
    b. 35.259
    c. 3,525.9
    d. not given

12. 32)18.176
    a. 0.056
    b. 5.68
    c. 0.568
    d. not given

13. 0.8
    × 16
    a. 1.28
    b. 12.8
    c. 1.028
    d. not given

14. The total mass of a banana, a grapefruit, and some cherry tomatoes is 1,138 g. How many kilograms is this?
    a. 1.138 kg
    b. 11.38 kg
    c. 0.1138 kg
    d. not given

CHAPTER 8

Geometry

MATH CONNECTIONS: ALGEBRA • PROBLEM SOLVING

1. What do you see in this picture?
2. What shapes do you see in the picture?
3. How can the tools in the picture be useful to you?
4. Write a problem about the picture.

DEVELOPING A CONCEPT
Geometry Around Us

Some basic geometric figures are suggested in this picture.

The source of the light from the lighthouse suggests a point. A **point** is an exact location in space.

	Read	Symbol
	point P	P

A light beam coming from the lighthouse suggests a ray. A **ray** is part of a line.

	Read	Symbol
	ray PR	$\overrightarrow{PR}$

The horizon suggests a line. A **line** is made up of points and goes on and on in both directions.

	Read	Symbol
	line AB or line BA	$\overleftrightarrow{AB}$ or $\overleftrightarrow{BA}$

The distance between the ships suggests a line segment. A **line segment** is part of a line.

	Read	Symbol
	line segment CD or line segment DC	$\overline{CD}$ or $\overline{DC}$

The surface of the ocean suggests a plane. A **plane** is a flat surface that goes on and on in all directions.

	Read	Symbol
	plane O	plane O

The pair of jet trails at the top left suggest intersecting lines. **Intersecting lines** cross each other. The pair of jet trails at the top right suggest parallel lines. **Parallel lines** are lines in the same plane that never intersect.

$\overleftrightarrow{EF}$ intersects $\overleftrightarrow{GH}$
$\overleftrightarrow{IJ}$ is parallel to $\overleftrightarrow{KL}$
$\overleftrightarrow{IJ} \parallel \overleftrightarrow{KL}$

SHARING IDEAS

Name two objects in the real world that suggest each geometric figure. Tell how they represent the geometric figure.

1. point
2. plane
3. line segment
4. line
5. parallel lines
6. intersecting lines

PRACTICE

Identify the figure. Then name it using symbols.

7. M • —— • N
8. ←—•——•—→ O P
9. • O
10. •——•—→ S   T

Write *intersecting* or *parallel* to describe the pair of lines.

11.
12.
13.
14.

Draw the figure.

15. $\overline{XY}$
16. ray AB
17. plane X
18. $\overleftrightarrow{GH} \parallel \overleftrightarrow{PQ}$

Use the diagram to answer the question.

19. Name the points.
20. Name the line segments.
21. Name the parallel lines.
22. Name a pair of intersecting lines.

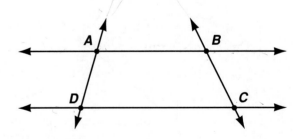

Solve. Use the drawing of the deck.

23. What geometric figures do the following suggest?
    a. the entire floor of the deck
    b. the length of one of the floor boards
    c. two boards that make up the floor
    d. the diagonal boards on the fence

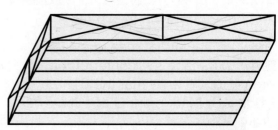

UNDERSTANDING A CONCEPT
Measuring Angles

A. The edges of some roofs meet on the side to form an angle. An **angle** is formed by two rays that start at the same endpoint.

vertex: S
sides: $\vec{SR}$, $\vec{ST}$
name: ∠S or ∠RST or ∠TSR

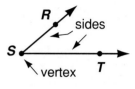

The **interior** of the angle is all points between the two rays. The **exterior** of the angle is all points not on the rays or in its interior.

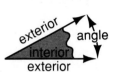

B. Angles are measured in **degrees (°).** You can use a protractor to measure an angle. Place the center of the protractor at the vertex of the angle. Make sure that the 0° mark is along one side of the angle.

Read the outer scale if the angle opens from the right.

Read the inner scale if the angle opens from the left.

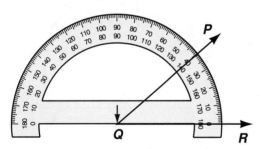

The measure of ∠PQR = 40°

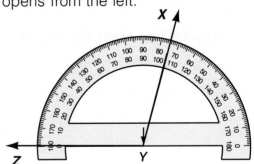

The measure of ∠XYZ = 105°

C. You can use a protractor to draw an angle to a given measure.

Draw an angle that measures 135°. Draw $\vec{AB}$. Place the center of the protractor on A so that the 0° mark is along $\vec{AB}$. Mark point C at 135°. Then draw $\vec{AC}$.

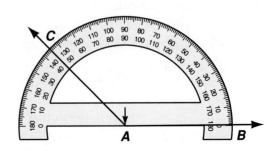

TRY OUT Name the angle. Then find the measure of the angle.

1.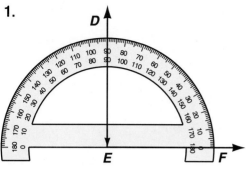

2.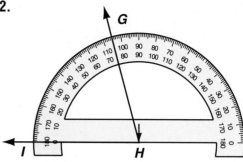

PRACTICE

Name the angle. Then find the measure of the angle.

3.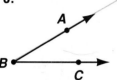

4.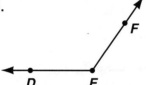

5.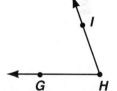

Write if point P is *on the angle,* in its *interior* or its *exterior.*

6.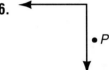

7.

8.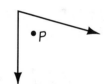

Use a protractor to draw the angle.

9. 60°   10. 150°   11. 45°   12. 165°   13. 90°   14. 100°

Mixed Applications Solve. You may need to use the Databank on page 520.

15. Maurice opened the front door to let his dog in. How wide is the door opening? Use the diagram at the right.

16. What kind of lines do the vertical lines of the floor suggest?

17. How much more expensive is the Traditional-style storm door than the Crossbuck-style door?

EXTRA Practice, page 340

Geometry 313

UNDERSTANDING A CONCEPT

Angles

A. You can identify angles according to their measure.

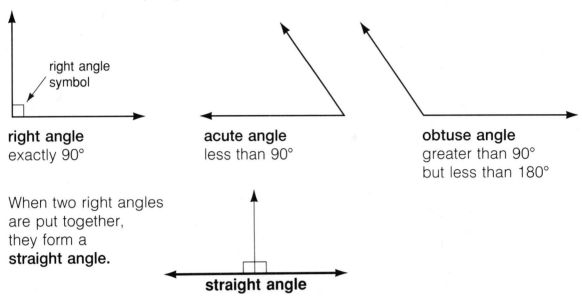

right angle
exactly 90°

acute angle
less than 90°

obtuse angle
greater than 90°
but less than 180°

When two right angles are put together, they form a **straight angle**.

straight angle

1. What is the measure of a straight angle? How do you know?
2. What real-world object suggests a right angle? an acute angle? an obtuse angle? a straight angle?

B. Perpendicular lines are lines that intersect to form right angles.

Read: Line *KL* is perpendicular to line *MN*.
Write: $\overleftrightarrow{KL} \perp \overleftrightarrow{MN}$

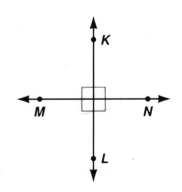

3. How many right angles do perpendicular lines form?
4. What real-world object suggests perpendicular lines?

TRY OUT Identify the angle as *right, acute,* or *obtuse.*
Identify whether the intersecting lines are perpendicular.

5. 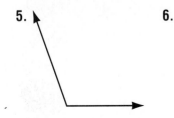   6. 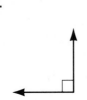   7. 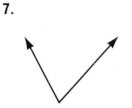   8.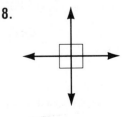

314   Lesson 8-3

PRACTICE

Write whether the angle is *acute, right,* or *obtuse.*

9.
10.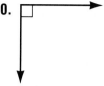
11.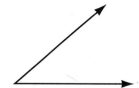

12.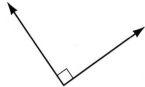
13.
14.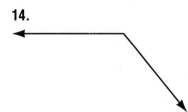

Are the lines perpendicular? Write *yes* or *no.* Use a protractor to measure if you need to.

15.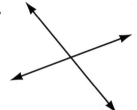
16.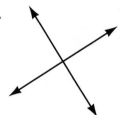
17.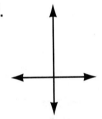

Draw the figure.

18. acute angle NOP   **19.** right angle X   **20.** obtuse angle BUS   **21.** $\overleftrightarrow{ST} \perp \overrightarrow{UV}$

Mixed Applications

22. Marilyn saw a lamp on the table at the hobby shop. What kind of angle does the arm of the lamp suggest? What kind of lines do the edges of the table suggest?

23. Marilyn bought a kite at the hobby shop. What kind of lines does the back of her kite suggest?

24. A spool of kite string costs $.59. Would you have enough money to buy 3 spools of kite string if you had $1.50? If not, how much more money would you need?

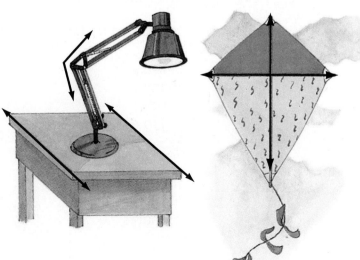

EXTRA Practice, page 340

Geometry **315**

UNDERSTANDING A CONCEPT
Plane Figures

The floor of the room above suggests a plane. The floor mats suggest **plane figures.**

A **polygon** is a closed plane figure the sides of which are line segments. The number of angles is the same as the number of sides.

A **regular polygon** has all sides the same length and all angles the same measure.

Some polygons are named by the number of sides or angles that they have.

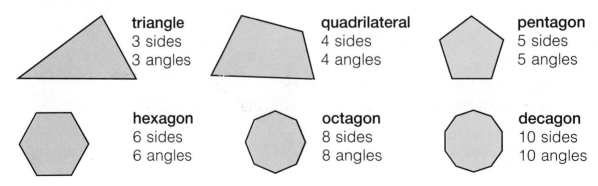

1. Which polygons above appear to be regular polygons?

TRY OUT   Match the name with the polygon.

2. pentagon

3. octagon

4. regular hexagon

a.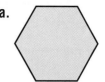
b.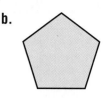
c.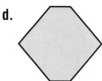
d.

PRACTICE

Identify the polygon. Tell how many sides and angles it has.

5.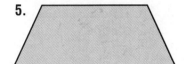
6.
7.
8.

9.
10.
11.
12.

13. Which polygons in Exercises 5–12 appear to be regular polygons?

Mixed Applications

Solve. Which method did you use?

14. A piece of rectangular carpeting is 7 ft long and 9.5 ft wide. What is the area of the carpet?

15. Mrs. Ribiero bought a rug that is a regular octagon. One side of the rug measures 2.75 ft. What is the perimeter of the rug?

16. It took Nino 25 minutes to vacuum, 15 minutes to wash the dishes, and 20 minutes to clean his room. Did he spend more than, less than, or exactly one hour straightening up the house?

ESTIMATION
MENTAL MATH
CALCULATOR
PAPER/PENCIL

EXTRA Practice, page 341; Practice *PLUS*, page 344

Geometry 317

DEVELOPING A CONCEPT

Triangles and Quadrilaterals

A. A triangle can be identified by the kind of angles it has.

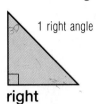
1 right angle
right

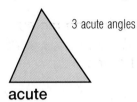
3 acute angles
acute

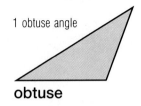
1 obtuse angle
obtuse

A triangle can also be identified according to the lengths of its sides.

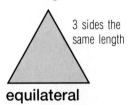
3 sides the same length
equilateral

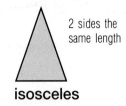
2 sides the same length
isosceles

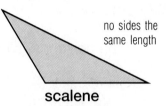
no sides the same length
scalene

B. Some quadrilaterals have special names.

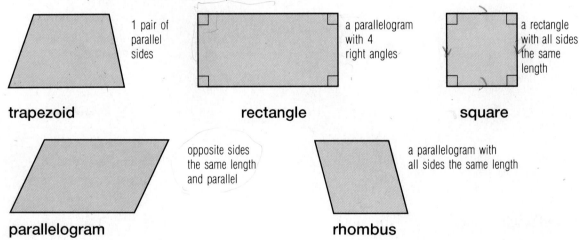

trapezoid — 1 pair of parallel sides

rectangle — a parallelogram with 4 right angles

square — a rectangle with all sides the same length

parallelogram — opposite sides the same length and parallel

rhombus — a parallelogram with all sides the same length

C. There is an interesting relationship among the sums of the measures of the angles of all triangles.

1. Use a protractor to measure the angles of the right triangle. What is the sum of the angle measures?

2. Find the sum of the angle measures in each of the other triangles. Do you see a pattern? If so, describe it.

3. Make up a rule for finding the sum of the angle measures in a triangle.

Lesson 8-5

SHARING IDEAS

4. If two angles of a triangle have a sum of 120°, what is the measure of the third angle?

5. What is the sum of the measures of the angles of a rectangle? How do you know?

6. What rule could you write for finding the sum of the angle measures of a quadrilateral? Use your protractor and the quadrilaterals on page 318 to test your rule.

PRACTICE

Write whether the triangle is *right, acute,* or *obtuse.*

7.   8.   9.   10.

Write whether the triangle is *equilateral, isosceles,* or *scalene.*

11.   12.   13. 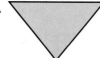 14.

Identify the quadrilateral.

15.   16.  17.   18.

Find the measure of the missing angle.

19. 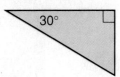 20. 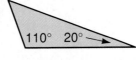 21.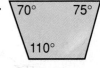

Critical Thinking Is the statement *true* or *false*? Give a reason to support your answer.

22. Every equilateral triangle is an isosceles triangle.

23. A square is also a rhombus.

24. Every quadrilateral with at least one right angle is a rectangle.

25. If a quadrilateral has three right angles, it must have four right angles.

EXTRA Practice, page 341

Geometry 319

PROBLEM SOLVING

Strategy: Drawing a Diagram

The scout group that Randal belongs to met at school on Saturday to go on a bicycling trip. The scouts started at the school and rode 3.5 kilometers east. Next they rode 2.5 kilometers south, 3.5 kilometers west, and then north, back to the school. How many kilometers long was their ride?

To understand the problem, Randal drew a diagram. He showed the school with a dot. To show the first part of the trip, he drew an arrow to the right. To show the next part, he drew an arrow going down. He labeled the arrow with the distances and directions.

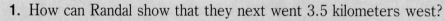

1. How can Randal show that they next went 3.5 kilometers west?

2. How can Randal show the final part of the trip?

3. How far north did they have to ride to get back to school? How do you know?

4. Show how the completed diagram should look. How can Randal find how long their ride was? How long was it?

5. What piece of information was given in the completed diagram that helped Randal solve the problem?

PRACTICE

Draw a diagram to help solve the problem.

6. The scouts collected newspapers for recycling. They started their route at Town Hall, went 5 miles north, 9 miles west, and 8 miles south. Next they went 9 miles east. How far and in what direction did they then have to go to get back to Town Hall?

7. The scouts helped a local day-care center build a fence around a rectangular play area. The fence was twice as long as it was wide. The width was 35 feet. Each foot of fencing cost $1.75. How much did the fencing cost in all?

8. From the picnic area, the scouts hiked 3 miles north, 4 miles east, 1 mile south, 6 miles west, and then 2 miles south. How far and in what direction will they have to go to get back to the picnic area?

9. At a scout picnic, Ben, Art, Dave, and Ed sat on the same side of a long table. Ben sat between Dave and Art. Dave sat between Ed and Ben. Which of the two boys sat on the ends?

Strategies and Skills Review

Solve. Use mental math, estimation, a calculator, or paper and pencil.

10. The scouts are collecting aluminum cans for recycling. For the month of October, Gary is asked to bring in 3 cans on October 1, 6 cans on October 3, 9 cans on October 5, and so on. How many cans do you predict he will be asked to bring in on October 15?

11. Mr. Parrella is the scout leader. He bought clay and paint for $12.90 for a meeting. The clay cost $3.00 more than the paint. How much did each item cost?

12. In September the scouts collected 203 pounds of newspapers. In October they collected 178 pounds of newspapers. How many more pounds did they collect in September than in October?

13. **Write a problem** that can be solved with the help of a diagram. Solve the problem. Ask other students to solve your problem.

Going Around in CIRCLES

Visual Reasoning

A. You will be drawing and coloring lots of circles for the activities below, so be sure you have a coin and pencils, crayons, or markers in four different colors.

Here are the rules:

- All circles are the same size.

- Circles may touch, but they may not overlap. O.K.   NOT O.K.

- Any two circles that touch must be different colors.

For example, in this arrangement of three circles, each circle touches at least one other and only two colors are needed:

1. Can you draw a different arrangement of three circles for which you need three colors?

2. Can you draw an arrangement of six circles where each circle touches at least two others, and you only need two colors? Can you find more than one way?

3. Can you draw an arrangement of six circles for which you need three colors? Can you find more than one way?

4. Next, try to draw an arrangement of ten circles for which you need three colors.

B. Copy the figures below.

5. Color each figure. Use as few colors as possible, but remember that any two circles that touch must have different colors.

   How many colors do you need for each?

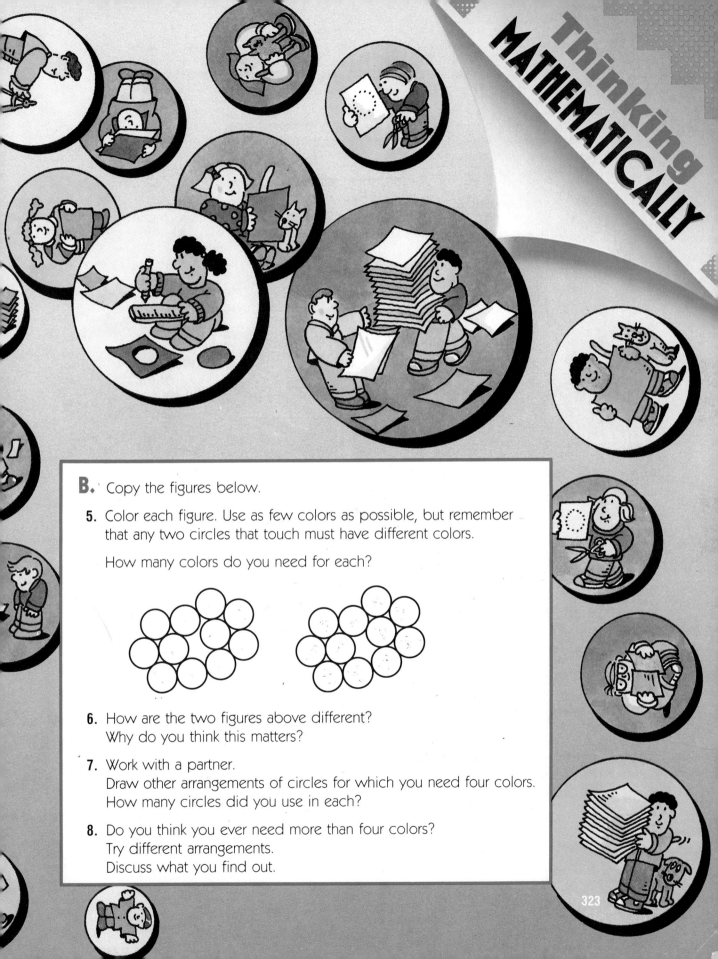

6. How are the two figures above different?
   Why do you think this matters?

7. Work with a partner.
   Draw other arrangements of circles for which you need four colors.
   How many circles did you use in each?

8. Do you think you ever need more than four colors?
   Try different arrangements.
   Discuss what you find out.

DEVELOPING A CONCEPT

Symmetry

Many buildings are examples of symmetric figures.

If you could fold this picture along the dashed line, the part of the building on one side of the line would exactly match the part on the other side. The dashed line is called a **line of symmetry.**

WORKING TOGETHER

Do the following activity to explore symmetry.

Step 1 Use a sheet of paper to make a square.

Step 2 Fold to find as many lines of symmetry as you can.

Step 3 Draw the lines of symmetry for the square.

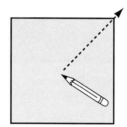
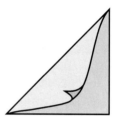

1. How many lines of symmetry does a square have?

2. Trace the trapezoid below and repeat the activity. How many lines of symmetry does the trapezoid have?

3. Trace the parallelogram and repeat the activity again. How many lines of symmetry does the parallelogram have?

SHARING IDEAS

4. Do all figures have a line of symmetry? If not, name such a figure.

5. Can a figure have more than one line of symmetry? If yes, name such a figure.

PRACTICE

Is the dashed line a line of symmetry? Write *yes* or *no*.

6.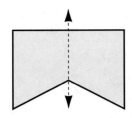
7.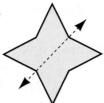
8.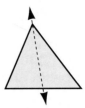

Trace the figure. Draw and then count all its lines of symmetry.

9.
10.
11.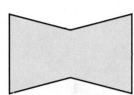

12.
13.
14.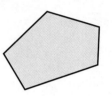

Trace and complete the figure so that the dashed line is a line of symmetry.

15.
16.
17.

Critical Thinking

18. Are $\overleftrightarrow{RS}$ and $\overleftrightarrow{XY}$ lines of symmetry for the circle? How many lines of symmetry can you draw for this circle?

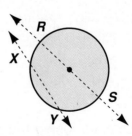

Mixed Review

Choose the appropriate metric unit of measure.

19. capacity of a pot   **a.** mL   **b.** L
20. mass of a speck of dust   **a.** mg   **b.** g   **c.** kg
21. mass of a pretzel   **a.** mg   **b.** g   **c.** kg

EXTRA Practice, page 342

DEVELOPING A CONCEPT

Slides, Flips, Turns

A. A skier skiing down a hill is an example of a **slide.**

A skier doing a somersault off a ski jump is an example of a **flip.**

An ice skater's spinning on one foot is an example of a **turn.**

B. You can move a figure by sliding it, flipping it, or turning it. Think of the movements of the skier and the skater as you do this activity.

Step 1 Copy the drawing at the right. Trace, label, and cut out each figure.

Step 2 Place the cutout of the triangle, label-side up, over the triangle. Slide it across the line. Trace to show the new position of the cutout. Label the move.

Step 3 Flip the trapezoid over the line. Trace and label the move.

Step 4 Turn the rectangle using the point as the center. Trace and label the move.

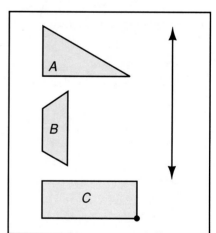

SHARING IDEAS

1. Describe the positions of the letters on the cutouts after the slide, flip, or turn.

2. Could you have reached these positions any other way? Try it and explain your results.

PRACTICE

Write whether a *slide*, *flip*, or *turn* was made.

3.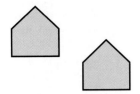

4.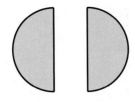

5.

6.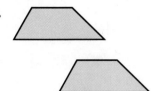

7.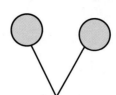

8.

9.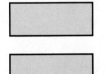

10.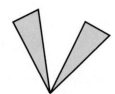

11.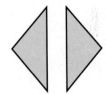

Critical Thinking

Is the movement made by a *slide*, a *flip*, or a *turn*? Tell why.

12. a merry-go-round spinning

13. a train moving on a track

14. opening the front door

VISUAL REASONING

You can create a geometric design by making slides, flips, and turns with a figure.

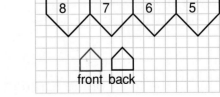

Tell whether a *slide*, *flip*, or *turn* of the figure was made to get to each step.

1. from step 1 to step 2
2. from step 2 to step 3
3. from step 3 to step 7
4. from step 6 to step 4
5. from step 5 to step 6
6. from step 2 to step 8

EXTRA Practice, page 342

UNDERSTANDING A CONCEPT

Congruent Figures

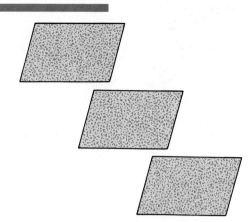

The two designs were made from the same stencil and suggest **congruent** figures.

Congruent segments have the same length.

A •————————————• B

C •————————————• D

Read: Line segment *AB* is congruent to line segment *CD*.

Write: $\overline{AB} \cong \overline{CD}$

Congruent angles have the same measure.

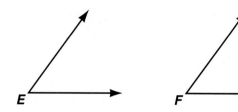

Read: Angle *E* is congruent to angle *F*.

Write: $\angle E \cong \angle F$

Congruent polygons have the same size and shape.

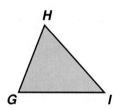 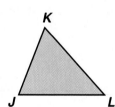

Read: Triangle *GHI* is congruent to triangle *JKL*.

Write: $\triangle GHI \cong \triangle JKL$

Matching, or **corresponding**, parts of congruent polygons are congruent.

Corresponding angles: $\angle G \cong \angle J$ Corresponding sides: $\overline{GH} \cong \overline{JK}$
                      $\angle H \cong \angle K$ $\overline{HI} \cong \overline{KL}$
                      $\angle I \cong \angle L$ $\overline{IG} \cong \overline{LJ}$

TRY OUT Write the letter of the figure that is congruent to the figure at the left.

1.   a. 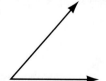   b.    c.

328   Lesson 8-10

PRACTICE

Is the figure congruent to the figure at the left? Write *yes* or *no*. Check by measuring or by making a tracing.

2.   a.  b.  c.

3. 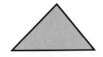 a.   b.   c.

4.  a.  b.  c.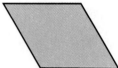

Quadrilateral *MNOP* ≅ Quadrilateral *QRST*.   TSRQ

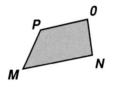
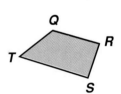

5. Use the symbol ≅ to tell which sides are congruent.

6. Use the symbol ≅ to tell which angles are congruent.

Critical Thinking

7. **What if** you made a slide, flip, or turn with a figure? Would the figure that you got be congruent to the original figure? Why or why not?

Mixed Applications

8. Miriam had an extra house key and mailbox key made at the locksmith's. Are the new keys congruent to each other? Why or why not?

9. Miriam raised the window three inches for some fresh air. Is this movement an example of a slide, flip, or turn?

Mixed Review

Find the answer. Which method did you use?

MENTAL MATH
CALCULATOR
PAPER/PENCIL

10. 12.8
   − 4.15

11.   879
    × 50

12. 6)647

13. $67.19
   + 33.00

EXTRA Practice, page 342

Geometry 329

UNDERSTANDING A CONCEPT
Similar Figures

Jeffrey has his school picture taken every year. The pictures come in three sizes. They suggest the idea of similar figures.

Similar figures have the same shape. They may or may not have the same size.

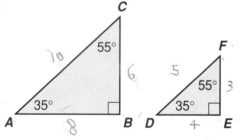

Read: Triangle ABC is similar to triangle DEF.

Write: △ABC ~ △DEF

1. Are the corresponding sides of the triangles congruent?

2. Are the corresponding angles congruent?

   In similar figures the corresponding angles are congruent.

   ∠A ≅ ∠D
   ∠B ≅ ∠E
   ∠C ≅ ∠F

TRY OUT Write the letters of the shapes that are similar.

3. a. 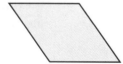   b.   c. 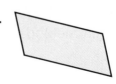   d.

PRACTICE

Are the figures similar? Write *yes* or *no*.

4.   5.    6.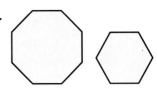

7. 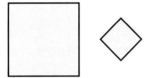   8.    9.

10.    11. 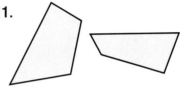   12.

13. Copy the triangle at the right on graph paper. Then draw a similar triangle with sides twice as long.

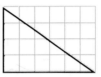

Critical Thinking

Is the statement *true* or *false*? Give a reason to support your answer.

14. All circles are similar.

15. All squares are similar.

Mixed Applications

16. The Copying Place charges $0.75 per enlargement for the first 10 enlargements and $0.35 for any additional ones. What would be the cost of 12 enlargements?

17. Brian needs an enlargement of a picture for a book report. Will the enlargement be similar to the original picture?

Mixed Review

Find the answer. Which method did you use?

MENTAL MATH
CALCULATOR
PAPER/PENCIL

18. $0.64 \div 8$

19. $10{,}000 - 475$

20. $6{,}392 + 838$

21. 239×54

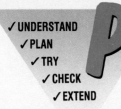

PROBLEM SOLVING

✓ UNDERSTAND
✓ PLAN
✓ TRY
✓ CHECK
✓ EXTEND

Strategies Review

You have used these problem-solving strategies to solve problems. Remember that sometimes you can use different strategies to solve the same problem.

- Drawing a Diagram
- Making a Table
- Working Backward
- Using Estimation
- Looking for a Pattern
- ~~Guess, Test, and Revise~~ Algebra
- Using Number Sense
- Solving a Multistep Problem

Solve. Tell which strategy you used.

1. A swimming pool is 21 meters wide and 50 meters long. It is divided by ropes into 8 lanes. The 6 inside lanes are each 2.5 meters wide. The 2 outside lanes have equal widths. What is the width of an outside lane?

2. Archie had a goal of scoring at least 200 points by the end of five games. He scored 34 points, 45 points, 46 points, 52 points, and 41 points. Did he reach his goal? Did you underestimate or overestimate?

3. In the 1976 Olympics, the winner of the platform dive received 600.51 points. That was 96.39 more points than the winner in 1972 received. The 1972 winner received 11.58 more than three times the number of points that the 1968 winner received. How many points did the 1968 winner receive?

4. Jay swam 130 laps in 3 days. The number of laps he swam on the second day was 9 less than the number he swam on the first day. The number of laps he swam on the last day was 10 more than the number he swam on the second day. How many laps did Jay swim each day?

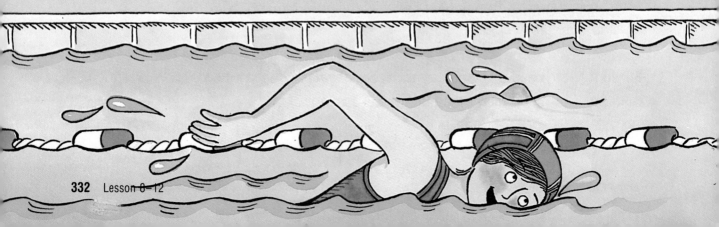

5. In the Olympics held from 1948 through 1968, the United States won 225 gold medals and 275 silver or bronze medals. The Soviet Union won 161 gold medals and 294 silver or bronze medals. If 3 points are awarded for gold medals and 1 point for other medals, which country scored more points?

6. Gwen wants to mount a poster of the 1988 Olympic Games on posterboard. The poster is 87 centimeters wide and 30 centimeters high. She wants 5 centimeters of posterboard to show at the bottom of the poster and 2 centimeters to show around the other 3 sides. What should be the size of the posterboard?

7. Ali bought a towel and swimming trunks for $29.96. The trunks cost $10 more than the towel. How much did each item cost?

8. In a track event, Olivia's time was just under Jean's. Linda finished ahead of Jean but behind Suzy. Which girl came in last?

9. Bert swam $\frac{1}{4}$ the distance that Arlene swam. Carolyn swam 15 meters more than Bert. Daryl swam 120 meters, which was twice the distance that Carolyn swam. How far did Arlene, Bert, and Carolyn swim?

10. Jill swims every 3 days and rides her bicycle every 5 days. She will swim on August 1 and ride her bicycle on August 2. What will be the first time in August when she will swim and ride on the same day?

11. Roy wants to increase the time he can hold his breath under water. For week 1 his goal is 20 seconds, for week 2 it is 24 seconds, for week 3 it is 28 seconds, for week 4 it is 32 seconds, and so on up to 60 seconds. In what week will he reach 60 seconds?

12. Juanita swam, ran, and rode a bicycle in a triathalon contest. She swam 3 miles. She ran twice as far as she swam. She rode the bicycle for 4 miles more than she ran. How many miles did she cover in the triathalon contest?

MANIPULATIVES

REASONING

DEVELOPING A CONCEPT

Space Figures

A. You can describe some **space figures** by their faces, edges, and vertices.

A rectangular prism has 6 flat surfaces called **faces**, 12 **edges**, and 8 **vertices**. A **cube** is a special kind of prism. It has six congruent square faces.

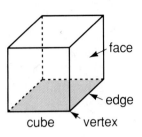

Other prisms are identified by the shape of their bases.

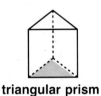
triangular prism

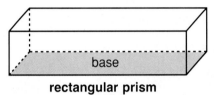
rectangular prism

hexagonal prism

1. What is the shape of the base of a rectangular prism? a triangular prism? a hexagonal prism?

B. A **pyramid** has triangular faces that meet at a common vertex. A pyramid is named by the shape of its base.

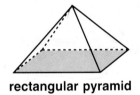
rectangular pyramid

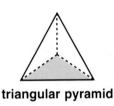
triangular pyramid

Trace and cut out the pattern at the right. Fold and tape it to make a space figure.

2. What name would you give this figure? Why?

3. How many faces, edges, and vertices does the figure have?

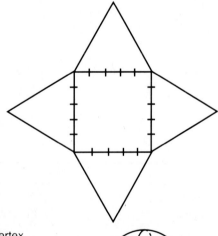

C. Here are some other space figures.

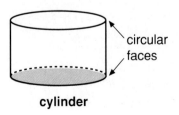
cylinder

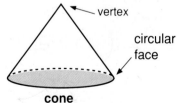
cone

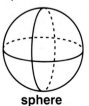
sphere

SHARING IDEAS

4. How is a triangular prism different from a triangular pyramid?

5. What would a hexagonal pyramid look like?

6. What space figure would this pattern fold into? How do you know? Check your answer by tracing, cutting, and folding the pattern.

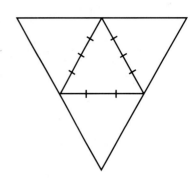

PRACTICE

What space figure does the object suggest?

7.    8.    9.    10.

Write the number of faces, edges, and vertices.

11. 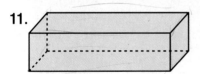   12.   13.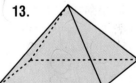

14.   15. 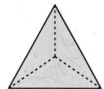   16.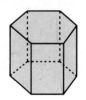

Mixed Applications

17. Find the volume of a rectangular prism with a length of 2.4 m, width of 1.3 m, and height of 1.5 m.

18. A floor tile measures 12 in. by 12 in. How much area do 25 of these tiles cover?

19. A regular hexagon has a perimeter of 72 m. What is the length of one of its sides?

20. Three angles of a trapezoid measure 90°, 90°, and 116°. What is the measure of the fourth angle?

EXTRA Practice, page 343; Practice **PLUS**, page 345

Geometry **335**

DECISION MAKING

Problem Solving: Paying for a School Newspaper

SITUATION

The students at Little Deer Elementary School want to start a school newspaper. Their teacher, Mr. Goodman, said that they would need to raise the money for printing the newspaper. Many local stores said that they would buy advertising space in the newspaper. The students also thought of selling the newspaper to cover costs.

PROBLEM

Should the students charge for the newspaper or sell ads?

DATA

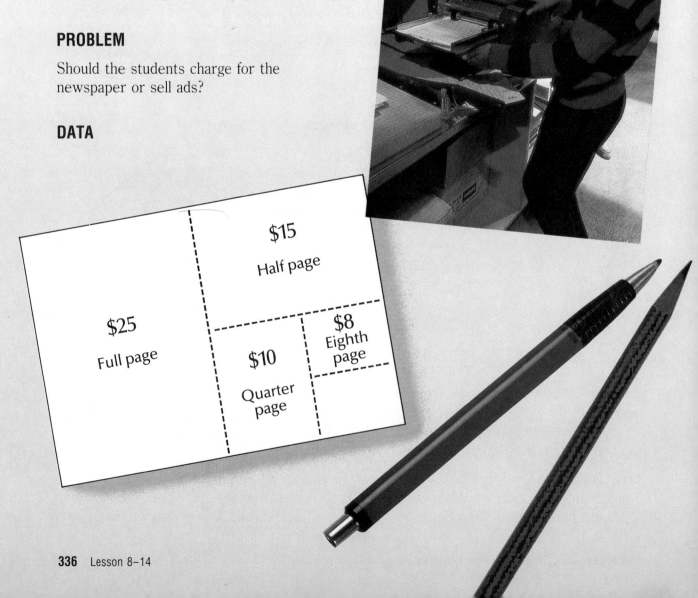

Top-Notch Printers

8-page newspaper, 1 fold
Estimated Cost:
100 copies $ 85.00
200 copies $100.00
500 copies $125.00

All bills must be paid within 30 days after order is delivered.

USING THE DATA

How much money could the students raise by selling the ads?

1. 4 quarter pages
2. 8 eighth pages
3. 2 half pages
4. 4 eighth pages
5. 1 half page and 2 quarter pages
6. 1 half page and 4 eighth pages
7. 2 quarter pages and 4 eighth pages
8. 1 half page, 1 quarter page, and 2 eighth pages

MAKING DECISIONS

9. **What if** the students thought their classmates would not buy the paper if it cost more than 25¢ a copy? Could they pay for the printing this way? Why or why not?

10. **What if** they decide to set aside $3\frac{1}{2}$ pages of space for ads? Will this be enough space to raise $100 needed to print 200 copies? Why?

11. What are the advantages of selling 2 half-page ads over selling 8 eighth-page ads? What are the disadvantages?

12. **What if** there are about 100 people who want a copy of the school newspaper? How many newspapers should the class have printed? List reasons for your answer.

13. **What if** the staff has time only to write the paper and does not have time to sell ads? How could the students pay for the printing?

14. Who should be given a free copy of a school newspaper? Tell why.

15. **Write a list** of the things the students must think about before deciding how to pay for the printing of the school newspaper.

16. How would you pay for the printing? Why?

Geometry 337

Curriculum Connection

Math and Art

The Hopi of northern Arizona make beautifully designed coiled baskets. Most Hopi baskets have colorful patterns. Sometimes the patterns are repeated irregularly around the basket. Other times the patterns may show *symmetry*. A figure has symmetry if it can be folded so that one part matches the other part exactly. The fold line of such a figure is a *line of symmetry*. Figures that do not have symmetry are *asymmetrical* as opposed to symmetrical.

Symmetrical

Asymmetrical

What if you want to use the design labeled *asymmetrical* as part of a symmetrical design? How could you use the asymmetrical figure in a symmetrical pattern?

Think: You need to position the design so that when it is folded in the middle, the two pieces match exactly. Although the figure itself is asymmetrical, you can repeat it to make a symmetrical design by turning the asymmetrical figure over as shown on the left.

ACTIVITIES

1. Design a basket. You may use graph paper and different color markers. Show your understanding of symmetry by making one part of your design symmetrical and another part asymmetrical. Label each.

2. The Navajo people of New Mexico have designed beautiful rugs as well as jewelry and other crafts. Research Navajo crafts for examples of rugs that have symmetrical and asymmetrical designs. Prepare an oral report for your class.

TECHNOLOGY

Computer: Exploring Angles

You can use Logo commands to draw different angles and discover an interesting relationship.

AT THE COMPUTER

When you input the number of degrees, the procedure TURN draws an angle with that measure. Enter the procedure.

```
TO TURN :DEGREES
RT :DEGREES
FD 50
BK 50
END
```

Enter FD 50 BK 50. To draw an angle of 50°, enter TURN 50.

Clear the screen and enter FD 50 BK 50 before drawing a new angle.

1. Draw angles that measure 110° and 165°. What commands did you enter?

2. Draw an angle that measures 60°. Repeat the procedure as many times as necessary to move the turtle around exactly one full turn (a circle). How many angles in all did you draw?

3. Repeat the steps in Problem 2 for angles that measure 120°, 90°, and 30°. Record your results in a table.

Angle	Number of Angles to Make a Circle
60°	
120°	
90°	
30°	

4. Study the results in your table. What relationship do you see between the measure of the angle and the number of angles it takes to make a circle?

5. What conclusion about a circle can you draw?

6. Can you find other pairs that work? Experiment on the computer and record your results.

Geometry

EXTRA PRACTICE

Geometry Around Us, page 311

Identify the figure. Then name it using symbols.

1.  2.   3.   4.

Write *intersecting* or *parallel* to describe the pair of lines.

5. 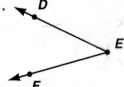 6. 7. 8.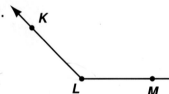

Draw the figure.

9. ray BC 10. $\overline{ST}$ 11. $\overleftrightarrow{DE} \parallel \overrightarrow{GH}$

Measuring Angles, page 313

Name the angle. Then find the measure of the angle.

1. 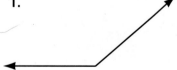 2.  3.

Use a protractor to draw the angle.

4. 45° 5. 30° 6. 120° 7. 90°

Identify Angles, page 315

Write whether the angle is *acute, right,* or *obtuse.*

1.   2.   3.

Are the lines perpendicular? Write *yes* or *no.* Use a protractor to measure if you need to.

4. 5. 6.

340 Chapter 8

Plane Figures, pages 317

Identify the polygon. Tell how many sides and angles it has.

1.
2.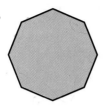
3.
4.

Triangles and Quadrilaterals, page 319

Write whether the triangle is *right, acute,* or *obtuse*.

1.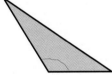
2.
3.

Write whether the triangle is *equilateral, isosceles,* or *scalene*.

4.
5.
6.

Identify the quadrilateral.

7.
8.
9.
10.

Problem-Solving Strategy: Drawing a Diagram, page 321

Draw a diagram to help solve the problem.

1. Ann and Liz went on a hike. They started at Ann's house and hiked 1.5 km west, then 2.1 km north, and 1.5 km east. They then went back south to Ann's house. How many kilometers did they hike?

2. Mike built a fence around a rectangular garden. The fence was three times as long as it was wide. The width was 18 feet. How much fencing did he use in all?

Geometry **341**

EXTRA PRACTICE

Symmetry, page 325

Is the dashed line a line of symmetry? Write *yes* or *no*.

1.
2.
3.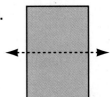

Trace the figure. Draw and then count all its lines of symmetry.

4.
5.
6.

Slides, Flips, Turns, page 327

Write whether a *slide, flip,* or *turn* was made.

1.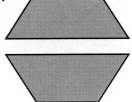
2.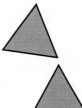
3.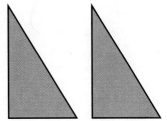

Congruent Figures, page 329

Is the figure congruent to the figure at the left? Write *yes* or *no*. Check by measuring or by making a tracing.

1.   a.   b.   c.

2.   a.   b.    c.

EXTRA PRACTICE

Similar Figures, page 331
Are the figures similar? Write *yes* or *no*.

1.
2.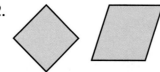
3.
4.
5.
6.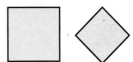

Problem Solving: Strategies Review, page 333
Solve. Tell which strategy you used.

1. A track is 32 m wide and 50 m long. It is divided into 6 lanes by chalk lines. The 4 inside lanes are each 3.5 m wide. The 2 outside lanes have equal widths. What is the width of an outside lane?

2. Judy jogged 17 mi in 3 days. The number of miles she jogged on the third day was 4 less than the number she jogged on the second day. The number of miles she jogged on the second day was 3 more than the number she jogged on the first day. How many miles did Judy jog each day?

Space Figures, page 335
What space figure does the object suggest?

1.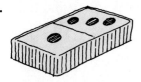
2.
3.

Write the number of faces, edges, and vertices.

4.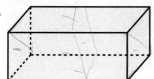
   - faces
   - edges
   - vertices

5.
   - faces
   - edges
   - vertices

Geometry

Practice PLUS

KEY SKILL: Triangles And Quadrilaterals (Use after page 319.)

Level A

Write whether the triangle is *right*, *acute*, or *obtuse*.

1.
2.

Write whether the triangle is *equilateral*, *isosceles*, or *scalene*.

3.
4.

Level B

Write whether the triangle is *right*, *acute*, or *obtuse*.

5.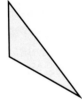
6.

Identify the quadrilateral.

7.
8.

Level C

Write whether the triangle is *equilateral*, *isosceles*, or *scalene*.

9.
10.
11.
12.

Identify the quadrilateral.

13.
14.
15.
16.

Practice PLUS

KEY SKILL: Space Figures (Use after page 335.)

Level A
What space figure does the object suggest?

1.
2.
3.

Level B
What space figure does the object suggest?

4.
5.
6.

Write the number of faces, edges, and vertices.

7.
   - faces
   - edges
   - vertices

8.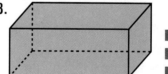
   - faces
   - edges
   - vertices

Level C
What space figure does the object suggest?

9.
10.
11.

Write the number of faces, edges, and vertices.

12.
    - faces
    - edges
    - vertices

13.
    - faces
    - edges
    - vertices

Geometry

Chapter Review

LANGUAGE AND MATHEMATICS

Complete the sentences. Use the words in the chart on the right.

1. A ■ is an exact location in space. *(page 310)*
2. A ■ is a closed plane figure whose sides are line segments. *(page 316)*
3. A ■ cuts a figure into two parts that match each other exactly. *(page 324)*
4. ■ segments have the same length. *(page 328)*
5. **Write a definition** or give an example of the words you did not use from the chart.

VOCABULARY
point
parallel lines
intersecting lines
similar
polygon
congruent
line of symmetry

CONCEPTS AND SKILLS

Complete. Use the diagram to answer. *(page 310)*

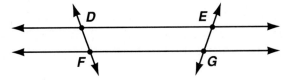

6. Name the points.
7. Name the line segments.
8. Name the parallel lines.
9. Name a pair of intersecting lines.

Use a protractor to draw the angle. *(page 312)*

10. 30° **11.** 130° **12.** 175° **13.** 90°

Write whether the angle is *acute, right,* or *obtuse.* *(page 314)*

14. 15. 16.

Name the polygon. *(page 316)*

17.   18.

Is the triangle *equilateral, isosceles,* or *scalene*? *(page 318)*

19.   20.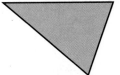

Write whether a *slide, flip,* or *turn* was made. *(page 326)*

21.
22.
23.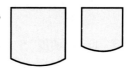

Are the figures similar? Write *yes* or *no*. *(page 330)*

24.
25.
26.

CRITICAL THINKING

27. How many lines of symmetry does a circle have? Explain your answer.

28. **What if** you made a slide, flip, or turn with a figure? Would the figure be congruent to the original figure? Why or why not?

MIXED APPLICATIONS

29. Barry rode 6 km east from school to the computer store. Then he rode 3 km north to his grandmother's house. From his grandmother's house he continued riding north to his house. He traveled 14 km in all. How far is Barry's house from the computer store? *(page 320)*

30. The members of the football team are choosing a team jersey. Their color choices are blue, orange, and silver. They can have emblems of the Warriors, the Bruins, and the Lions. How many different ways can they design their team jersey? *(page 320)*

31. John made a model of a skyscraper. It has 6 flat surfaces, 12 edges, and 8 vertices. What shape is the model? *(page 334)*

32. A knitted scarf shows this pattern: one pink row, one white row, then one pink and two white, then one pink and three white, and so on. It ends with six white and one pink. How many rows is the pattern? *(page 332)*

33. Carrie made a model of a globe. It has no face or edges. What shape is the model? *(page 334)*

Chapter Test

Use the diagram to answer Questions 1–6.

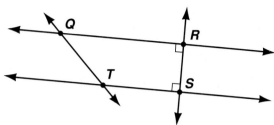

1. Name the points.
2. Name the line segments.
3. Name the parallel lines.
4. Name the intersecting lines.
5. Name the perpendicular lines.
6. Find the measure of ∠QTS. Identify it as *acute*, *obtuse*, or *right*.

Name the polygon.   Tell how many lines of symmetry.

7.   8.   9.   10.

Write whether the figures are *congruent* or *similar*.

11. 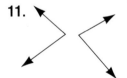   12.    13.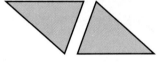

Write whether a slide, flip, or turn was made.

14. 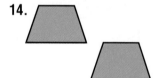   15.   16.   17.

Write the name of the figure. Then write the number of faces, edges, and vertices.

18.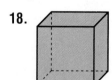
■ faces
■ edges
■ vertices

19.
■ faces
■ edges
■ vertices

Solve. Draw a diagram to help you solve the problem.

20. John has a 17-link gold chain that he wants to cut into two pieces. He wants one piece to have 5 more links than the other. How many links will be on each piece?

Enrichment For All

INFORMAL ALGEBRA: VENN DIAGRAMS

Venn diagrams show relationships between groups. Look at this diagram. Read the labels carefully.

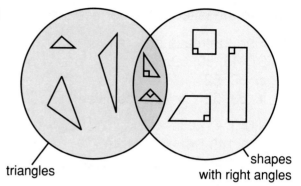

The left circle contains only triangles. The right circle contains only shapes with right angles. The space where the circles overlap contains triangles that have right angles.

Where would the shape go? Write *triangles, shapes with right angles,* or *both*.

1.   **2.** 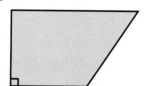   **3.** 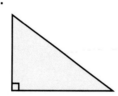   **4.**

If two groups have no common members, the circles do not overlap.

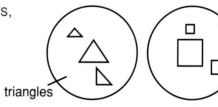

5. Sometimes one circle is entirely inside another circle. What does this mean?

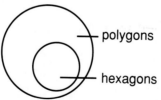

Draw and label a Venn diagram for the pair of groups.

6. circles and hexagons

7. quadrilaterals and shapes with congruent sides

8. prisms and cubes

9. right triangles and isosceles triangles

Geometry   **349**

Cumulative Review

Choose the letter of the correct answer.

1. Find the mean and range:
   37, 45, 89, 62, 82
   a. 62; 52   c. 62; 63
   b. 63; 52   d. not given

2. 8,021 ÷ 7
   a. 1,145 R6   c. 1,145
   b. 1,146 R5   d. not given

3. 0.007
   ×   8
   ─────
   a. 0.56   c. 0.056
   b. 0.5   d. not given

4. Name the angle.

   a. ∠C   c. ∠ABC
   b. ∠BCA   d. not given

5. Estimate by rounding to the nearest whole number.
   12.36
   ×   6
   ─────
   a. 6   c. 72
   b. 60   d. not given

6. $70.38 ÷ 23
   a. $30.60   c. $3.06
   b. $3.60   d. not given

7. Identify the figure.

   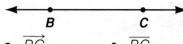

   a. $\overrightarrow{BC}$ c. $\overline{BC}$
   b. $\overleftrightarrow{BC}$ d. not given

8. $15,000 ÷ 25
   a. $600   c. $6,000
   b. $60   d. not given

9. Identify the angle.

   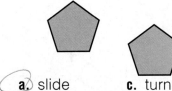

   a. right   c. obtuse
   b. acute   d. not given

10. Identify the transformation.

    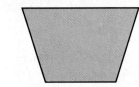

    a. slide   c. turn
    b. flip   d. not given

11. Name the quadrilateral.

    a. rhombus   c. parallelogram
    b. trapezoid   d. not given

12. How many lines of symmetry?

    a. 1   c. 3
    b. 2   d. not given

13. Name the polygon.

    a. pentagon   c. octagon
    b. decagon   d. not given

Understanding Fractions and Mixed Numbers

CHAPTER 9

MATH CONNECTIONS: MEASUREMENT
• LENGTH • PROBLEM SOLVING

1. What information do you see in this picture?
2. What does the information tell you?
3. How can the information be useful to you?
4. Write a problem about the picture.

UNDERSTANDING A CONCEPT
Fractions

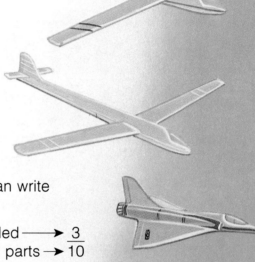

A. Martin has 4 gliders in his collection of 9 model airplanes. What fraction of his collection is gliders?

You can write a **fraction** to show how many of the airplanes are gliders.

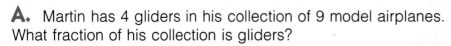

Read: four-ninths

So $\frac{4}{9}$ of Martin's collection is gliders.

1. What fraction would you write to describe the airplanes that are not gliders?

B. When a figure is divided into equal parts, you can write a fraction to show what part is shaded.

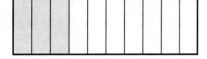   number of parts shaded ⟶ $\frac{3}{10}$
total number of equal parts ⟶

2. **What if** all the parts of the rectangle were shaded? What fraction would you write? What whole number is this equal to? Why?

3. What fraction would you write if none of the parts were shaded? What whole number is this equal to? Why?

4. **What if** you shaded 4 parts? What fraction would you write? What decimal would you write?

C. You can write a fraction to name a point on a number line.

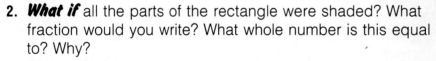

$\frac{3}{8}$ and $\frac{5}{8}$ name points A and B.

5. What fraction names point C?

352 Lesson 9-1

TRY OUT Find the answer.

6. What part of the set of squares is shaded?

7. What part of the rectangle is shaded?

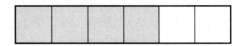

8. What fraction names point D?

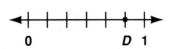

PRACTICE

Write the fraction for the part that is shaded.

9. ★★★★
★★★★
★★★★

10. ●●○
●●○
●○○

11. (circle divided into tenths, partially shaded)

12. (rectangle divided into horizontal strips, partially shaded)

Write the fraction for the point.

13.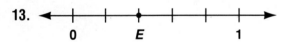

14. (number line from 0 to 1 with point F marked)

Write the fraction.

15. one-half
16. three-sevenths
17. seven-twelfths
18. nine-tenths
19. ten-elevenths
20. seventeen-twentieths
21. 0.8
22. 0.93
23. 0.415
24. The numerator is 2. The denominator is 9.
25. The numerator is 8. The denominator is 15.

Write the decimal.

26. $\frac{9}{10}$ **27.** $\frac{12}{100}$ **28.** $\frac{2}{10}$ **29.** $\frac{9}{100}$ **30.** $\frac{56}{100}$

Solve.

31. There are 25 students in Martin's class. Eight of them belong to the model airplane club. What fraction of the class are club members?

32. Write a fraction to describe:
a. the number of boys in your class.
b. the number of girls in your class.

DEVELOPING A CONCEPT
Equivalent Fractions

A. Ayad drew a design for a poster. He divided the design into 12 equal parts. He painted $\frac{1}{4}$ of the design blue. How many parts did he paint blue?

Look at these fraction strips.

1. How many $\frac{1}{12}$s are equal to $\frac{1}{4}$?

Equivalent fractions name the same amount.

2. Complete to name an equivalent fraction for $\frac{1}{4}$: $\frac{1}{4} = \frac{\blacksquare}{12}$.

3. How many parts of the design did Ayad paint blue?

4. **What if** Ayad had divided his design into 8 equal parts and painted $\frac{1}{4}$ of it blue? How many parts would he have painted blue? Use fraction strips to find out. Write another equivalent fraction for $\frac{1}{4}$.

B. You can multiply or divide the numerator and the denominator by the same nonzero number to find equivalent fractions.

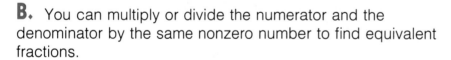

$\frac{1}{4} = \frac{1 \times 2}{4 \times 2} = \frac{2}{8}$ $\frac{1}{4} = \frac{1 \times 3}{4 \times 3} = \frac{3}{12}$ $\frac{12}{30} = \frac{12 \div 2}{30 \div 2} = \frac{6}{15}$ $\frac{12}{30} = \frac{12 \div 6}{30 \div 6} = \frac{2}{5}$

So $\frac{1}{4} = \frac{2}{8} = \frac{3}{12}$.   So $\frac{12}{30} = \frac{6}{15} = \frac{2}{5}$.

5. How can you find the missing numerator in $\frac{2}{3} = \frac{\blacksquare}{6}$? What is it?

6. How can you find the missing numerator in $\frac{8}{12} = \frac{\blacksquare}{3}$? What is it?

SHARING IDEAS

7. Why can you multiply or divide the numerator and denominator of a fraction by the same number without changing its value?

PRACTICE

Complete.

8. $\frac{1}{6} = \frac{1 \times 5}{6 \times 5} = \blacksquare$

9. $\frac{3}{5} = \frac{3 \times 3}{5 \times 3} = \blacksquare$

10. $\frac{6}{8} = \frac{6 \div 2}{8 \div 2} = \blacksquare$

Write the next three equivalent fractions.

11. $\frac{2}{8}, \frac{4}{16}, \frac{6}{24}, \blacksquare, \blacksquare, \blacksquare$

12. $\frac{1}{5}, \frac{2}{10}, \frac{3}{15}, \blacksquare, \blacksquare, \blacksquare$

13. $\frac{2}{7}, \frac{4}{14}, \frac{6}{21}, \blacksquare, \blacksquare, \blacksquare$

14. $\frac{2}{3}, \frac{4}{6}, \frac{6}{9}, \blacksquare, \blacksquare, \blacksquare$

15. $\frac{2}{9}, \frac{4}{18}, \frac{6}{27}, \blacksquare, \blacksquare, \blacksquare$

16. $\frac{3}{10}, \frac{6}{20}, \frac{9}{30}, \blacksquare, \blacksquare, \blacksquare$

Complete.

17. $\frac{3}{9} = \frac{\blacksquare}{3}$

18. $\frac{5}{20} = \frac{1}{\blacksquare}$

19. $\frac{10}{25} = \frac{\blacksquare}{5}$

20. $\frac{9}{21} = \frac{\blacksquare}{7}$

21. $\frac{24}{32} = \frac{\blacksquare}{4}$

22. $\frac{3}{4} = \frac{\blacksquare}{8}$

23. $\frac{5}{6} = \frac{\blacksquare}{12}$

24. $\frac{8}{11} = \frac{16}{\blacksquare}$

25. $\frac{2}{3} = \frac{\blacksquare}{18}$

26. $\frac{4}{9} = \frac{8}{\blacksquare}$

27. $\frac{1}{7} = \frac{2}{\blacksquare}$

28. $\frac{2}{16} = \frac{\blacksquare}{8}$

29. $\frac{7}{12} = \frac{14}{\blacksquare}$

30. $\frac{2}{5} = \frac{\blacksquare}{30}$

31. $\frac{2}{3} = \frac{6}{\blacksquare}$

Mixed Applications

32. Ayad painted $\frac{2}{6}$ of his design yellow and $\frac{4}{12}$ of it orange. Did he paint the same amount in each color? Why or why not?

33. There are 24 students in Mina's class. Eighteen of the students made posters. What fraction of the students made posters?

34. It took Ayad 3 hours 20 minutes to design and paint his poster. He spent 1 hour 45 minutes designing the poster. How much time did he spend painting it?

35. Count the number of letters in your full name. Write a fraction to show what part of the letters are vowels. Find another student who has written an equivalent fraction.

Mixed Review

Find the answer. Which method did you use?

MENTAL MATH / CALCULATOR / PAPER/PENCIL

36. 576 + 235

37. 3,500 − 400

38. 251 × 60

39. 143 + 257

40. 458 + 97

41. 26 × 45

42. 1,565 − 287

43. 47)1,645

UNDERSTANDING A CONCEPT
Greatest Common Factor

A. Sonia is arranging 24 apples in equal rows on a shelf. She can arrange the apples in four different ways.

Look at the arrangements and the related multiplication sentences. They show all the **factors** of 24.

The factors of 24 are 1, 2, 3, 4, 6, 8, 12, and 24.

1. What are the factors of 36?

B. A group of numbers can have common factors. **Common factors** are factors that are the same for each number.

Step 1
List the factors of each number.

Factors of 12: 1, 2, 3, 4, 6, 12
Factors of 18: 1, 2, 3, 6, 9, 18
Factors of 24: 1, 2, 3, 4, 6, 8, 12, 24

Step 2
List the common factors.

1, 2, 3, 6

The **greatest common factor (GCF)** of a group of numbers is the greatest of the common factors. So the greatest common factor of 12, 18, and 24 is 6.

2. What is the GCF of 28 and 36?

TRY OUT
Find the common factors. Then find the GCF.

3. 12 and 16
4. 6 and 18
5. 15 and 20
6. 9, 27, and 45

PRACTICE
Find the common factors. Then find the GCF.

7. 4 and 8
8. 3 and 15
9. 10 and 40
10. 18 and 36
11. 24 and 36
12. 45 and 60
13. 18 and 48
14. 19 and 57
15. 9 and 24
16. 3 and 21
17. 12 and 72
18. 24 and 32
19. 2, 10, and 30
20. 4, 8, and 16
21. 5, 15, and 40
22. 12 and 15

Critical Thinking
23. What number is a common factor of every set of numbers? Why?
24. Can zero ever be a common factor? Give an example to support your answer.

Mixed Applications
25. Sonia has 12 oranges to arrange in the grocery store window. How many ways can she arrange them in rows with an equal number of oranges in each? Name the ways.
26. In three weekends Danielle earned $15.75, $20.25, and $18.50. How much did she earn all together?

CHALLENGE
A **prime number** has exactly two factors, itself and 1.
A **composite number** has more than two factors.

List all the factors of 7.
Think: $1 \times 7 = 7$
Factors of 7: 1, 7
So 7 is a prime number.

List all the factors of 8.
Think: $1 \times 8 = 8$ $2 \times 4 = 8$
Factors of 8: 1, 2, 4, 8
So 8 is a composite number.

List all the factors of the number. Is the number *prime* or *composite*?

1. 6
2. 13
3. 15
4. 12
5. 21
6. 31

UNDERSTANDING A CONCEPT

Simplifying Fractions

A. Of the 12 trees in Mr. Kent's backyard, 4 are apple. Write the number of apple trees as a fraction of all 12 trees in simplest form.

A fraction is in **simplest form** when the numerator and the denominator have no common factors greater than 1.

To write a fraction in simplest form, divide the numerator and denominator by a common factor.

A common factor of 4 and 12 is 2.

$\frac{4 \div 2}{12 \div 2} = \frac{2}{6}$ ← not in simplest form

Divide again until the numerator and the denominator have no common factor greater than 1.

$\frac{2}{6} = \frac{2 \div 2}{6 \div 2} = \frac{1}{3}$ ← simplest form

So $\frac{1}{3}$ of Mr. Kent's trees are apple trees.

1. Are $\frac{4}{12}$, $\frac{2}{6}$, and $\frac{1}{3}$ equivalent fractions? Why or why not?

B. You can also write a fraction in simplest form by dividing the numerator and the denominator by the GCF.

Find the simplest form of $\frac{18}{27}$.

Think: The GCF of 18 and 27 is 9.   So $\frac{18}{27} = \frac{18 \div 9}{27 \div 9} = \frac{2}{3}$.

2. Why is simplifying by dividing with the GCF a quicker method than dividing with a smaller factor?

TRY OUT
Write the letter of the correct answer.
Find the equivalent fraction in simplest form.

3. $\frac{12}{18}$ **a.** $\frac{1}{3}$ **b.** $\frac{4}{6}$ **c.** $\frac{6}{9}$ **d.** $\frac{2}{3}$

4. $\frac{8}{40}$ **a.** $\frac{1}{5}$ **b.** $\frac{2}{10}$ **c.** $\frac{2}{20}$ **d.** $\frac{4}{20}$

Lesson 9-4

PRACTICE

Complete.

5. $\frac{3}{6} = \frac{\blacksquare}{2}$
6. $\frac{6}{8} = \frac{3}{\blacksquare}$
7. $\frac{3}{9} = \frac{1}{\blacksquare}$
8. $\frac{6}{24} = \frac{\blacksquare}{4}$
9. $\frac{7}{14} = \frac{1}{\blacksquare}$
10. $\frac{10}{14} = \frac{5}{\blacksquare}$
11. $\frac{8}{12} = \frac{\blacksquare}{3}$
12. $\frac{5}{15} = \frac{1}{\blacksquare}$
13. $\frac{18}{24} = \frac{\blacksquare}{4}$
14. $\frac{21}{28} = \frac{3}{\blacksquare}$
15. $\frac{8}{16} = \frac{1}{\blacksquare}$
16. $\frac{15}{24} = \frac{\blacksquare}{8}$
17. $\frac{2}{18} = \frac{1}{\blacksquare}$
18. $\frac{3}{12} = \frac{1}{\blacksquare}$
19. $\frac{2}{10} = \frac{\blacksquare}{5}$

Write in simplest form.

20. $\frac{8}{10}$
21. $\frac{2}{12}$
22. $\frac{4}{16}$
23. $\frac{10}{25}$
24. $\frac{6}{27}$
25. $\frac{7}{21}$
26. $\frac{9}{18}$
27. $\frac{6}{30}$
28. $\frac{3}{21}$
29. $\frac{3}{24}$
30. $\frac{9}{54}$
31. $\frac{35}{40}$
32. $\frac{2}{8}$
33. $\frac{3}{15}$
34. $\frac{18}{20}$
35. $\frac{12}{36}$
36. $\frac{14}{28}$
37. $\frac{9}{63}$
38. $\frac{20}{24}$
39. $\frac{18}{27}$
40. $\frac{18}{45}$
41. $\frac{4}{32}$
42. $\frac{10}{40}$
43. $\frac{16}{32}$

Mixed Applications

44. Mr. Kent has 24 pear tree seedlings in his nursery. He will plant 5 in the ground and 3 in large tubs. What fraction of the seedlings will be planted? Give the answer in simplest form.

45. Ed Kent helps his father trim trees. Mr. Kent trimmed 27 out of 36 apple trees, and Ed trimmed 18 out of 24 pear trees. Did they both trim an equal fraction of their trees? Why or why not?

46. A nursery had 162 customers this spring. Of these, 27 came to the nursery in March and 18 came in April. What fraction of all the customers came in March and April?

47. The Kents' garden is divided into 8 equal parts. If $\frac{3}{4}$ of the garden is planted with vegetables, how many eighths are planted with vegetables?

Mixed Review

Find the answer. Which method did you use?

MENTAL MATH
CALCULATOR
PAPER/PENCIL

48. 2.85 + 6.15
49. 3.57 − 1.09
50. 3.25 × 4
51. 3)4.08

52. 4.08 − 2.95
53. 8.561 × 10
54. 13.85 + 4.193
55. 10)5.897

UNDERSTANDING A CONCEPT
Mixed Numbers

A. Keshia helped her mother pack 17 mugs in 3 boxes. Each box had 6 compartments. $\frac{17}{6}$ or $2\frac{5}{6}$ boxes held the mugs.

$\frac{17}{6}$ is an **improper fraction**. An **improper fraction** has a numerator that is equal to or greater than its denominator.

$2\frac{5}{6}$ is a **mixed number**. A **mixed number** is made up of a whole number and a fraction.

B. You can use multiplication and addition to rename a mixed number as an improper fraction.

Write $2\frac{5}{6}$ as a fraction.

Step 1	Step 2	Step 3
Multiply the whole number by the denominator.	Add the numerator to the product.	Write the sum over the denominator.
$2\frac{5}{6} \rightarrow 6 \times 2 = 12$	$12 + 5 = 17$	$\frac{17}{6}$

C. You can use division to rename an improper fraction as a mixed number.

Write $\frac{17}{6}$ as a mixed number.

Step 1	Step 2
Divide the numerator by the denominator.	Write the quotient as the whole number part. Write the remainder over the divisor.
$\begin{array}{r} 2 \\ 6\overline{)17} \\ -12 \\ \hline 5 \end{array}$	$\frac{17}{6} = 2\frac{5}{6}$

TRY OUT Find the answer.

1. Write $3\frac{3}{8}$ as an improper fraction.
2. Complete: $7 = \frac{\blacksquare}{3}$
3. Write $\frac{14}{4}$ as a mixed number.
4. Write $\frac{28}{7}$ as a whole number.

PRACTICE

Write an improper fraction and a whole number or mixed number for the shaded part or the point.

5.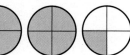
6.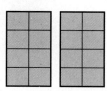
7.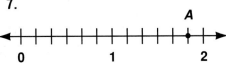

Write as a whole number or mixed number in simplest form.

8. $\frac{9}{2}$
9. $\frac{15}{4}$
10. $\frac{12}{3}$
11. $\frac{16}{5}$
12. $\frac{10}{7}$
13. $\frac{18}{9}$
14. $\frac{22}{5}$
15. $\frac{27}{4}$
16. $\frac{13}{6}$
17. $\frac{21}{8}$
18. $\frac{15}{10}$
19. $\frac{30}{6}$
20. $\frac{17}{3}$
21. $\frac{18}{6}$
22. $\frac{25}{4}$
23. $\frac{11}{5}$
24. $\frac{28}{7}$
25. $\frac{32}{9}$

Write as an improper fraction.

26. $2\frac{1}{2}$
27. $4\frac{3}{7}$
28. $1\frac{8}{9}$
29. $8\frac{2}{3}$
30. $7\frac{1}{8}$
31. $2\frac{5}{7}$
32. $5\frac{1}{4}$
33. $3\frac{2}{5}$
34. $6\frac{5}{9}$
35. $3\frac{7}{8}$
36. $9\frac{1}{6}$
37. $4\frac{3}{10}$
38. $3\frac{3}{4}$
39. $2\frac{1}{3}$
40. $5\frac{5}{8}$
41. $1\frac{2}{9}$
42. $4\frac{5}{6}$
43. $8\frac{1}{2}$

Mixed Applications

44. Eva's desk has 8 same-size drawers. Each drawer has 4 compartments. If Eva fills 21 of the compartments, what fraction of the drawers does she fill? What mixed number also describes this amount?

45. Write the lengths or heights of four mammals as improper fractions. You may need to use the Databank on page 521.

EXPLORING A CONCEPT
Renaming Mixed Numbers

Sally baked 4 pans of brownies to share with her friends. She cut one pan into eighths, and then ate one piece to see how it tasted. How much of the brownies did she have to share?

WORKING TOGETHER

1. Explore the problem by making a paper-folding model.

Sally had $3\frac{7}{8}$ pans of brownies to share.

Record the model and the answer on a chart like this.

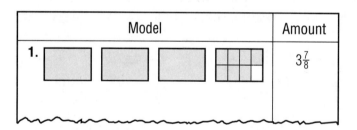

Use paper-folding models to answer these questions. Record your models and answers on the chart.

2. **What if** Sally cut a second pan of brownies into eighths? What mixed number would you write to show how much she has to share now?

3. **What if** Sally cut a third pan of brownies into eighths? What mixed number would you write?

4. **What if** Sally cut the fourth pan of brownies into eighths? Write an improper fraction for the amount she would have to share.

SHARING IDEAS

5. Look at your chart. Tell what happens to the whole number part when you rename the mixed number. What happens to the numerator?

6. **What if** Sally had *not* eaten a piece of brownie and had four full pans of brownies to share? What would be your models and answers to Problems 2, 3, and 4? Use paper-folding models to find out. Record your models and answers on the chart.

7. Are $3\frac{8}{8}$, $2\frac{16}{8}$, $1\frac{24}{8}$, and $\frac{32}{8}$ equivalent? Why or why not?

ON YOUR OWN

Solve using a model. Explain how you solved the problem.

8. After the class picnic Manuel found that there were $1\frac{3}{8}$ pizzas left. Show another way he can write this amount.

9. Mark has 3 pans of lasagna to share. He cuts each pan into tenths. Show the different ways he can write this number.

10. Erin has $2\frac{1}{3}$ cups of raisins. Show the ways she can write the amount of raisins she has.

11. Cheryl has $4\frac{3}{4}$ apples. Show the different ways she can write this number.

12. Joe has $2\frac{5}{6}$ cups of milk, Billy has $1\frac{10}{6}$ cups, and John has $\frac{17}{6}$ cups. Do they have the same amount of milk? Why or why not?

13. Margo has $\frac{16}{4}$ yards of fabric, Betty has $2\frac{8}{4}$ yards, while Jean has $1\frac{12}{4}$ yards. Do they have the same amount of fabric? Why or why not?

Understanding Fractions and Mixed Numbers

UNDERSTANDING A CONCEPT

Rounding Mixed Numbers

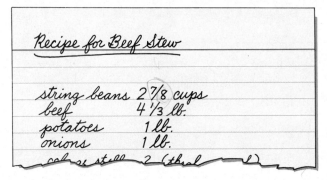

Molly is making a beef stew for a large gathering of friends. She found this recipe in a cookbook and is collecting all the ingredients. About how many cups of string beans does she need?

To find about how many cups, you can round $2\frac{7}{8}$ to the nearest whole number.

To round a mixed number, compare the fraction part to $\frac{1}{2}$.

If less than $\frac{1}{2}$, round down.
If greater than or equal to $\frac{1}{2}$, round up.

Think: $\frac{7}{8} > \frac{1}{2}$
Round up.
$2\frac{7}{8} \rightarrow 3$

Molly needs about 3 cups of string beans.

1. Why is $\frac{7}{8}$ greater than $\frac{1}{2}$?
2. How can you tell whether a fraction is greater or less than $\frac{1}{2}$?
3. About how many pounds of beef are needed?

TRY OUT Round to the nearest whole number.

4. $3\frac{1}{8}$
5. $7\frac{7}{12}$
6. $2\frac{3}{5}$
7. $6\frac{1}{4}$

PRACTICE

Round to the nearest whole number.

8. $5\frac{2}{3}$ 9. $9\frac{2}{11}$ 10. $6\frac{1}{5}$ 11. $2\frac{4}{5}$ 12. $12\frac{8}{9}$

13. $17\frac{1}{3}$ 14. $10\frac{2}{9}$ 15. $21\frac{7}{16}$ 16. $36\frac{4}{9}$ 17. $27\frac{8}{13}$

18. $5\frac{5}{8}$ 19. $10\frac{9}{16}$ 20. $9\frac{14}{16}$ 21. $3\frac{4}{13}$ 22. $12\frac{12}{23}$

23. $3\frac{7}{12}$ 24. $11\frac{2}{16}$ 25. $32\frac{7}{9}$ 26. $12\frac{3}{7}$ 27. $25\frac{3}{13}$

28. $18\frac{6}{16}$ 29. $11\frac{6}{14}$ 30. $27\frac{13}{20}$ 31. $18\frac{11}{23}$ 32. $7\frac{3}{8}$

Mixed Applications

33. Sam chopped $2\frac{3}{4}$ cups of vegetables for soup. He has a 1-cup measure. About how many times can he fill it with the chopped vegetables?

34. Molly and Sam like green beans and peas. They added $1\frac{3}{4}$ cups beans and $1\frac{6}{8}$ cups peas to their soup. Did they use the same amount of both vegetables? Why or why not?

35. Molly and Sam spent $\frac{7}{8}$ hour shopping and $1\frac{3}{8}$ hours making the stew. Did they spend the same amount of time on each activity? Explain.

MENTAL MATH

You can use these rules to tell whether a fraction is close to 0, $\frac{1}{2}$, or 1.

If the numerator is about the same as the denominator, the fraction is close to 1.

If the numerator is very small compared to the denominator, the fraction is close to zero.

If the denominator is about twice as large as the numerator, the fraction is close to $\frac{1}{2}$.

Tell whether the fraction is close to 0, $\frac{1}{2}$, or 1.

1. $\frac{1}{5}$ 2. $\frac{4}{7}$ 3. $\frac{18}{21}$ 4. $\frac{2}{13}$ 5. $\frac{1}{6}$ 6. $\frac{4}{5}$ 7. $\frac{5}{8}$

EXTRA Practice, page 385   Understanding Fractions and Mixed Numbers

PROBLEM SOLVING

Strategy: Making an Organized List

Ada, Tara, and Laurie want to have a group photo taken. In how many different ways can the three girls arrange themselves in a line from left to right for the photo?

Ada started to make an organized list to help solve this problem.
She used *A* for Ada, *T* for Tara, and *L* for Laurie.

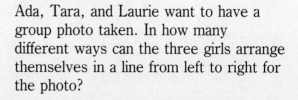

Left	Middle	Right
A	T	L

1. In what other way can you arrange the girls with Ada at the left end of the line? How would you put this arrangement in Ada's list?

2. Now list all the possible arrangements with Tara at the left end of the line.

3. List all the possible arrangements with Laurie at the left end of the line.

4. Have all the possible arrangements been listed? How do you know?

5. In how many different ways can the girls be arranged in a line?

6. **What if** Ellen joined the girls in taking the group photo? Copy and complete this organized list to show the different ways the four girls could be arranged in a line.

Left	Left Middle	Right Middle	Right
A	T	L	E
A	L	T	E

7. Tell how making an organized list helped you solve these problems.

PRACTICE

Make an organized list to solve the problem.

8. Dave, Juan, and Sam are going to play "Follow the Leader." List all the possible ways they can line up. Use *D* for Dave, *J* for Juan, and *S* for Sam. In how many ways can they line up so that Dave is first? is second? is third?

9. Brett wants an after-school snack. He can have a banana, an orange, or an apple to eat. He can have milk or juice to drink. List all the possible fruit and drink combinations that Brett can have. In all, how many are there?

10. List all the possible 3-digit numbers that can be written using each of the digits 2, 3, and 5 exactly once. How many 3-digit numbers did you write?

11. In how many ways is it possible to give a person 25 cents in coins? List them.

Strategies and Skills Review

Solve. Use mental math, estimation, a calculator, or paper and pencil.

12. Tara wants to buy a camera that costs $35.75 and two rolls of film that cost $3.60 each. She has $50. Does she have enough money to buy all these items?

13. Ada got 2 regular-size copies of each of 24 pictures. She got enlargements of her 3 favorite pictures. The regular-size prints cost 12¢ each. Do you have enough information to find how much she paid in all? If you answer no, tell what other information you need.

14. Ada got some pictures developed at the photo shop. It cost $3.85. She paid with a $10 bill. How much did she have left for new rolls of film?

15. Ellen bought some rolls of 24-exposure film and some rolls of 36-exposure film. The film she bought was enough for 108 pictures. How many rolls of each kind of film did she buy?

16. Ada took 3 rolls of pictures while she was on vacation. Each roll had 36 exposures. Tara took 12 more pictures than Ada. How many pictures did Tara take?

17. **Write a problem** that can be solved by making an organized list. Solve the problem. Ask other students to solve the problem.

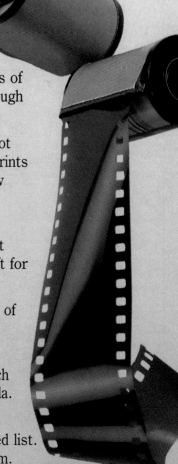

SOME WEIGHTY PROBLEMS

Logical Reasoning

A. Look at the pictures.

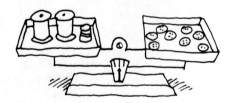

1. How many buttons will balance the spool of thread?
   How did you find your answer?

2. Can you find another way to solve the problem? Which way do you prefer? Why?

B. A full container of flour balances a container that is $\frac{2}{3}$ full plus a 2-lb weight.

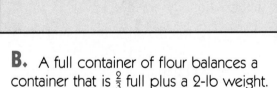

3. How much does the full container weigh? How do you know?

Lesson 9-9

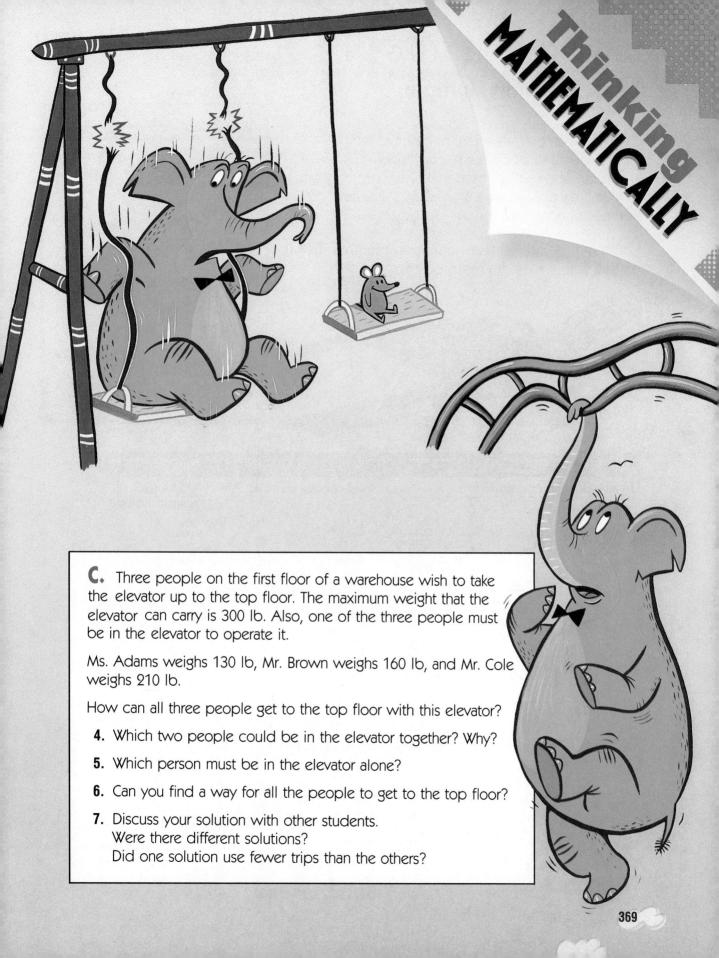

Thinking MATHEMATICALLY

C. Three people on the first floor of a warehouse wish to take the elevator up to the top floor. The maximum weight that the elevator can carry is 300 lb. Also, one of the three people must be in the elevator to operate it.

Ms. Adams weighs 130 lb, Mr. Brown weighs 160 lb, and Mr. Cole weighs 210 lb.

How can all three people get to the top floor with this elevator?

4. Which two people could be in the elevator together? Why?

5. Which person must be in the elevator alone?

6. Can you find a way for all the people to get to the top floor?

7. Discuss your solution with other students.
   Were there different solutions?
   Did one solution use fewer trips than the others?

UNDERSTANDING A CONCEPT

Least Common Multiple

Ken wants to buy an equal number of each kind of decal to decorate his model planes. The chart shows that the decals are sold only in multiples of 3, 4, and 6. What is the least number of each kind he should buy?

You need to find the **least common multiple (LCM)** of 3, 4, and 6 to solve the problem.

Decal	Price
⚡	6 for $1.50
▬	4 for $1.25
★	3 for $1.00

Step 1
List multiples of each number.

Multiples of 3: 0, 3, 6, 9, 12, 15, 18, 21, 24, . . .
Multiples of 4: 0, 4, 8, 12, 16, 20, 24, 28, 32, . . .
Multiples of 6: 0, 6, 12, 18, 24, 30, 36, . . .

Think:
You can find multiples of a number by multiplying by 0, 1, 2, 3, and so on.

Step 2
List common multiples.

0, 12, 24

Think:
Look for numbers common to all lists.

Step 3
Choose the least common multiple.

12

Think:
Look for the smallest nonzero number.

Ken should buy 12 of each kind of decal.

1. What number is a common multiple for all numbers? Why?

TRY OUT Write the letter of the correct answer.
Find the least common multiple.

2. 2 and 6
   a. 2   **c.** 12
   b. 6   **d.** 18

3. 9 and 15
   a. 3   **c.** 30
   b. 5   **d.** 45

4. 4, 8, and 12
   a. 2   **c.** 12
   b. 4   **d.** 24

PRACTICE

Find the LCM.

5. 2 and 5
6. 3 and 7
7. 5 and 3
8. 2 and 9
9. 12 and 18
10. 2 and 7
11. 6 and 8
12. 4 and 9
13. 4 and 7
14. 8 and 20
15. 9 and 12
16. 3 and 11
17. 20 and 25
18. 3, 6, and 9
19. 5, 12, and 15
20. 4, 8, and 10

Find the answer.

21. The least common multiple of two numbers is 24. If you add the two numbers, the sum is 11. What are the numbers?

22. The least common multiple of three numbers in a row is 60. The sum of the numbers is 15. What are the numbers?

Critical Thinking

23. Can you ever find the greatest common multiple for a pair of numbers? Why or why not?

Mixed Applications

24. There are lighthouses at two points of Pelican Island. The beacon at the Point Lookout lighthouse flashes every 4 seconds. The beacon at the Lands End lighthouse flashes every 5 seconds. If both beacons flash at 10 PM, when will they flash together again?

25. Use the chart on page 370 to find how much money Ken spent to buy 12 star decals. How much did he spend buying 12 of each kind of decal?

26. Red paper sells in multiples of 5 sheets, green in multiples of 10. What is the least number of sheets of each color Erin could buy to have the same number of each color?

Mixed Review

Find the measure of the angle. Tell whether it is acute, right, or obtuse.

27.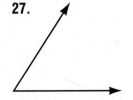
28.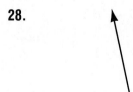
29.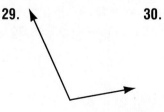
30.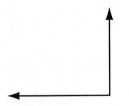

UNDERSTANDING A CONCEPT
Comparing Fractions and Mixed Numbers

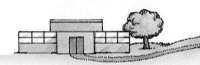

A. Beth walked $\frac{3}{8}$ mile to school each day. Jerry walked $\frac{5}{8}$ mile. Who walked the greater distance?

Compare: $\frac{3}{8}$ and $\frac{5}{8}$

You can use a number line.

When fractions have like denominators, you can compare the numerators.

$\frac{3}{8} \bullet \frac{5}{8}$ **Think:** $3 < 5$ $\frac{3}{8} < \frac{5}{8}$

$\frac{5}{8}$ is to the right of $\frac{3}{8}$, so $\frac{5}{8} > \frac{3}{8}$.

Jerry walked the greater distance.

B. You can compare mixed numbers and fractions with unlike denominators by using equivalent fractions.

Compare: $\frac{3}{4}$ and $\frac{5}{6}$

Step 1
Write equivalent fractions with like denominators.

$\frac{3}{4} = \frac{3 \times 3}{4 \times 3} = \frac{9}{12}$

$\frac{5}{6} = \frac{5 \times 2}{6 \times 2} = \frac{10}{12}$

Step 2
Compare the numerators.

$9 < 10$

So $\frac{9}{12} < \frac{10}{12}$ and $\frac{3}{4} < \frac{5}{6}$.

1. What other denominators could have been used?
2. Why do you think 12 was used?

C. You can order $\frac{5}{6}$, $\frac{7}{8}$, and $\frac{2}{3}$ by comparing.

$\frac{5}{6} = \frac{5 \times 4}{6 \times 4} = \frac{20}{24}$ $\frac{7}{8} = \frac{7 \times 3}{8 \times 3} = \frac{21}{24}$ $\frac{2}{3} = \frac{2 \times 8}{3 \times 8} = \frac{16}{24}$

Think: $\frac{16}{24} < \frac{20}{24} < \frac{21}{24}$

From least to greatest: $\frac{2}{3}, \frac{5}{6}, \frac{7}{8}$ From greatest to least: $\frac{7}{8}, \frac{5}{6}, \frac{2}{3}$

Try Out

Compare. Write >, <, or =.

3. $\frac{5}{9} \bullet \frac{8}{9}$
4. $4\frac{7}{12} \bullet 4\frac{2}{3}$
5. $6\frac{1}{5} \bullet 5\frac{7}{9}$

Order from least to greatest.

6. $\frac{2}{3}, \frac{5}{8}, \frac{7}{12}$
7. $2\frac{3}{5}, 1\frac{1}{3}, 1\frac{1}{2}$

Practice

Compare. Write >, <, or =.

8. $\frac{3}{4} \bullet \frac{1}{4}$
9. $\frac{7}{8} \bullet \frac{3}{8}$
10. $\frac{1}{3} \bullet \frac{2}{5}$
11. $\frac{5}{10} \bullet \frac{1}{2}$
12. $\frac{3}{4} \bullet \frac{7}{16}$
13. $4\frac{2}{3} \bullet 5\frac{1}{9}$
14. $2\frac{8}{9} \bullet 2\frac{2}{3}$
15. $3\frac{3}{15} \bullet 3\frac{1}{5}$
16. $2\frac{2}{3} \bullet 2\frac{1}{5}$
17. $5\frac{1}{4} \bullet 5\frac{1}{3}$
18. $2\frac{1}{2} \bullet 1\frac{1}{2}$
19. $2\frac{1}{5} \bullet 2\frac{1}{5}$
20. $\frac{11}{12} \bullet \frac{3}{4}$
21. $\frac{5}{7} \bullet \frac{7}{9}$
22. $\frac{3}{10} \bullet \frac{5}{20}$
23. $4\frac{2}{5} \bullet 4\frac{1}{3}$
24. $7 \bullet 6\frac{5}{9}$
25. $\frac{8}{3} \bullet 2\frac{2}{3}$
26. $1\frac{3}{8} \bullet \frac{13}{8}$
27. $\frac{9}{8} \bullet 1\frac{1}{4}$

Write in order from least to greatest.

28. $\frac{2}{5}, \frac{3}{4}, \frac{3}{10}$
29. $\frac{4}{9}, \frac{3}{6}, \frac{15}{18}$
30. $\frac{11}{12}, \frac{7}{8}, \frac{5}{6}$
31. $\frac{5}{8}, \frac{3}{16}, \frac{3}{4}$
32. $1\frac{2}{3}, 1\frac{3}{5}, 2\frac{1}{8}$
33. $2\frac{5}{6}, 2\frac{3}{4}, 2\frac{7}{8}$
34. $4\frac{5}{8}, 4\frac{3}{5}, 4\frac{3}{4}$
35. $1\frac{2}{9}, \frac{5}{6}, 1\frac{1}{3}$

Write in order from greatest to least.

36. $\frac{2}{5}, \frac{1}{4}, \frac{3}{10}$
37. $\frac{5}{6}, \frac{3}{4}, \frac{7}{12}$
38. $\frac{1}{2}, \frac{3}{4}, \frac{2}{3}$
39. $\frac{7}{9}, \frac{5}{6}, \frac{2}{3}$
40. $1\frac{4}{7}, 2\frac{1}{5}, 2\frac{3}{10}$
41. $3\frac{1}{5}, 3\frac{2}{3}, 2\frac{4}{5}$
42. $5\frac{3}{8}, 5\frac{3}{4}, 5\frac{1}{2}$
43. $2\frac{2}{5}, 2\frac{1}{3}, \frac{27}{9}$

Mixed Applications

44. One day Beth walked $1\frac{7}{8}$ miles and Jerry walked $1\frac{9}{16}$ miles. Which distance was greater?

45. Jerry wants to ride his bicycle 3.5 miles each day for 10 days. How many miles will he ride in all?

46. On three different days Beth swam $\frac{5}{16}$ mile, $\frac{7}{8}$ mile, and $\frac{3}{4}$ mile. Order the distances from greatest to least.

47. **Write a problem** comparing two mixed numbers. Ask others to solve your problem.

PROBLEM SOLVING

✓ UNDERSTAND
✓ PLAN
✓ TRY
✓ CHECK
✓ EXTEND

Interpreting the Quotient and Remainder

Jane works in the Margaretfield General Store. She has 139 flashlights to package in boxes of 15 each. How many boxes does she need?

$$15\overline{)139} \begin{array}{r} 9 \\ -135 \\ \hline 4 \end{array}$$

Jane divides and finds that she needs 10 boxes for the flashlights.

1. How many full boxes will Jane have?

2. Will Jane have a box that is only part full? If so, how many flashlights will there be for this box?

3. **What if** Jane put 13 flashlights in each box? Would 10 boxes be enough then? Why or why not?

4. **What if** Jane wanted to place the store's 21 cameras as evenly as possible along 4 shelves? How might she do it?

PRACTICE

Solve by interpreting a quotient and a remainder.

5. Jane is setting up 255 postcards in a display case. Each slot in the case holds 21 postcards. How many slots does she need?

6. The stationery section of the store sells 12 greeting cards for $2.93. How much would you expect to pay if you bought just one greeting card?

7. Jane had 1,098 pencils in stock. She put them in boxes of 25 pencils each and sold them at a special price. How many boxes will she completely fill? How many loose pencils will she have?

8. The store received a shipment of 368 video movies. Jane put the videos on shelves that each held 18 videos. How many shelves did she completely fill?

Strategies and Skills Review

Solve. Use mental math, estimation, a calculator, or paper and pencil.

9. The store sold camera film for $2.69 a roll. Marty bought 12 rolls. How much did he spend?

10. Jane sells yarn for $3.75 a ball. The tax is $.23 on each ball. What is the total price for 78 balls?

11. The first year it was open, the store sold 150 cameras. The second year it sold 165 cameras. This year, the third year, it sold 180 cameras. How many cameras did it sell in the first half of this year? Is there enough information to solve the problem? If not, what else do you need to know?

12. A child's jump rope costs $1.95. A flashlight costs $4.95, and a radio costs $19.95. What is the total cost of the three items?

13. Shana wants to buy a vacuum cleaner for $129.95, a radio for $19.95, and a toaster for $14.95. Is $150 enough to cover the cost of these items?

14. ***Write a problem*** that can be solved by dividing and then interpreting the quotient and remainder. Solve the problem. Give your problem to other students to solve.

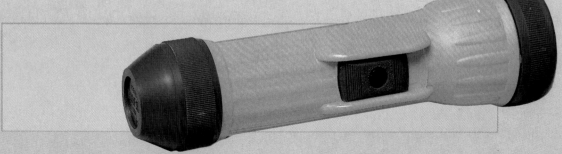

EXTRA Practice, page 387   Understanding Fractions and Mixed Numbers

DEVELOPING A CONCEPT

Estimating Length

The **inch (in.)**, **foot (ft)**, **yard (yd)**, and **mile (mi)** are customary units of length.

```
12 in. = 1 ft
 3 ft = 1 yd
5,280 ft = 1 mi
```

about 1 in. wide.   about 1 ft long.   about 1 yd wide.

A mile is about the distance a person can walk in 15 minutes.

WORKING TOGETHER

Step 1 Choose the following objects to measure.

- three objects that are longer than 1 in. but less than 1 ft
- three objects that are between 1 ft and 1 yd long
- three objects that are longer than 1 yd

Step 2 Estimate the length of each object. Then use an inch ruler or measuring tape to measure. Record your answers in a table like this.

Object	Estimate	Measure

SHARING IDEAS

1. Compare your table with those of others. What unit of measure did you use for lengths between 1 inch and 1 foot? between 1 foot and 1 yard? longer than 1 yard?

2. How does each estimate in your table compare with the actual measure?

3. What unit would you use to measure the length of an eraser? Why?

4. What unit would you use to measure the height of a desk? Why?

5. What unit would you use to measure the length of the classroom? Why?

6. What unit would you use to measure the distance between Dallas and Houston? Why?

Lesson 9-13

PRACTICE

Which unit would you use to measure? Write *inch, foot, yard* or *mile*.

7. the length of a truck
8. the width of a notebook
9. the height of a room
10. the length of a road

Which is the most reasonable estimate?

11. the height of a 10-year-old   a. 4 in   b. 4 ft   c. 4 yd
12. the width of a TV screen   a. 15 in.   b. 15 ft   c. 15 yd
13. the distance between two towns   a. 10 ft   b. 10 yd   c. 10 mi
14. the height of a table   a. $2\frac{1}{2}$ in.   b. $2\frac{1}{2}$ ft   c. $2\frac{1}{2}$ yd
15. the distance from first to second base on a baseball field   a. 30 in.   b. 30 ft   c. 30 yd
16. the width of a camera   a. 8 in.   b. 8 ft   c. 8 yd
17. the distance from New York to Boston   a. 225 ft   b. 225 yd   c. 225 mi

Choose the object that is close to the given length.

18. 9 in.   a. stamp   b. scissors   c. stepladder
19. 20 yd   a. rake   b. bookcase   c. garden hose
20. 5 mi   a. soccer field   b. hiking path   c. school
21. 6 ft   a. bed   b. spoon   c. airplane

Solve.

22. Chen plans to hang a basketball hoop above his garage door, 10 yd from the ground. Is he using the appropriate unit of measure? If not, what unit should he use?

23. Lisa needs a piece of plywood to use as the top of a Ping-Pong table. Will a piece that measures 9 yd by 5 yd be about the right size? Why?

24. Rick estimated the distance he had to walk from the school to his house as 1.4 yd. Would this be a reasonable estimate? Why or why not?

25. **Write a problem** involving length in which you leave out all the units of measure. Ask others to solve your problem.

UNDERSTANDING A CONCEPT
Measuring Length

A. Phil measured a strip of balsa wood to use in building a model plane. The piece is between 4 and 5 in. long.

Phil can measure to get a more precise measurement. To the nearest $\frac{1}{2}$ in., the piece of wood is $4\frac{0}{2}$ in. long. To the nearest $\frac{1}{4}$ in., the piece of wood is $4\frac{1}{4}$ in. long.

1. What is the length of the wood to the nearest $\frac{1}{8}$ in.?

2. Why is measuring to the nearest $\frac{1}{8}$ in. more precise than measuring to the nearest $\frac{1}{4}$ or $\frac{1}{2}$ in.?

3. Measuring to what fraction of an inch would given an even more precise measurement?

4. When might you need a very precise measurement of something? Give some examples.

B. Paula drew a line segment that is $2\frac{1}{2}$ in. long to the nearest $\frac{1}{2}$ in.

5. What is the shortest length to the nearest $\frac{1}{8}$ in. that is still $2\frac{1}{2}$ in. to the nearest $\frac{1}{2}$ in.? What is the longest length to the nearest $\frac{1}{8}$ in.

TRY OUT Measure to the nearest inch, ½ inch, ¼ inch, and ⅛ inch.

6.   7.

PRACTICE

Measure to the nearest inch, ½ inch, ¼ inch, and ⅛ inch.

8.   9.

10.    11.

12. _____   13. _____
14. _____   15. _____
16. _____   17. _____

Draw a line with the given length.

18. $4\frac{1}{4}$ in.   **19.** $2\frac{5}{8}$ in.   **20.** $\frac{3}{4}$ ft   **21.** $3\frac{7}{8}$ in.

Critical Thinking
22. What is the measure of the line segment that when measured to the nearest inch, ½, ¼, and ⅛ inch is 3 in.?

Mixed Applications
23. Phil has a piece of wood 19 in. long. He needs three 6-in. pieces for airplane parts. Does he have enough? Why or why not?

24. Phil has a piece of wire $1\frac{1}{2}$ ft long. He wants to cut it into pieces $\frac{1}{2}$ ft long. How many pieces will he have? Draw a picture to help you.

25. Terry has to cut a strip of felt $6\frac{1}{4}$ in. long and $\frac{7}{8}$ in. wide. Draw a pattern for her to use.

COOPERATIVE LEARNING

DECISION MAKING

Problem Solving: Building a Model Airplane

SITUATION

Howard Nelson enjoys model airplanes. Howard would like to make a model of one of his favorite airplanes. He has the information about the size of each part of the airplane and how it should look. There is also an airplane kit of the plane that Howard can buy.

PROBLEM

Should Howard build his model from a kit, or should he buy the individual parts and materials?

DATA

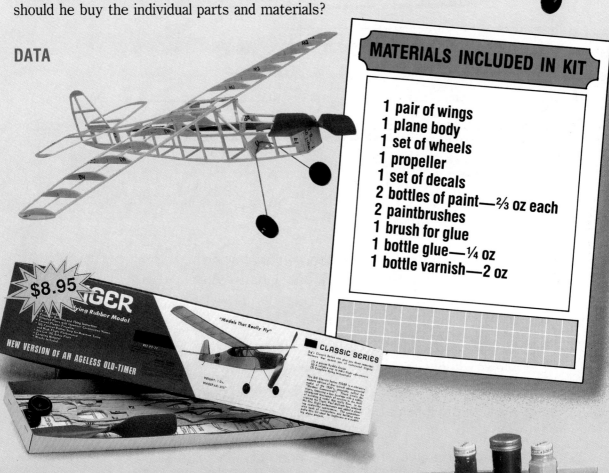

MATERIALS INCLUDED IN KIT

- 1 pair of wings
- 1 plane body
- 1 set of wheels
- 1 propeller
- 1 set of decals
- 2 bottles of paint—⅔ oz each
- 2 paintbrushes
- 1 brush for glue
- 1 bottle glue—¼ oz
- 1 bottle varnish—2 oz

$8.95

THINGS I NEED TO BUILD MODEL AIRPLANE

ITEM	SIZE	PRICE	NUMBER NEEDED
Materials:			
Balsa wood	8 in. by 10 in.	$.69	4
Wheels	small set	$.75	1
Propeller	small	$.50	1
Glue with brush	2/3 oz	$1.50	1
Paintbrush	Thin	$1.75	2
	Thick	$2.50	1
Paint	2 3/4 oz	$1.75	2
Varnish	5 1/2 oz	$2.39	1
Tools:			
Balsa-wood saw	small	$1.89	1
Wood file	small	$2.95	1

USING THE DATA

Find the cost.

1. 4 pieces of balsa wood
2. 2 bottles of paint
3. 2 thin paint brushes
4. to build the airplane from scratch

MAKING DECISIONS

5. Is it less expensive to buy the airplane kit or to build the plane from scratch?

6. What are some of the advantages of buying the ready-to-assemble kit?

7. **What if** Howard had never made a model airplane before? Would he be better off buying the kit or buying individual parts and materials? Tell why.

8. **What if** Howard is very familiar with assembling a model airplane and plans to assemble many more? Would he be better off building the plane from scratch or buying the airplane kit? Tell why.

9. **Write a list** of other factors Howard should consider before deciding whether to build a model airplane from scratch or from a kit.

10. What would you do, buy the airplane kit or build the airplane from scratch? Why?

Understanding Fractions and Mixed Numbers

CURRICULUM CONNECTION

Math and Social Studies

In 1958, a group of students at the Massachusetts Institute of Technology had some fun with measurements. They measured the Harvard Bridge using chalk and a classmate, Oliver Reed Smoot, Jr. Their human measure, Smoot, was 5 feet 7 inches. By using him as a ruler, they found the bridge to measure 364.4 Smoots plus one ear. Twice a year since then, the bridge has been re-marked, with paint marks every 10 Smoots. When the bridge was rebuilt, lines scored the new sidewalk in even 5-foot 7-inch intervals.

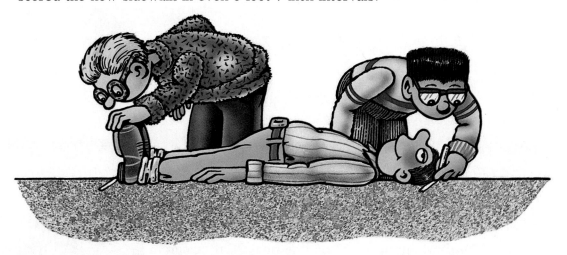

What if you want to know the length of the Harvard Bridge in feet? The only figure you have is in Smoots.

Think: One Smoot is 5 feet 7 in. = 67 in.   The bridge = 364.4 Smoots.
67 in. × 364.4 = 24,414.8 in.
24,414.8 ÷ 12 = 2,034.57 ft
The Harvard Bridge is about 2,035 ft long.

ACTIVITIES

1. Choose an object to use as a standard of measurement. Measure several things with this new ruler. Make a chart showing the things you measured in your new measure and in standard measurements.

2. Read about the history of the yard as a unit of measure. Share what you learn with your class.

TECHNOLOGY

Computer: Exploring Quadrilaterals

Recall that a parallelogram is a quadrilateral whose opposite sides are the same length and parallel. You can use Logo commands to see how parallelograms can change into other special quadrilaterals.

AT THE COMPUTER

When you input the measure of one angle and the lengths of the two sides, the procedure PARA draws the parallelogram.

Enter the procedure.

```
TO PARA :ANGLE :SIDE1 :SIDE2
RT 90 - :ANGLE
REPEAT 2 FD :SIDE1 RT :ANGLE
FD :SIDE2 RT 180 - :ANGLE
END
```

To draw a parallelogram with an angle measuring 50° and sides measuring 36 units and 24 units, enter PARA 50 36 24.

Clear the screen each time you use the procedure.

1. Using a 50° angle, change the lengths of the sides so that your parallelogram is short and wide. Then redraw the parallelogram to be tall and narrow. What commands did you enter?

2. Draw some rectangles. What commands did you enter to draw the rectangles? Which part of the command must be the same for drawing all rectangles?

3. Draw a rhombus with a 35° angle and one side 25 units long. What commands did you enter?

4. Change the rhombus you drew in Problem 3 into a square with sides 25 units long. Which part of the command did you change? Why?

5. Can you use the PARA procedure to draw a quadrilateral with sides 10 units, 15 units, 20 units, and 25 units long? Explain.

6. Draw a parallelogram with an angle of 60° and sides 15 units and 24 units long. Draw another parallelogram with an angle of 60° and sides 25 units and 40 units long. Are the parallelograms congruent or similar?

Understanding Fractions and Mixed Numbers

EXTRA PRACTICE

Fractions, page 353

Write the fraction for the part that is shaded.

1.    2.   3.

Write the fraction.

4. four-sixths   5. six-sevenths   6. eight-twelfths
7. 0.2   8. 0.85   9. 0.317

Write the decimal.

10. $\frac{8}{10}$ 11. $\frac{14}{100}$ 12. $\frac{3}{10}$ 13. $\frac{8}{100}$ 14. $\frac{38}{100}$

Equivalent Fractions, page 355

Write the next three equivalent fractions.

1. $\frac{1}{4}, \frac{2}{8}, \frac{3}{12}, \blacksquare, \blacksquare, \blacksquare$ 2. $\frac{2}{9}, \frac{4}{18}, \frac{6}{27}, \blacksquare, \blacksquare, \blacksquare$

3. $\frac{4}{10}, \frac{8}{20}, \frac{12}{30}, \blacksquare, \blacksquare, \blacksquare$ 4. $\frac{1}{3}, \frac{2}{6}, \frac{3}{9}, \blacksquare, \blacksquare, \blacksquare$

Complete.

5. $\frac{2}{4} = \frac{\blacksquare}{2}$ 6. $\frac{4}{8} = \frac{1}{\blacksquare}$ 7. $\frac{7}{21} = \frac{\blacksquare}{3}$ 8. $\frac{2}{3} = \frac{\blacksquare}{6}$ 9. $\frac{7}{11} = \frac{14}{\blacksquare}$

10. $\frac{8}{24} = \frac{1}{\blacksquare}$ 11. $\frac{25}{50} = \frac{1}{\blacksquare}$ 12. $\frac{4}{20} = \frac{1}{\blacksquare}$ 13. $\frac{3}{8} = \frac{\blacksquare}{24}$ 14. $\frac{7}{16} = \frac{14}{\blacksquare}$

Greatest Common Factor, page 357

Find the common factors. Then find the GCF.

1. 5 and 25   2. 2 and 4   3. 10 and 30
4. 18 and 24   5. 7 and 21   6. 7 and 35
7. 3, 9, and 36   8. 2, 8, and 16   9. 5, 20, and 40

EXTRA PRACTICE

Simplifying Fractions, page 359

Complete.

1. $\frac{4}{8} = \frac{\blacksquare}{2}$
2. $\frac{9}{12} = \frac{3}{\blacksquare}$
3. $\frac{5}{25} = \frac{1}{\blacksquare}$
4. $\frac{6}{12} = \frac{3}{\blacksquare}$
5. $\frac{15}{21} = \frac{5}{\blacksquare}$
6. $\frac{21}{28} = \frac{3}{\blacksquare}$
7. $\frac{15}{25} = \frac{\blacksquare}{5}$
8. $\frac{3}{18} = \frac{1}{\blacksquare}$

Write in simplest form.

9. $\frac{4}{6}$
10. $\frac{2}{14}$
11. $\frac{3}{24}$
12. $\frac{14}{28}$
13. $\frac{10}{30}$
14. $\frac{2}{22}$
15. $\frac{6}{36}$
16. $\frac{10}{50}$
17. $\frac{8}{64}$
18. $\frac{6}{12}$
19. $\frac{7}{63}$
20. $\frac{3}{21}$
21. $\frac{45}{54}$
22. $\frac{18}{36}$
23. $\frac{18}{32}$

Mixed Numbers, page 361

Write as a whole number or mixed number.

1. $\frac{18}{2}$
2. $\frac{14}{5}$
3. $\frac{15}{7}$
4. $\frac{31}{6}$
5. $\frac{36}{5}$
6. $\frac{11}{7}$
7. $\frac{15}{3}$
8. $\frac{27}{5}$
9. $\frac{30}{7}$
10. $\frac{19}{6}$
11. $\frac{28}{3}$
12. $\frac{33}{8}$
13. $\frac{16}{10}$
14. $\frac{21}{7}$
15. $\frac{19}{9}$

Write as an improper fraction.

16. $3\frac{1}{2}$
17. $5\frac{3}{7}$
18. $2\frac{7}{8}$
19. $8\frac{1}{4}$
20. $6\frac{2}{3}$
21. $2\frac{3}{7}$
22. $4\frac{1}{3}$
23. $1\frac{4}{9}$
24. $6\frac{5}{6}$
25. $8\frac{1}{2}$
26. $3\frac{1}{3}$
27. $5\frac{7}{10}$
28. $3\frac{2}{5}$
29. $4\frac{9}{10}$
30. $2\frac{7}{10}$

Rounding Mixed Numbers, page 365

Round to the nearest whole number.

1. $4\frac{1}{2}$
2. $6\frac{7}{8}$
3. $5\frac{2}{3}$
4. $3\frac{1}{2}$
5. $7\frac{7}{10}$
6. $8\frac{9}{10}$
7. $16\frac{11}{25}$
8. $10\frac{3}{7}$
9. $18\frac{4}{9}$
10. $3\frac{6}{7}$
11. $21\frac{5}{12}$
12. $18\frac{1}{3}$
13. $9\frac{15}{16}$
14. $25\frac{3}{14}$
15. $8\frac{3}{8}$

Understanding Fractions and Mixed Numbers

EXTRA PRACTICE

Problem-Solving Strategy: Making an Organized List, page 367
Make an organized list to solve the problem.

1. List all the possible 3-digit numbers that can be written using each of the digits 3, 6, and 9 exactly once. How many 3-digit numbers did you write?

2. In a cash register there are $10 bills, $5 bills, and $1 bills. How many different ways could you give someone change for a $20 bill using no coins and no more than five $1 bills? List them.

Least Common Multiple, page 371
Find the first three common multiples. Then find the LCM.

1. 4 and 8
2. 2 and 9
3. 8 and 12
4. 4 and 5
5. 6 and 12
6. 12 and 18
7. 5 and 3
8. 4, 6, and 8
9. 5, 6, and 10
10. 3 and 9
11. 7 and 14
12. 2, 4, and 6

Comparing Fractions and Mixed Numbers, page 373
Compare. Write >, <, or =.

1. $\frac{1}{3} \bullet \frac{1}{4}$
2. $\frac{5}{8} \bullet \frac{7}{8}$
3. $\frac{1}{5} \bullet \frac{1}{3}$
4. $\frac{7}{10} \bullet \frac{1}{2}$
5. $\frac{3}{6} \bullet \frac{4}{8}$
6. $\frac{7}{8} \bullet \frac{15}{16}$
7. $\frac{6}{12} \bullet \frac{3}{6}$
8. $\frac{3}{5} \bullet \frac{6}{15}$
9. $5\frac{1}{2} \bullet 6\frac{1}{4}$
10. $2\frac{1}{4} \bullet 2\frac{2}{8}$
11. $3\frac{1}{3} \bullet 4\frac{1}{2}$
12. $6\frac{2}{3} \bullet 7\frac{1}{4}$
13. $2\frac{3}{9} \bullet 2\frac{2}{3}$
14. $\frac{7}{2} \bullet 3\frac{1}{2}$
15. $4\frac{1}{3} \bullet 4\frac{1}{4}$

Write in order from least to greatest.

16. $\frac{1}{4}, \frac{5}{12}, \frac{2}{3}$
17. $\frac{7}{24}, \frac{1}{3}, \frac{3}{8}$
18. $1\frac{1}{3}, 1\frac{5}{6}, 1\frac{2}{9}$
19. $1\frac{1}{3}, 1\frac{2}{5}, 2\frac{1}{15}$
20. $3\frac{1}{6}, 2\frac{2}{3}, 2\frac{3}{4}$
21. $\frac{1}{2}, 1\frac{1}{5}, 1\frac{3}{10}$

Write in order from greatest to least.

22. $\frac{3}{5}, \frac{1}{3}, \frac{7}{15}$
23. $\frac{7}{8}, \frac{3}{4}, \frac{1}{6}$
24. $\frac{1}{2}, \frac{1}{3}, \frac{3}{4}$
25. $2\frac{3}{10}, 2\frac{1}{2}, 1\frac{1}{5}$
26. $5\frac{3}{5}, 2\frac{2}{7}, 5\frac{1}{2}$
27. $1\frac{1}{5}, 1\frac{3}{10}, 1\frac{5}{12}$

Chapter 9

EXTRA PRACTICE

Problem Solving: Interpreting the Quotient and Remainder, page 375
Solve by interpreting a quotient and a remainder.

1. The office-supply store sells pens in boxes of 24 for $15.25. How much would you expect to pay if you bought just one pen?

2. The office-supply store wants to display 216 boxes of stationery. Each shelf can hold 15 boxes. How many shelves will they completely fill?

Estimating Length, page 377
Which unit would you use to measure? Write *inch, foot, yard,* or *mile.*

1. the length of a mobile home
2. the height of a fence
3. the width of a shoe
4. the length of a highway

Which is the most reasonable estimate?

5. the height of a woman   a. 5 in.   b. 5 ft   c. 5 yd
6. the width of a painting   a. 20 in.   b. 20 ft   c. 20 yd
7. the distance from Denver to Cheyenne   a. 120 ft   b. 120 yd   c. 120 mi

Choose the object that is close to the given length.

8. 7 in.   a. bracelet   b. ladder   c. TV screen
9. 10 yd   a. stamp   b. garden   c. football field
10. 15 mi   a. bike path   b. skyscraper   c. ladder
11. 6 ft   a. rake   b. necklace   c. man

Measuring Length, page 379
Measure to the nearest inch, $\frac{1}{2}$ inch, $\frac{1}{4}$ inch, and $\frac{1}{8}$ inch.

1.    2.    3.

Draw a line with the given length.

4. $3\frac{1}{2}$ in.   5. $2\frac{3}{4}$ in.   6. $\frac{1}{2}$ ft   7. $4\frac{5}{8}$ in.

Understanding Fractions and Mixed Numbers   **387**

Practice PLUS

KEY SKILL: Equivalent Fractions (Use after page 355.)

Level A
Complete.

1. $\dfrac{4}{6} = \dfrac{\blacksquare}{3}$

2. $\dfrac{2}{4} = \dfrac{\blacksquare}{2}$

3. $\dfrac{4}{5} = \dfrac{\blacksquare}{10}$

Level B
Complete.

4. $\dfrac{2}{5} = \dfrac{2 \times 3}{5 \times 3} = \dfrac{\blacksquare}{\blacksquare}$

5. $\dfrac{1}{10} = \dfrac{1 \times 5}{10 \times 5} = \dfrac{\blacksquare}{\blacksquare}$

6. $\dfrac{3}{8} = \dfrac{3 \times 4}{8 \times 4} = \dfrac{\blacksquare}{\blacksquare}$

7. $\dfrac{5}{8} = \dfrac{5 \times 2}{8 \times 2} = \dfrac{\blacksquare}{\blacksquare}$

8. $\dfrac{1}{4} = \dfrac{1 \times 6}{4 \times 6} = \dfrac{\blacksquare}{\blacksquare}$

9. $\dfrac{2}{3} = \dfrac{2 \times 5}{3 \times 5} = \dfrac{\blacksquare}{\blacksquare}$

10. $\dfrac{5}{6} = \dfrac{5 \times 3}{6 \times 3} = \dfrac{\blacksquare}{\blacksquare}$

11. $\dfrac{1}{3} = \dfrac{1 \times 3}{3 \times 3} = \dfrac{\blacksquare}{\blacksquare}$

12. $\dfrac{3}{6} = \dfrac{3 \times 2}{6 \times 2} = \dfrac{\blacksquare}{\blacksquare}$

Level C
Write the next three equivalent fractions.

13. $\dfrac{2}{3}, \dfrac{4}{6}, \dfrac{6}{9}, \dfrac{\blacksquare}{\blacksquare}, \dfrac{\blacksquare}{\blacksquare}, \dfrac{\blacksquare}{\blacksquare}$

14. $\dfrac{7}{10}, \dfrac{14}{20}, \dfrac{21}{30}, \dfrac{\blacksquare}{\blacksquare}, \dfrac{\blacksquare}{\blacksquare}, \dfrac{\blacksquare}{\blacksquare}$

15. $\dfrac{3}{12}, \dfrac{6}{24}, \dfrac{9}{36}, \dfrac{\blacksquare}{\blacksquare}, \dfrac{\blacksquare}{\blacksquare}, \dfrac{\blacksquare}{\blacksquare}$

16. $\dfrac{4}{7}, \dfrac{8}{14}, \dfrac{12}{21}, \dfrac{\blacksquare}{\blacksquare}, \dfrac{\blacksquare}{\blacksquare}, \dfrac{\blacksquare}{\blacksquare}$

17. $\dfrac{1}{3}, \dfrac{2}{6}, \dfrac{3}{9}, \dfrac{\blacksquare}{\blacksquare}, \dfrac{\blacksquare}{\blacksquare}, \dfrac{\blacksquare}{\blacksquare}$

18. $\dfrac{3}{5}, \dfrac{6}{10}, \dfrac{9}{15}, \dfrac{\blacksquare}{\blacksquare}, \dfrac{\blacksquare}{\blacksquare}, \dfrac{\blacksquare}{\blacksquare}$

Complete.

19. $\dfrac{2}{3} = \dfrac{\blacksquare}{15}$

20. $\dfrac{3}{10} = \dfrac{9}{\blacksquare}$

21. $\dfrac{18}{24} = \dfrac{\blacksquare}{4}$

22. $\dfrac{4}{20} = \dfrac{\blacksquare}{5}$

23. $\dfrac{16}{32} = \dfrac{\blacksquare}{8}$

24. $\dfrac{4}{9} = \dfrac{\blacksquare}{36}$

25. $\dfrac{11}{33} = \dfrac{\blacksquare}{3}$

26. $\dfrac{4}{11} = \dfrac{24}{\blacksquare}$

27. $\dfrac{21}{27} = \dfrac{\blacksquare}{9}$

28. $\dfrac{8}{32} = \dfrac{1}{\blacksquare}$

29. $\dfrac{2}{6} = \dfrac{\blacksquare}{3}$

30. $\dfrac{15}{25} = \dfrac{3}{\blacksquare}$

31. $\dfrac{1}{4} = \dfrac{\blacksquare}{8}$

32. $\dfrac{4}{12} = \dfrac{1}{\blacksquare}$

33. $\dfrac{2}{9} = \dfrac{\blacksquare}{81}$

34. $\dfrac{5}{6} = \dfrac{25}{\blacksquare}$

Practice PLUS

KEY SKILL: Mixed Numbers (Use after page 361.)

Level A

Write an improper fraction and a whole number or mixed number for the shaded part or the point.

1.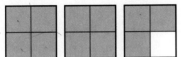
2.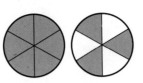
3.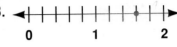

4.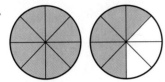
5.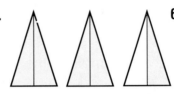
6.

7.
8.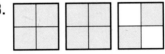
9.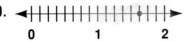

Level B

Complete.

10. $\frac{10}{3} = 3\frac{\blacksquare}{3}$
11. $\frac{18}{5} = 3\frac{\blacksquare}{5}$
12. $\frac{17}{4} = 4\frac{\blacksquare}{4}$
13. $\frac{23}{8} = 2\frac{\blacksquare}{8}$
14. $\frac{17}{6} = \blacksquare\frac{5}{6}$
15. $1\frac{3}{4} = \frac{\blacksquare}{4}$
16. $3\frac{2}{3} = \frac{\blacksquare}{3}$
17. $3\frac{3}{8} = \frac{\blacksquare}{8}$
18. $4\frac{1}{6} = \frac{25}{\blacksquare}$
19. $6\frac{4}{5} = \frac{\blacksquare}{5}$

Write as an improper fraction.

20. $4\frac{3}{5}$
21. $1\frac{7}{9}$
22. $5\frac{3}{8}$
23. $3\frac{5}{6}$
24. $2\frac{3}{7}$
25. $6\frac{9}{10}$

Level C

Write as a whole number or a mixed number.

26. $\frac{58}{7}$
27. $\frac{35}{6}$
28. $\frac{400}{100}$
29. $\frac{89}{10}$
30. $\frac{63}{9}$
31. $\frac{72}{4}$

Write as an improper fraction.

32. $4\frac{7}{12}$
33. $3\frac{3}{10}$
34. $6\frac{1}{6}$
35. $9\frac{5}{8}$
36. $5\frac{4}{5}$
37. $2\frac{8}{11}$

Chapter Review

LANGUAGE AND MATHEMATICS

Complete the sentences. Use the words in the chart on the right.

1. A ■ may name a part of a whole or a part of a set. *(page 352)*

2. Fractions that name the same amount are called ■. *(page 354)*

3. Factors that are the same for a group of numbers are called ■. *(page 356)*

4. A number that is made up of a whole number and a fraction is a ■ *(page 360)*

5. ***Write a definition*** or give an example of the words you did not use from the chart.

VOCABULARY
fraction
mixed number
proper fraction
common factors
equivalent fractions
improper fraction

CONCEPTS AND SKILLS

Write the fraction. *(page 352)*

6. seven-eighths   7. four-fifths   8. one-fourth   9. eleven-fifteenths

Write the decimal. *(page 352)*

10. $\frac{4}{10}$ 11. $\frac{37}{100}$ 12. $\frac{3}{100}$ 13. $\frac{7}{10}$ 14. $\frac{71}{100}$

Write the next three equivalent fractions. *(page 354)*

15. $\frac{1}{2}, \frac{2}{4}, \frac{3}{6}, \blacksquare, \blacksquare, \blacksquare$ 16. $\frac{8}{12}, \frac{10}{15}, \frac{12}{18}, \blacksquare, \blacksquare, \blacksquare$ 17. $\frac{4}{20}, \frac{5}{25}, \frac{6}{30}, \blacksquare, \blacksquare, \blacksquare$

Complete. *(page 354)*

18. $\frac{2}{8} = \frac{\blacksquare}{4}$ 19. $\frac{9}{18} = \frac{1}{\blacksquare}$ 20. $\frac{9}{12} = \frac{\blacksquare}{4}$ 21. $\frac{27}{36} = \frac{3}{\blacksquare}$

Find the common factors. Then find the GCF. *(page 356)*

22. 16 and 20   23. 27 and 63   24. 9 and 6   25. 9 and 18

26. 2, 8, and 16   27. 4, 12, and 24   28. 18, 27, and 45   29. 15, 30, and 45

Write in simplest form. *(page 358)*

30. $\frac{12}{16}$ 31. $\frac{20}{25}$ 32. $\frac{6}{14}$ 33. $\frac{9}{24}$ 34. $\frac{24}{27}$

Write as a whole number or a mixed number. *(page 360)*

35. $\frac{11}{4}$ **36.** $\frac{27}{7}$ **37.** $\frac{33}{9}$ **38.** $\frac{25}{15}$ **39.** $\frac{18}{6}$

Write as an improper fraction. *(page 360)*

40. $3\frac{1}{8}$ **41.** $2\frac{2}{5}$ **42.** $4\frac{3}{7}$ **43.** $6\frac{2}{3}$ **44.** $9\frac{6}{8}$ **45.** $5\frac{7}{10}$

Round to the nearest whole number. *(page 364)*

46. $12\frac{1}{4}$ **47.** $9\frac{5}{8}$ **48.** $2\frac{1}{2}$ **49.** $36\frac{15}{16}$ **50.** $21\frac{1}{7}$

Find the first three common multiples. Then find the LCM. *(page 370)*

51. 3 and 4   **52.** 2 and 6   **53.** 4 and 9   **54.** 8 and 12

55. 2, 3, and 8   **56.** 5, 15, and 30   **57.** 3, 9, and 18   **58.** 4, 8, and 9

Compare. Write >, <, or =. *(page 372)*   **Write in order from greatest to least.** *(page 372)*

59. $\frac{1}{2}$ ● $\frac{1}{8}$ **60.** $2\frac{18}{20}$ ● $2\frac{9}{10}$ **61.** $\frac{2}{3}, \frac{7}{12}, \frac{5}{8}$ **62.** $\frac{7}{8}, 1\frac{1}{4}, 1\frac{2}{9}$

Which is the most reasonable estimate? *(page 376)*

63. the height of a house   **a.** 6 in.   **b.** 6 ft   **c.** 6 yd

64. length of swimming pool   **a.** 25 in.   **b.** 25 yd   **c.** 25 mi

65. the width of a calculator   **a.** 3 in.   **b.** 3 ft   **c.** 3 yd

Draw a line with the given length. *(page 378)*

66. $3\frac{1}{4}$ in.   **67.** $\frac{1}{2}$ ft   **68.** $5\frac{7}{8}$ in.   **69.** $2\frac{3}{4}$ in.   **70.** $4\frac{3}{8}$ in.

CRITICAL THINKING

71. Is zero a common factor to *every* set of numbers? Give an example to support your answer.

MIXED APPLICATIONS

72. Ben, José, and Art buy tickets for a soccer game. List all the different ways the friends can sit next to each other at the game. *(page 366)*

73. Mariana has 1,018 stamps in her collection. Each page in her album holds 25 stamps. How many pages are full? How many stamps are on the partly filled page? *(page 374)*

Understanding Fractions and Mixed Numbers

Chapter Test

Write the fraction for the part that is shaded.

1.
2.

Complete.

3. $\frac{4}{8} = \frac{\blacksquare}{2}$

4. $\frac{5}{30} = \frac{1}{\blacksquare}$

5. $\frac{9}{12} = \frac{\blacksquare}{4}$

Find the GCF.

6. 4 and 5

7. 10 and 30

Find the LCM.

8. 3 and 5

9. 12 and 8

Write as a mixed number. Then round to the nearest whole number.

10. $\frac{15}{10}$

11. $\frac{42}{5}$

12. $\frac{19}{9}$

13. $\frac{60}{8}$

Compare. Write >, <, or =.

14. $\frac{3}{4} \bullet \frac{5}{16}$

15. $2\frac{7}{8} \bullet 3\frac{7}{9}$

16. $\frac{9}{5} \bullet 1\frac{4}{5}$

17. $2\frac{3}{4} \bullet 2\frac{6}{8}$

Write in order from greatest to least.

18. $1\frac{1}{3}, 1\frac{3}{5}, 2\frac{1}{7}$

19. $6\frac{3}{8}, 6\frac{3}{4}, 4\frac{1}{2}$

20. $\frac{2}{5}, \frac{3}{10}, \frac{1}{2}$

Which is the most reasonable estimate?

21. the length of a suitcase   **a.** 4 in.   **b.** 4 ft   **c.** 4 mi

22. the height of a flagpole   **a.** 20 in.   **b.** 20 ft   **c.** 20 yd

Measure to the nearest $\frac{1}{2}$ inch, $\frac{1}{4}$ inch, and $\frac{1}{8}$ inch.

23. _____

Solve.

24. There are 198 people going to the zoo. Each bus holds 48 passengers. How many buses will they need?

25. Arnold is buying a new sofa. He can choose from 3 different colors and 3 different manufacturers. How many choices does he have?

ENRICHMENT FOR ALL

PRIME FACTORIZATION

A **prime number** has exactly two factors, itself and 1. The factors of 5 are 1 and 5.

  1. What are the factors of 3?

  2. Why is the number 1 not a prime number?

A **composite number** has more than two factors. The factors of 10 are 1, 2, 5, and 10.

  3. Is 12 a composite number? How do you know?

Every composite number can be written as the product of prime numbers. This is called the **prime factorization** of the number. The prime factors of a number are always the same. The order of the factors is not important.

You can use a factor tree to find the prime factors of a composite number.

Here are two factor trees for 24:

```
        24                              24
       /  \                            /  \
      6  ×  4   ← 6 and 4             2 × 12   ← 12 is not
     /\    /\     are not prime.         /\      prime.
    2×3  2×2  ← prime               2 × 2 × 6   ← 6 is not
                 factors               /  /  /\    prime.
                                    2 × 2 × 2 × 3  ← prime
                                                     factors
```

Copy and complete.

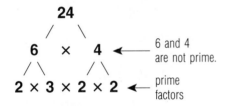

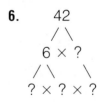

Draw a factor tree to show the prime factorization of the number.

  7. 12   **8.** 18   **9.** 28   **10.** 32

  11. 36   **12.** 40   **13.** 44   **14.** 54

  15. 60   **16.** 75   **17.** 90   **18.** 100

Understanding Fractions and Mixed Numbers 393

Cumulative Review

Choose the letter of the correct answer.

1. 32.873
   × 9
   a. 295.857   c. 2,958.57
   b. 29.5857   d. not given

2. 63 d = ■ wk
   a. 6   c. 9
   b. 7   d. not given

3. Find the volume of a rectangular prism with length = 9 cm, width = 5 cm, height = 6 cm.
   a. 30   c. 20 cm³
   b. 30 cm³   d. not given

4. $\frac{2}{3} = \frac{\blacksquare}{18}$
   a. 3   c. 12
   b. 6   d. not given

5. Name the polygon.

   a. parallelogram   c. pentagon
   b. quadrilateral   d. not given

6. Identify the triangle.

   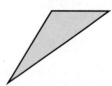

   a. right   c. obtuse
   b. acute   d. not given

7. 7)$58.52
   a. $80.36   c. $8.36
   b. $8.26   d. not given

8. What is the fraction for the part that is shaded?

   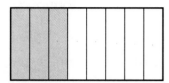

   a. $\frac{3}{4}$ c. $\frac{1}{2}$
   b. $\frac{3}{8}$ d. not given

9. Find the GCF of 36 and 48.
   a. 12   c. 8
   b. 6   d. not given

10. What is $\frac{5}{30}$ in simplest form?
    a. $\frac{1}{5}$ c. $\frac{1}{10}$
    b. $\frac{1}{6}$ d. not given

11. What is the mixed number for $\frac{22}{9}$?
    a. $2\frac{1}{9}$ c. $2\frac{4}{9}$
    b. $2\frac{3}{9}$ d. not given

12. Round to the nearest whole number. $10\frac{9}{16}$
    a. 10   c. 12
    b. 11   d. not given

13. Find the LCM of 3 and 11.
    a. 3   c. 33
    b. 11   d. not given

14. One day Margie walked $2\frac{7}{8}$ miles and Miguel walked $2\frac{11}{16}$ miles. Which distance was greater?
    a. $2\frac{7}{8}$ c. both are equal
    b. $2\frac{11}{16}$ d. not given

394 Chapter 9

Adding and Subtracting Fractions

CHAPTER 10

MATH CONNECTION: PROBLEM SOLVING

1. What information do you see in this picture?
2. How can the information be useful to you?
3. Which two patterns are the same amount of fabric?
4. Write a problem using the information.

EXPLORING A CONCEPT
Sums and Differences of Fractions

Vicki needs to buy ribbon to trim a T-shirt. She needs a piece $\frac{1}{2}$ yd long and another piece $\frac{3}{8}$ yd long. About how much ribbon does she need?

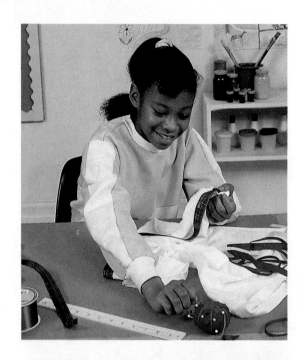

WORKING TOGETHER

Explore the problem by making or using a model. Then answer the questions below.

1. How does your model show $\frac{1}{2}$ yd?
2. How does it show $\frac{3}{8}$ yd?
3. How does the sum compare to one whole yd?

Vicki uses fraction strips to solve the problem.

She uses the unit strip to represent 1 yd.

She uses the $\frac{1}{2}$ strip to represent $\frac{1}{2}$ yd.

She uses the $\frac{1}{8}$ strips to represent $\frac{3}{8}$ yd.

4. What does her model tell her?

Vicki uses the fraction strips to find exactly how much ribbon she needs.

Add: $\frac{1}{2} + \frac{3}{8}$

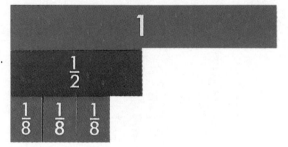

5. What does Vicki rename in order to add?
6. How does the renaming help find the sum? What is the sum?

What if you had a piece of ribbon $\frac{5}{6}$ yd long and you cut from it a piece $\frac{2}{3}$ yd long? About how much ribbon would you have left? Exactly how much ribbon would you have left? Use a model to find out.

7. How does the amount of ribbon left compare to $\frac{1}{2}$?

8. What is the exact amount of ribbon left?

SHARING IDEAS

9. Compare your model and results with those of others. How are they the same? How are they different?

10. How did you find the sum or difference of two fractions with unlike denominators?

ON YOUR OWN

Solve using a model.

11. Matt dyed a T-shirt purple by mixing $\frac{1}{4}$ cup of blue dye and $\frac{1}{3}$ cup of red dye. About how much dye did he mix in all? Exactly how much dye did he mix?

12. Jenny bought $1\frac{1}{8}$ yd of fabric to make a T-shirt. She used $\frac{3}{4}$ yd. Did she have more or less than 1 yd of fabric left? Exactly how much fabric did she have left?

13. Ali had $1\frac{5}{6}$ cups of orange dye. He used $\frac{1}{4}$ cup to tie-dye two T-shirts. About how much orange dye did he have left: less than 1 cup, about 1 cup, or more than 1 cup? Exactly how much orange dye did he have left?

14. Eva used $\frac{1}{6}$ yd of braid on one T-shirt and $\frac{1}{3}$ yd on another. How much braid did she use in all? Is this amount the same as $\frac{1}{2}$ yd?

Adding and Subtracting Fractions **397**

DEVELOPING A CONCEPT
Adding Fractions: Like Denominators

A. Franklin made a design that is in 8 equal parts. He painted $\frac{3}{8}$ of the design green and $\frac{1}{8}$ of it orange. How much of the design did Franklin paint?

Add: $\frac{3}{8} + \frac{1}{8}$

You can use a diagram to solve the problem.

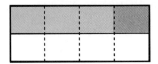

Think: three-eighths + one-eighth = four-eighths
four-eighths = one-half

Franklin painted $\frac{1}{2}$ of the design.

Here is a way to add fractions without models.

Step 1	Step 2	Step 3
Add the numerators.	Use the common denominator.	Write the sum in simplest form.
$\frac{3}{8} + \frac{1}{8} = \frac{4}{}$	$\frac{3}{8} + \frac{1}{8} = \frac{4}{8}$	$\frac{4}{8} = \frac{1}{2}$

B. You can use the same steps to add three fractions.

Add: $\frac{3}{4} + \frac{3}{4} + \frac{1}{4}$

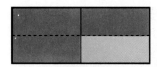
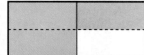

$\frac{3}{4} + \frac{3}{4} + \frac{1}{4} = \frac{7}{4}$

$\frac{7}{4} = 1\frac{3}{4}$

SHARING IDEAS

1. **What if** you had painted $\frac{3}{7}$ of a rectangle red and $\frac{4}{7}$ of it blue? How much of the rectangle would you have painted? Why?

2. Add $\frac{1}{3} + \frac{3}{3} + \frac{2}{3}$. Does it matter in what order you add the numerators? Why?

398 Lesson 10-2

PRACTICE

Write an addition sentence to show how much is shaded.

3.   4. 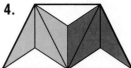   5.

Add. Write the sum in simplest form.

6. $\frac{1}{4} + \frac{1}{4}$ 7. $\frac{3}{8} + \frac{5}{8}$ 8. $\frac{5}{16} + \frac{3}{16}$ 9. $\frac{1}{9} + \frac{5}{9}$ 10. $\frac{1}{12} + \frac{7}{12}$

11. $\frac{1}{5} + \frac{3}{5} + \frac{2}{5}$ 12. $\frac{1}{4} + \frac{3}{4} + \frac{2}{4}$ 13. $\frac{5}{9} + \frac{2}{9} + \frac{7}{9}$ 14. $\frac{9}{14} + \frac{1}{14} + \frac{3}{14}$

15. $\frac{1}{4} + \frac{3}{4}$ 16. $\frac{5}{6} + \frac{5}{6}$ 17. $\frac{1}{9} + \frac{4}{9}$ 18. $\frac{7}{8} + \frac{5}{8}$ 19. $\frac{3}{10} + \frac{9}{10}$

20. $\frac{5}{9} + \frac{7}{9}$ 21. $\frac{1}{10} + \frac{7}{10}$ 22. $\frac{1}{12} + \frac{1}{12}$ 23. $\frac{7}{15} + \frac{8}{15}$ 24. $\frac{4}{9} + \frac{2}{9}$

Mixed Applications

25. June spent $\frac{3}{4}$ hour drawing her design and $\frac{3}{4}$ hour painting it on her T-shirt. How much time did she spend working on her shirt?

26. The 28 fifth graders in art class each contributed $3.55 to pay for supplies. How much did the class spend on T-shirts and paint?

27. Mark drew a circle and divided it into 12 equal sections. He painted 3 sections red, 4 sections yellow, and 5 sections blue. How much of the circle did he paint?

28. **Write a problem** using addition for a fraction design in which you have painted several parts in different colors. Be sure the design is in equal parts.

Mixed Review

Find the greatest common factor.

29. 6 and 8   30. 14 and 21   31. 9, 18, and 36   32. 8, 12, and 16

Find the least common multiple.

33. 2 and 3   34. 6 and 9   35. 3, 4, and 6   36. 2, 4, and 5

EXTRA Practice, page 428   Adding and Subtracting Fractions   399

UNDERSTANDING A CONCEPT
Adding Fractions: Unlike Denominators

A. Marisa decided to add words to her T-shirt design. She plans to allow $\frac{7}{8}$ in. space for each letter plus $\frac{1}{4}$ in. space between letters. How much space should she allow for each letter and space she uses?

Add: $\frac{7}{8} + \frac{1}{4}$

Marisa used this diagram of a model to find the sum of $\frac{7}{8} + \frac{1}{4}$.

 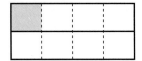

Marisa should allow $1\frac{1}{8}$ in. for each letter and space.

1. How did Marisa rename $\frac{1}{4}$ in order to add?

B. Here is a way to add fractions with unlike denominators without models or diagrams.

Step 1	Step 2	Step 3
Find a common denominator.	**Write equivalent fractions.**	**Add. Write the sum in simplest form.**
$\frac{7}{8}$ $+\frac{1}{4}$	$\frac{7}{8} = \frac{7}{8}$ $+\frac{1 \times 2}{4 \times 2} = \frac{2}{8}$	$\frac{7}{8} = \frac{7}{8}$ $+\frac{1 \times 2}{4 \times 2} = \frac{2}{8}$ $\phantom{+\frac{1 \times 2}{4 \times 2} = {}}\frac{9}{8} = 1\frac{1}{8}$

Think: 8 is a common denominator.

2. Why must you find a common denominator before you can add $\frac{7}{8}$ and $\frac{1}{4}$?

3. Find the sum using 16 as the common denominator. Is the sum the same? What extra step did you have to take?

4. What if you had to add $\frac{1}{3} + \frac{1}{6} + \frac{6}{12}$? What denominator would you use? Why? Find the sum.

TRY OUT
Add. Write the sum in simplest form.

5. $\dfrac{1}{5} + \dfrac{3}{10}$

6. $\dfrac{1}{3} + \dfrac{5}{6}$

7. $\dfrac{1}{2} + \dfrac{1}{4} + \dfrac{3}{4}$

8. $\dfrac{1}{3} + \dfrac{1}{6} + \dfrac{1}{12}$

PRACTICE
Add. Write the sum in simplest form.

9. $\dfrac{1}{4} + \dfrac{1}{8}$

10. $\dfrac{1}{2} + \dfrac{3}{4}$

11. $\dfrac{2}{3} + \dfrac{5}{6}$

12. $\dfrac{5}{8} + \dfrac{1}{2}$

13. $\dfrac{4}{5} + \dfrac{1}{10}$

14. $\dfrac{5}{6} + \dfrac{1}{2}$

15. $\dfrac{3}{4} + \dfrac{1}{12}$

16. $\dfrac{1}{8} + \dfrac{3}{4}$

17. $\dfrac{2}{3} + \dfrac{4}{9}$

18. $\dfrac{5}{12} + \dfrac{1}{3}$

19. $\dfrac{2}{3} + \dfrac{5}{6} + \dfrac{1}{3}$

20. $\dfrac{1}{4} + \dfrac{7}{8} + \dfrac{1}{2}$

21. $\dfrac{1}{2} + \dfrac{5}{6} + \dfrac{1}{12}$

22. $\dfrac{5}{18} + \dfrac{5}{9} + \dfrac{1}{3}$

23. $\dfrac{3}{4} + \dfrac{5}{8} + \dfrac{3}{4}$

24. $\dfrac{1}{6} + \dfrac{2}{3}$

25. $\dfrac{3}{4} + \dfrac{7}{12}$

26. $\dfrac{1}{8} + \dfrac{3}{4}$

27. $\dfrac{1}{3} + \dfrac{8}{9}$

28. $\dfrac{3}{10} + \dfrac{2}{5}$

29. $\dfrac{5}{8} + \dfrac{1}{2}$

30. $\dfrac{1}{2} + \dfrac{1}{6} + \dfrac{5}{12}$

31. $\dfrac{3}{4} + \dfrac{1}{2} + \dfrac{1}{8}$

32. the sum of one-half and one-fifth

33. four-ninths plus five-eighteenths

Mixed Applications

34. Jaime plans to draw letters $\dfrac{9}{10}$ in. wide and allow $\dfrac{2}{5}$ in. space between them. How much space does he need for one letter and one space?

35. The stencil Jaime will use to draw letters includes letters with widths of $\dfrac{5}{8}$ in., $\dfrac{3}{16}$ in., $\dfrac{3}{4}$ in., $\dfrac{7}{8}$ in., $\dfrac{9}{16}$ in., and $\dfrac{15}{16}$ in. Which size letter is the widest? Which is narrowest?

36. Sandy wants to decorate the letters on her T-shirt with metal dots. The dots come in a package of 360. She wants to decorate 28 letters. How many metal dots can she use on each letter?

37. Andy has one decal that is $\dfrac{3}{4}$ in. wide and another that is $\dfrac{7}{8}$ in. wide. He plans to put them on a T-shirt with $\dfrac{3}{4}$ in. space between them. How much space will he need in all?

PROBLEM SOLVING

✓ UNDERSTAND
✓ PLAN
✓ TRY
✓ CHECK
✓ EXTEND

Strategy: Using Number Sense

At the grocery deli, Sam asked the clerk to slice $\frac{5}{8}$ pound of Swiss cheese, $\frac{3}{4}$ pound of cheddar, $\frac{9}{16}$ pound of Muenster, and $\frac{7}{16}$ pound of provolone. Did the customer buy more or less than 2 pounds of cheese?

You can solve this problem by using number sense. Part of using number sense is *examining* the numbers in the problem to see if you notice anything special.

1. Which two fractions in this problem have the same denominator? What do you get if you add these fractions?

2. Look at the other two fractions in the problem. Are they less than, equal to, or greater than $\frac{1}{2}$? Is their sum less than, equal to, or greater than 1?

3. Is the total amount of cheese the customer bought more than 2 pounds? How do you know?

4. **What if** the customer had bought all except the Swiss cheese? Would the amount bought have been more or less than 2 pounds? How do you know?

402 Lesson 10–4

PRACTICE

Use number sense to solve the problem.

5. Mrs. Franklin bought $\frac{1}{4}$ pound of black olives and $\frac{3}{8}$ pound of green olives. Did she buy more or less than 1 pound of olives? How do you know?

6. Ben bought $\frac{1}{2}$ pound of bananas and $\frac{2}{3}$ pound of pears. Did he buy more or less than 1 pound of fruit? How do you know?

7. Mr. Ortiz bought $\frac{2}{3}$ quart of fresh orange juice and $\frac{3}{4}$ quart of fresh grapefruit juice. Mrs. Peng bought $\frac{1}{2}$ quart of fresh orange juice and $\frac{2}{3}$ quart of fresh grapefruit juice. Who bought more juice? How do you know?

8. Mañuel had 2 cups of milk. He used $\frac{1}{3}$ cup in the morning and another $\frac{1}{3}$ cup in the afternoon. Did he have more or less than 1 cup of milk left?

Strategies and Skills Review

Solve. Use mental math, estimation, a calculator, or paper and pencil.

9. Mark and Bernice each work 20 hours a week at the store. This week each worked an additional 3 hours. They earned $5.20 for each additional hour they worked. How much extra money did each earn this week?

10. The owner of the grocery store built this display of canned goods.

    Then he decided to make it 7 cans tall. If he continued the pattern, how many cans did he use in the bottom row?

11. The shelves in the cereal section can each hold 24 boxes. There are 175 boxes of cereal on the shelves. All the shelves are full except one. How many shelves are there in the cereal section?

12. **Write a problem** that involves fractions and can be solved using number sense. Solve the problem. Ask others to solve the problem.

DEVELOPING A CONCEPT
Subtracting Fractions

A. Tony had $\frac{3}{4}$ hour before his trumpet lesson. He spent $\frac{1}{4}$ hour talking to his teacher. How much time did he have to practice before his lesson?

Subtract: $\frac{3}{4} - \frac{1}{4}$

You can use this picture of a model to solve the problem.

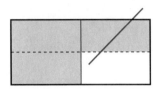

Think: three-fourths − one-fourth = two-fourths
two-fourths = one-half

Tony has $\frac{1}{2}$ hour to practice.

Here is a way to subtract fractions without models.

Step 1	Step 2	Step 3
Subtract the numerators.	Use the common denominator.	Write the difference in simplest form.
$\frac{3}{4} - \frac{1}{4} = \frac{2}{}$	$\frac{3}{4} - \frac{1}{4} = \frac{2}{4}$	$\frac{2}{4} = \frac{1}{2}$

B. Sometimes you have to subtract fractions with unlike denominators.

Subtract: $\frac{3}{4} - \frac{3}{8}$

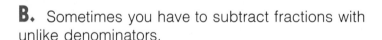

Step 1	Step 2	Step 3
Find a common denominator.	Write equivalent fractions.	Subtract. Write the difference in simplest form.
$\begin{array}{r}\frac{3}{4}\\-\frac{3}{8}\\\hline\end{array}$	$\frac{3 \times 2}{4 \times 2} = \frac{6}{8}$ $-\frac{3}{8} = \frac{3}{8}$	$\frac{3 \times 2}{4 \times 2} = \frac{6}{8}$ $-\frac{3}{8} = \frac{3}{8}$ $\phantom{-\frac{3}{8} = }\frac{3}{8}$

Think: 8 is a common denominator.

SHARING IDEAS

1. What denominator would you use for each subtraction? Why?

   a. $\frac{5}{6} - \frac{1}{3}$ **b.** $\frac{7}{10} - \frac{2}{5}$ **c.** $\frac{1}{2} - \frac{3}{8}$

2. How would you find the difference between $\frac{7}{9}$ and $\frac{5}{6}$?

PRACTICE

Subtract. Write the difference in simplest form.

3. $\frac{4}{5} - \frac{2}{5}$
4. $\frac{8}{9} - \frac{2}{9}$
5. $\frac{7}{8} - \frac{5}{8}$
6. $\frac{5}{6} - \frac{1}{6}$
7. $\frac{6}{7} - \frac{3}{7}$

8. $\frac{5}{6} - \frac{2}{3}$
9. $\frac{5}{8} - \frac{1}{2}$
10. $\frac{9}{10} - \frac{4}{5}$
11. $\frac{4}{9} - \frac{1}{3}$
12. $\frac{1}{2} - \frac{1}{10}$

13. $\frac{3}{5} - \frac{1}{5}$
14. $\frac{5}{6} - \frac{1}{2}$
15. $\frac{1}{3} - \frac{2}{9}$
16. $\frac{5}{7} - \frac{2}{7}$
17. $\frac{2}{3} - \frac{1}{6}$

18. $\frac{7}{12} - \frac{1}{12}$
19. $\frac{4}{5} - \frac{3}{10}$
20. $\frac{5}{12} - \frac{1}{3}$
21. $\frac{7}{8} - \frac{1}{2}$

22. $\frac{3}{4} - \frac{5}{12}$
23. $\frac{7}{8} - \frac{5}{8}$
24. $\frac{8}{9} - \frac{2}{3}$
25. $\frac{7}{10} - \frac{1}{2}$

Mixed Applications

26. Cal's tuba lesson lasted $\frac{11}{12}$ hour. He spent only $\frac{7}{12}$ hour playing his tuba. How much time did he not play his tuba?

27. Jan practiced the piano for $\frac{5}{6}$ hour on Monday and $\frac{2}{3}$ hour on Tuesday. How long did she practice in all?

28. On Thursday, Laura practiced on her guitar for $\frac{7}{12}$ hour. On Friday, she practiced $\frac{3}{4}$ hour, and on Saturday she practiced $\frac{2}{6}$ hour. On which day did she practice the most?

29. Tony played his trumpet with the school band for $\frac{3}{4}$ hour during a concert. He had a $\frac{1}{12}$-hour solo part. How much longer was his band part than his solo part?

ANY WAY YOU LOOK AT IT

Visual Reasoning

A. A stack of cubes is pictured at the right. The top view, front view, and a side view of the stack are shown below.

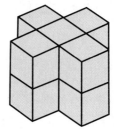

Top view

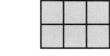
Front view

Side view

Use centimeter cubes to build each stack of cubes. Then draw the three views for each. Assume that the only hidden cubes are those that support others.

1.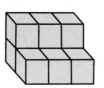

2.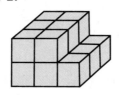

3.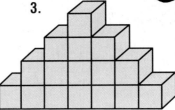

4.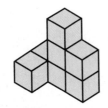

5.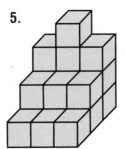

6.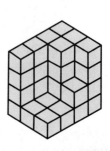

Lesson 10-6

Thinking MATHEMATICALLY

B. Top views, front views, and side views of stacks of cubes are given below.

7. Use the three views to build the stack. Tell how many cubes you used.

   Top   Front   Side

8. Build the stack. What is the least number of cubes you need? What is the greatest number you need?

   Top   Front   Side

9. Build this stack using 48 cubes. Then build it using only 30 cubes.

   Top   Front   Side

UNDERSTANDING A CONCEPT
Estimating Sums and Differences

A. The Washington Street School art club needs modeling clay for its papier-mâché workshop. Lindsay brings in $2\frac{1}{4}$ lb of clay; Glenn brings $1\frac{1}{8}$ lb; and Darryl brings $1\frac{5}{6}$ lb. About how much clay will the club have to work with?

To find about how much clay, you can estimate the sum of $2\frac{1}{4}$, $1\frac{1}{8}$, and $1\frac{5}{6}$.

Here are two methods.

Rounding
Round each number to the nearest whole number. Then add.

$2\frac{1}{4} + 1\frac{1}{8} + 1\frac{5}{6}$
↓   ↓   ↓
$2 + 1 + 2 = 5$

Front-End Estimation
Add the whole number parts of each mixed number.

$2\frac{1}{4} + 1\frac{1}{8} + 1\frac{5}{6}$
↓   ↓   ↓
$2 + 1 + 1 = 4$

The club will have about 4 or 5 lb of clay to work with.

1. Why is the rounded estimate greater than the front-end estimate?

B. You can also estimate differences of mixed numbers.

Estimate: $5\frac{1}{2} - 1\frac{2}{5}$

Rounding
Round each number to the nearest whole number. Then subtract.

$5\frac{1}{2} - 1\frac{2}{5}$
↓   ↓
$6 - 1 = 5$

Front-End Estimation
Subtract the whole number parts of each mixed number.

$5\frac{1}{2} - 1\frac{2}{5}$
↓   ↓
$5 - 1 = 4$

2. Will the exact answer be greater or less than 4? Why?

TRY OUT   Estimate the sum or difference.

Use rounding.

3. $2\frac{1}{5} + 5\frac{3}{8}$ **4.** $4\frac{9}{10} - 1\frac{1}{3}$

Use front-end estimation.

5. $4\frac{5}{8} + 1\frac{7}{9}$ **6.** $8\frac{3}{4} - 2\frac{7}{8}$

PRACTICE

Estimate the sum or difference. Which method did you use?

7. $1\frac{1}{8} + 2\frac{1}{3}$ **8.** $4\frac{2}{3} + 1\frac{4}{5}$ **9.** $3\frac{7}{9} + 3\frac{3}{10}$ **10.** $1\frac{6}{7} + 1\frac{1}{4}$ **11.** $1\frac{3}{4} + 3\frac{3}{5}$

12. $4\frac{4}{5} - 1\frac{3}{7}$ **13.** $2\frac{3}{5} - 1\frac{1}{4}$ **14.** $3\frac{5}{12} - 2\frac{5}{6}$ **15.** $5\frac{2}{3} - 2\frac{2}{9}$ **16.** $3\frac{5}{7} - 1\frac{3}{8}$

17. $4\frac{2}{3} + 1\frac{1}{3}$ **18.** $4\frac{6}{7} - 2\frac{3}{4}$ **19.** $3\frac{9}{10} - 1\frac{1}{4}$ **20.** $2\frac{1}{6} + 1\frac{4}{5}$ **21.** $2\frac{1}{4} + 5\frac{9}{10}$

Estimate to compare. Use > or <.

22. $3\frac{2}{9} + 1\frac{3}{8}$ ■ 5   **23.** $8\frac{7}{10} - 6\frac{2}{3}$ ■ 3   **24.** $5\frac{1}{8} + 2\frac{6}{7}$ ■ 7   **25.** $3\frac{5}{6} - 2\frac{1}{4}$ ■ 1

26. $4\frac{3}{4} + 5\frac{1}{3}$ ■ 9   **27.** $7\frac{1}{8} - 5\frac{5}{12}$ ■ 2   **28.** $9\frac{5}{6} + 2\frac{3}{8}$ ■ 12   **29.** $8\frac{2}{3} - 2\frac{3}{8}$ ■ 6

Critical Thinking

30. Which estimation method would give a more reasonable estimate to $9\frac{1}{4} - 8\frac{5}{8}$? Why?

Mixed Applications

Solve. Which method did you use?

31. Lindsay spent $1\frac{1}{3}$ hr cutting newspaper for the papier-mâché workshop and another $2\frac{1}{4}$ hr making her clay model. About how many hours in all did Lindsay work?

32. Darryl started with $1\frac{5}{6}$ lb of clay. After making his model, he had $\frac{2}{3}$ lb left. About how much clay did Darryl use in his model?

33. The art club decided to sell their papier-mâché figures for $5.95 each. How much money did they make selling 7 figures?

34. The art club bought 5 jars of poster paints with $10.50 of the money they made by selling the papier-mâché figures. How much did each jar of paint cost?

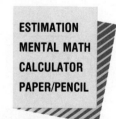

MANIPULATIVES

EXPLORING A CONCEPT
Adding, Subtracting Mixed Numbers

Jacob and his father are building shelves for a cabinet. One shelf will be $4\frac{3}{8}$ ft long. The other shelf will be $3\frac{1}{4}$ ft long. How long a board do Jacob and his father need for both shelves?

WORKING TOGETHER

Explore the problem by making a model. Then answer the questions below.

1. How does your model show $4\frac{3}{8}$ ft? How does it show $3\frac{1}{4}$ ft?

2. How did you combine the models to show the sum? What is the sum?

Jacob solves the problem using a graph-paper model.

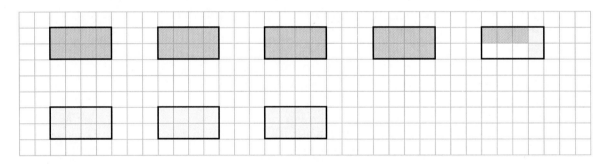

Each rectangle represents 1 whole, and each square in the rectangle represents $\frac{1}{8}$.

3. How does Jacob show $4\frac{3}{8}$? $3\frac{1}{4}$?

4. How does Jacob rename $3\frac{1}{4}$?

5. What is the sum of $4\frac{3}{8}$ and $3\frac{1}{4}$? How long a board do Jacob and his father need for the shelves?

6. **What if** the shelves were to be $4\frac{5}{8}$ ft and $3\frac{7}{8}$ ft long? How long a board would Jacob and his father need? Make a model to solve.

410   Lesson 10-8

Jacob's father had a piece of board that was $1\frac{3}{4}$ ft long and used $1\frac{1}{2}$ ft to make a shelf. How much of the board would he have left?

Explore the problem by making a model. Then answer the questions.

7. How does your model show $1\frac{3}{4}$ ft?

8. How did you show the subtraction of $1\frac{1}{2}$ ft?

9. How much of the board would he have left? What did you have to rename?

10. **What if** the board were $1\frac{3}{4}$ ft long and Jacob's father used $1\frac{1}{4}$ ft? How much of the board would he have left? Make a model to solve the problem.

SHARING IDEAS

11. Compare your models and results to those of others. How are they the same? How are they different?

12. How did you find the sum or difference of two mixed numbers with like denominators? with unlike denominators?

ON YOUR OWN

Solve using a model. Explain how you solved the problem.

13. Jacob used $2\frac{1}{3}$ pt of paint for part of the shelves. His father used $3\frac{2}{3}$ pt. How much paint did they use in all?

14. Jacob's father spent $6\frac{5}{6}$ hours working on the cabinet. Jacob spent $3\frac{1}{2}$ hours. How much longer did his father work?

15. The space between shelves is $9\frac{1}{2}$ in. If Jacob's radio is $6\frac{1}{4}$ in. high, how much space will be above it when he puts it on the shelf?

16. There is a $9\frac{1}{2}$-in. space between two shelves. Each shelf is $\frac{7}{8}$ in. thick. How much space is this in all?

DEVELOPING A CONCEPT
Adding Mixed Numbers

A. Eileen bought two new music boxes. One music box is $2\frac{1}{8}$ in. wide, and the other is $2\frac{5}{8}$ in. wide. How much space does she need to put them side by side?

Add: $2\frac{1}{8} + 2\frac{5}{8}$

Eileen needs $4\frac{3}{4}$ in. of space.

Here is a way to add mixed numbers without models.

Step 1	Step 2	Step 3
Add the fractions.	Add the whole numbers.	Write the sum in simplest form.
$2\frac{1}{8}$ $+\ 2\frac{5}{8}$ $\overline{\quad\ \frac{6}{8}}$	$2\frac{1}{8}$ $+\ 2\frac{5}{8}$ $\overline{\quad 4\frac{6}{8}}$	$4\frac{6}{8} = 4\frac{3}{4}$

B. Sometimes you have to add mixed numbers with unlike denominators.

Add: $6\frac{1}{2} + 3\frac{1}{6}$

Step 1	Step 2	Step 3
Find equivalent fractions with a common denominator.	Add the fractions. Add the whole numbers.	Write the sum in simplest form.
$6\frac{1}{2} = 6\frac{3}{6}$ $+\ 3\frac{1}{6} = 3\frac{1}{6}$	$6\frac{1}{2} = 6\frac{3}{6}$ $+\ 3\frac{1}{6} = 3\frac{1}{6}$ $\overline{\quad\quad\ \ 9\frac{4}{6}}$	$9\frac{4}{6} = 9\frac{2}{3}$

1. Is $9\frac{2}{3}$ a reasonable answer? Why?

SHARING IDEAS

2. How is adding mixed numbers similar to adding fractions? How is it different?

PRACTICE

Add. Write the sum in simplest form.

3. $3\frac{1}{8} + 1\frac{5}{8}$
4. $9\frac{1}{12} + 1\frac{5}{12}$
5. $2\frac{2}{9} + 3\frac{1}{9}$
6. $2\frac{1}{6} + 3\frac{1}{6}$
7. $6\frac{1}{10} + 3\frac{3}{10}$

8. $3\frac{1}{2} + 2\frac{1}{4}$
9. $6\frac{1}{5} + 3\frac{5}{10}$
10. $1\frac{1}{6} + 2\frac{1}{3}$
11. $4\frac{2}{5} + 3\frac{1}{15}$
12. $7\frac{1}{4} + 3\frac{1}{8}$

13. $3 + 5\frac{1}{7}$
14. $2\frac{1}{9} + 8\frac{5}{9}$
15. $4\frac{7}{8} + 2\frac{1}{16}$
16. $4\frac{7}{8} + 3$
17. $5\frac{7}{10} + 2\frac{3}{100}$

18. $3\frac{7}{16} + 2\frac{1}{8} + 5\frac{1}{4}$
19. $1\frac{3}{5} + 4\frac{1}{4} + 2\frac{1}{4}$
20. $6\frac{3}{5} + 2\frac{1}{15} + 1\frac{2}{15}$
21. $4\frac{1}{2} + 1\frac{1}{8} + 2\frac{1}{4}$
22. $1\frac{2}{7} + 5\frac{3}{14} + 4\frac{5}{14}$

23. $3\frac{2}{3} + 2\frac{1}{9}$
24. $5\frac{1}{6} + 3\frac{1}{18}$
25. $4\frac{1}{8} + 3\frac{3}{4}$
26. $1\frac{2}{5} + 3\frac{1}{10} + 2\frac{1}{5}$

Mixed Applications

27. Eileen's bedroom window is $42\frac{1}{4}$ in. wide. She wants her curtains to extend $4\frac{1}{8}$ in. beyond each side of the window. How wide should the curtain rod be?

28. Mrs. Casey ordered 21 yd of fabric at $7.98 per yd. How much will she spend for the fabric?

29. Eileen's bed measures $39\frac{1}{2}$ in. wide. Her night table is $18\frac{1}{6}$ in. wide. Will they fit in a space 65 in. wide? Why or why not?

UNDERSTANDING A CONCEPT
Adding Mixed Numbers with Renaming

A. David plans to hang a cork strip to use as a bulletin board along two walls of his room. The two walls measure $10\frac{7}{8}$ ft and $12\frac{3}{4}$ ft. How long should David make the cork strip?

Add: $10\frac{7}{8} + 12\frac{3}{4}$

Step 1	Step 2	Step 3
Find equivalent fractions with a common denominator.	Add the fractions. Add the whole numbers.	Write the sum in simplest form.
$10\frac{7}{8} = 10\frac{7}{8}$ $+ 12\frac{3}{4} = 12\frac{6}{8}$	$10\frac{7}{8} = 10\frac{7}{8}$ $+ 12\frac{3}{4} = 12\frac{6}{8}$ $\overline{22\frac{13}{8}}$	**Think:** $\frac{13}{8} = 1\frac{5}{8}$ $22 + 1\frac{5}{8} = 23\frac{5}{8}$ $22\frac{13}{8} = 23\frac{5}{8}$

David should make the cork strip $23\frac{5}{8}$ ft long.

1. Is $23\frac{5}{8}$ a reasonable answer? Why?
2. Why did $22\frac{13}{8}$ have to be renamed?
3. Find the sum: $4\frac{2}{5} + 3\frac{1}{5} + 7\frac{2}{5}$. Is the sum a mixed number or a whole number? Why?

B. Sometimes you can add mixed numbers mentally. Look for fractional parts that add to 1.

Add: $2\frac{3}{8} + 4\frac{1}{2} + 3\frac{5}{8}$

$\begin{array}{r} 2\frac{3}{8} \\ 4\frac{1}{2} \\ + 3\frac{5}{8} \end{array}$ **Think:** $\frac{3}{8} + \frac{5}{8} = \frac{8}{8} = 1$

So $2\frac{3}{8} + 3\frac{5}{8} = 6$.

$\begin{array}{r} 6 \\ + 4\frac{1}{2} \\ \hline 10\frac{1}{2} \end{array}$

4. **What if** you had to add $3\frac{1}{4} + 1\frac{1}{2} + 2\frac{1}{4}$? How would you find the sum mentally? What is the sum?

414 Lesson 10-10

TRY OUT Add. Write the sum in simplest form.

5. $3\frac{4}{9}$
   $+ 2\frac{7}{9}$

6. $5\frac{3}{8}$
   $+ 4\frac{3}{4}$

7. $2\frac{1}{5}$
   $+ 3\frac{4}{5}$

8. $1\frac{2}{3} + 3\frac{1}{6} + 2\frac{7}{12}$

PRACTICE

Add. Write the sum in simplest form.

9. $1\frac{1}{4}$
   $+ 2\frac{3}{4}$

10. $2\frac{4}{5}$
    $+ 5\frac{3}{10}$

11. $6\frac{5}{9}$
    $+ 2\frac{2}{3}$

12. $5\frac{3}{4}$
    $+ 2\frac{5}{12}$

13. $3\frac{7}{9}$
    $+ 2\frac{2}{9}$

14. $3\frac{9}{10}$
    $+ 4\frac{2}{5}$

15. $3\frac{2}{3}$
    $+ 5\frac{8}{9}$

16. $7\frac{3}{7}$
    $+ 3\frac{5}{7}$

17. $3\frac{7}{8}$
    $+ 5\frac{1}{4}$

18. $4\frac{1}{2}$
    $+ 3\frac{1}{2}$

19. $3\frac{1}{5}$
    $1\frac{2}{5}$
    $+ 4\frac{9}{10}$

20. $5\frac{4}{9}$
    $3\frac{2}{3}$
    $+ 2\frac{2}{9}$

21. $1\frac{3}{4}$
    $6\frac{3}{8}$
    $+ 2\frac{5}{8}$

22. $6\frac{1}{8}$
    $3\frac{3}{4}$
    $+ 4\frac{1}{2}$

23. $2\frac{5}{6}$
    $1\frac{1}{3}$
    $+ 3\frac{1}{12}$

24. $5\frac{1}{2} + 3\frac{3}{4}$

25. $5\frac{7}{9} + 7\frac{1}{3}$

26. $3\frac{3}{5} + 6\frac{2}{5}$

27. $3\frac{5}{8} + 2\frac{5}{8} + 1\frac{1}{4}$

Critical Thinking Write *always true*, *sometimes true*, or *never true*.

28. When you add more than two mixed numbers, the fractional part of the sum will be more than 1. Give examples to support your answer.

Mixed Applications

29. David needs $23\frac{5}{8}$ ft of corkboard. To the nearest foot, how many feet of corkboard should he buy? If the corkboard sells for $1.19 per ft, how much will it cost?

30. Karen wants to frame a rectangular map of the United States. The map measures $3\frac{1}{4}$ ft by $2\frac{3}{8}$ ft. How many feet of frame will she need?

31. Sabra spent $142.50 on wallpaper for her room. The wallpaper cost $9.50 per roll. How many rolls did she buy?

32. **Write a problem** involving addition of mixed numbers. Use the Databank on page 521. Ask others to solve your problem.

PROBLEM SOLVING

✓ UNDERSTAND
✓ PLAN
✓ TRY
✓ CHECK
✓ EXTEND

Strategy: Solving a Simpler Problem

Donna is a fashion designer. She needs $119\frac{2}{3}$ yards of navy corduroy and $127\frac{1}{4}$ yards of green corduroy to manufacture slacks she has designed. How many yards of corduroy does she need all together?

Sometimes solving a simpler problem helps you understand the problem better. Try using lesser numbers in the original problem.

1. **What if** Donna needs only 3 yards of navy and 4 yards of green corduroy? Which operation could you use to find how many yards of corduroy she needs all together? How many yards would she need?

2. Which operation can you use to solve the original problem? How many yards of corduroy does she need?

3. How do you know that your answer is reasonable?

4. About how much more green corduroy than navy corduroy does Donna need?

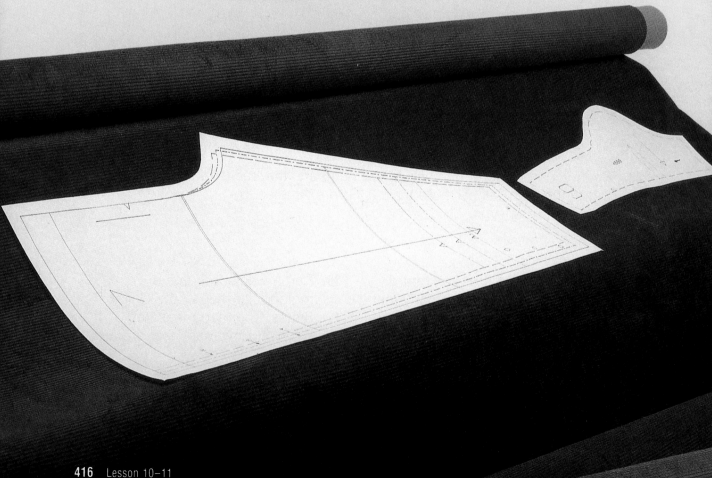

PRACTICE

Think about solving a simpler problem using lesser numbers. Then solve the original problem.

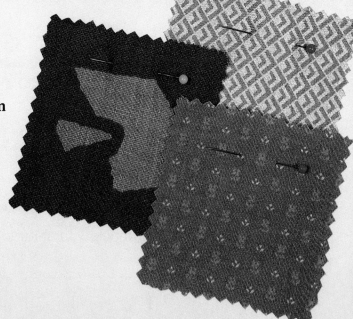

5. Donna needs $203\frac{1}{3}$ yards of green silk and $196\frac{3}{8}$ yards of white silk to fill an order for blouses. How many yards of silk does she need all together?

6. Donna had $117\frac{1}{8}$ yards of wool fabric. She used $98\frac{2}{9}$ yards. About how much wool fabric did she have left?

7. Donna bought some fabric from the BKT Company. She bought $43\frac{1}{2}$ yards of cotton, $58\frac{2}{3}$ yards of wool, and $63\frac{1}{6}$ yards of rayon. How many yards of fabric did Donna buy in all?

8. Tina makes dress samples. Last week she worked for $37\frac{2}{3}$ hours. This week she worked for $45\frac{1}{3}$ hours. About how much longer did she work this week than last week?

Strategies and Skills Review

Solve. Use mental math, estimation, a calculator, or paper and pencil.

9. Janice is buying a dress and a blouse for $70.50. The dress is $30.50 more than the blouse. How much is the blouse?

10. The ATM Fashion Company sold 325 suits at $69.95 each. How much did the company take in from the sale of the suits?

11. Ann bought $163\frac{2}{3}$ yards of lace and $129\frac{4}{5}$ yards of velvet. To the nearest yard, how many more yards of lace did she buy?

12. Anna has $70 to spend on supplies for the ATM Fashion Company. Can she buy tape measures for $8.98, scissors for $33.49, thread for $18.98, and pins for $7.50? Did you underestimate or overestimate?

13. Greta has won twice as many fashion awards as Bonnie. Rita has won 3 fewer awards than Greta. Rita has won 11 awards. How many awards has Bonnie won?

14. **Write a problem** that you can solve by first solving a simpler problem using lesser numbers. Solve the problem. Ask others to solve the problem.

EXTRA Practice, page 430

Adding and Subtracting Fractions   **417**

DEVELOPING A CONCEPT

Subtracting Mixed Numbers

A. Mrs. Arnold had $2\frac{7}{8}$ pans of lasagna. She served $1\frac{1}{8}$ pans at lunch. How much lasagna did she have left?

Subtract: $2\frac{7}{8} - 1\frac{1}{8}$

You can use a model to help you subtract.

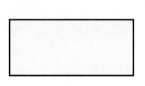
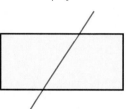
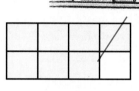

Mrs. Arnold had $1\frac{3}{4}$ pans of lasagna left.

Here is a way to subtract without using models.

Step 1	Step 2	Step 3
Subtract the fractions.	Subtract the whole numbers.	Write the difference in simplest form.
$\begin{array}{r} 2\frac{7}{8} \\ -\,1\frac{1}{8} \\ \hline \frac{6}{8} \end{array}$	$\begin{array}{r} 2\frac{7}{8} \\ -\,1\frac{1}{8} \\ \hline 1\frac{6}{8} \end{array}$	$1\frac{6}{8} = 1\frac{3}{4}$

B. Sometimes you need to subtract mixed numbers with unlike denominators.

Subtract: $3\frac{7}{12} - 1\frac{1}{4}$

Step 1	Step 2	Step 3
Find equivalent fractions with a common denominator.	Subtract the fractions. Subtract the whole numbers.	Write the difference in simplest form.
$\begin{array}{r} 3\frac{7}{12} = 3\frac{7}{12} \\ -\,1\frac{1}{4} = 1\frac{3}{12} \\ \hline \end{array}$	$\begin{array}{r} 3\frac{7}{12} = 3\frac{7}{12} \\ -\,1\frac{1}{4} = 1\frac{3}{12} \\ \hline 2\frac{4}{12} \end{array}$	$2\frac{4}{12} = 2\frac{1}{3}$

Lesson 10-12

SHARING IDEAS

1. How is subtracting mixed numbers similar to subtracting fractions? How is it different?
2. Why is it easy to find the difference of $5\frac{2}{3}$ and 4?

PRACTICE

Subtract. Write the difference in simplest form.

3. $5\frac{5}{6} - 1\frac{1}{6}$
4. $3\frac{9}{10} - 2\frac{1}{10}$
5. $6\frac{5}{7} - 1\frac{3}{7}$
6. $3\frac{9}{10} - 1\frac{7}{10}$
7. $4\frac{11}{12} - 2\frac{1}{12}$

8. $3\frac{3}{4} - 1\frac{1}{2}$
9. $6\frac{5}{12} - 2\frac{1}{3}$
10. $7\frac{4}{5} - 2\frac{1}{10}$
11. $3\frac{5}{6} - 1\frac{1}{3}$
12. $6\frac{5}{7} - 2\frac{3}{28}$

13. $13\frac{8}{9} - 4$
14. $1\frac{5}{8} - \frac{3}{8}$
15. $5\frac{7}{8} - 1\frac{3}{4}$
16. $12\frac{1}{5} - 6$
17. $9\frac{5}{6} - 2\frac{1}{2}$

18. $8\frac{7}{9} - 7\frac{1}{3}$
19. $10\frac{7}{8} - 4\frac{1}{4}$
20. $9\frac{4}{7} - 3$
21. $3\frac{5}{6} - 1\frac{1}{2}$
22. $5\frac{5}{12} - 3\frac{1}{4}$

23. $8\frac{2}{3} - 5\frac{1}{6}$
24. $7\frac{4}{5} - 3\frac{2}{5}$
25. $4\frac{5}{9} - 1\frac{1}{3}$
26. $6\frac{5}{12} - 2\frac{1}{6}$

Mixed Applications

27. Hana had $3\frac{1}{2}$ loaves of bread. She used $1\frac{1}{8}$ loaves to make stuffing for a turkey. How much bread did she have left?

28. Hana and Renée combined $\frac{1}{2}$ cup of white flour, $\frac{3}{8}$ cup of whole-wheat flour, and $\frac{1}{4}$ cup of rye flour to make bread. How much flour was used?

Mixed Review

Find the answer. Which method did you use?

**MENTAL MATH
CALCULATOR
PAPER/PENCIL**

29. $2.27 + 5.8$
30. $5,608 - 4,225$
31. $\$8.62 \times 6$
32. $32\overline{)6.496}$

33. $1,110 \times 7$
34. $87.1 - 60.4$
35. 49.5×0.09
36. $16.32 + 25.328$

EXTRA Practice, page 430

Adding and Subtracting Fractions

DEVELOPING A CONCEPT
Renaming to Subtract Mixed Numbers

A. Valerie has $3\frac{1}{3}$ sheets of blue paper. She cuts one of the sheets into $\frac{1}{3}$ strips. How many whole sheets does she have now? How many $\frac{1}{3}$ strips does she have? Write a mixed number to show how much blue paper Valerie has.

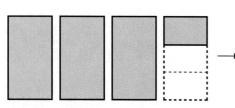

Valerie now has two whole sheets of blue paper and four $\frac{1}{3}$ strips. So she has $2\frac{4}{3}$ sheets of blue paper.

1. **What if** Valerie cut another of the $2\frac{4}{3}$ sheets into $\frac{1}{3}$ strips? How many whole sheets would she have then? How many $\frac{1}{3}$ strips would she have? What mixed number could you write to show how much blue paper she would have?

2. Tom has 4 sheets of green paper. He cuts one of the sheets into $\frac{1}{6}$ strips. Write a mixed number to show how much green paper Tom has.

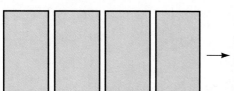

B. You can rename any whole number or mixed number by thinking of different names for 1.

$5\frac{2}{5} = 4\frac{\blacksquare}{5}$ $7 = 6\frac{\blacksquare}{8}$

Think: $5\frac{2}{5} = 4 + 1 + \frac{2}{5}$ **Think:** $7 = 6 + 1$
$\qquad\qquad = 4 + \frac{5}{5} + \frac{2}{5}$ $\qquad\qquad = 6 + \frac{8}{8}$
$\qquad\qquad = 4 + \frac{7}{5}$ $\qquad\qquad = 6\frac{8}{8}$
$\qquad\qquad = 4\frac{7}{5}$

SHARING IDEAS

3. Find the mixed number in each group that is not equivalent to the others.

   a. $3\frac{1}{4}, 2\frac{5}{4}, 1\frac{8}{4}$ b. $6\frac{3}{8}, 5\frac{10}{8}, 4\frac{19}{8}$

4. What happens to the numerator during each regrouping?

PRACTICE

Find the missing number.

5. $2\frac{1}{3} = 1\frac{\blacksquare}{3}$ 6. $8\frac{1}{4} = 7\frac{\blacksquare}{4}$ 7. $9 = 8\frac{\blacksquare}{5}$ 8. $12 = 11\frac{\blacksquare}{7}$ 9. $15\frac{5}{6} = 14\frac{\blacksquare}{6}$

10. $6 = 5\frac{\blacksquare}{8}$ 11. $3\frac{1}{4} = 2\frac{\blacksquare}{4}$ 12. $4\frac{3}{8} = 3\frac{\blacksquare}{8}$ 13. $3 = 2\frac{\blacksquare}{10}$ 14. $5\frac{1}{2} = 4\frac{\blacksquare}{2}$

15. $5 = 4\frac{3}{\blacksquare}$ 16. $7 = \blacksquare\frac{2}{2}$ 17. $4\frac{3}{10} = 3\frac{13}{\blacksquare}$ 18. $12\frac{5}{9} = \blacksquare\frac{14}{9}$ 19. $18 = 17\frac{\blacksquare}{3}$

20. $6\frac{3}{5} = 4\frac{\blacksquare}{5}$ 21. $9\frac{1}{6} = 6\frac{\blacksquare}{6}$ 22. $16\frac{5}{8} = 14\frac{21}{\blacksquare}$ 23. $10\frac{6}{7} = 9\frac{\blacksquare}{7}$ 24. $15\frac{3}{4} = \blacksquare\frac{11}{4}$

25. $21 = \blacksquare\frac{18}{6}$ 26. $7 = 6\frac{\blacksquare}{10}$ 27. $17\frac{2}{9} = 16\frac{\blacksquare}{9}$ 28. $4\frac{5}{12} = 3\frac{\blacksquare}{12}$ 29. $5\frac{3}{5} = 3\frac{\blacksquare}{5}$

Mixed Applications

30. Tom has 5 strips of wood, each 1 yd long. He cut one strip into $\frac{1}{4}$-yd pieces. Write a mixed number for the amount of wood Tom has.

31. Valerie mixed $1\frac{1}{4}$ c of red paint and $2\frac{1}{2}$ c of yellow paint to make orange paint. How many cups of orange paint did she make?

32. Look through newspapers and magazines to find three mixed numbers. Then rename each mixed number as another mixed number.

33. Charlie earns $20.75 each week doing yard work. How much will he have earned in 9 weeks?

VISUAL REASONING

Window Problem
Trace this picture of the window. Color exactly half of the large square, so that the uncolored half is still a square.

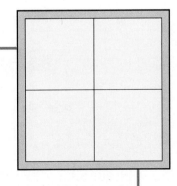

UNDERSTANDING A CONCEPT
Subtracting Mixed Numbers with Renaming

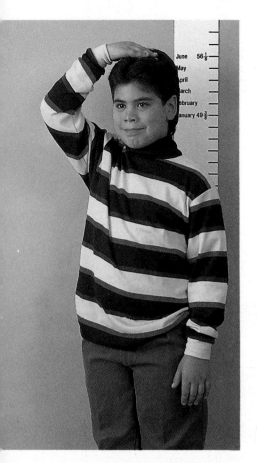

A. Julio records his height in inches every six months. How many inches did he grow in the period from January to June?

Subtract: $56\frac{1}{8} - 49\frac{3}{8}$

Step 1	Step 2
Rename fractions where necessary.	Subtract the fractions. Subtract the whole numbers. Write the difference in simplest form.
$56\frac{1}{8} = 55\frac{9}{8}$ $-\ 49\frac{3}{8} = 49\frac{3}{8}$	$56\frac{1}{8} = 55\frac{9}{8}$ $-\ 49\frac{3}{8} = 49\frac{3}{8}$ $\overline{ 6\frac{6}{8} = 6\frac{3}{4}}$

Think: $56\frac{1}{8} = 55 + 1\frac{1}{8}$
$\phantom{Think: 56\frac{1}{8}} = 55 + \frac{8}{8} + \frac{1}{8}$
$\phantom{Think: 56\frac{1}{8}} = 55\frac{9}{8}$

Julio grew $6\frac{3}{4}$ in. from January to June.

1. Why was it necessary to rename $56\frac{1}{8}$ as $55\frac{9}{8}$?

B. Subtract: $56\frac{1}{2} - 52\frac{3}{4}$

Step 1	Step 2	Step 3
Find equivalent fractions with a common denominator.	Rename if necessary.	Subtract the fractions. Subtract the whole numbers. Write the difference in simplest form.
$56\frac{1}{2} = 56\frac{2}{4}$ $-\ 52\frac{3}{4} = 52\frac{3}{4}$	$56\frac{1}{2} = 56\frac{2}{4} = 55\frac{6}{4}$ $-\ 52\frac{3}{4} = 52\frac{3}{4} = 52\frac{3}{4}$	$56\frac{1}{2} = 56\frac{2}{4} = 55\frac{6}{4}$ $-\ 52\frac{3}{4} = 52\frac{3}{4} = 52\frac{3}{4}$ $\overline{ 3\frac{3}{4}}$

Think: $56\frac{2}{4} = 55 + 1\frac{2}{4}$
$\phantom{Think: 56\frac{2}{4}} = 55 + \frac{4}{4} + \frac{2}{4}$
$\phantom{Think: 56\frac{2}{4}} = 55\frac{6}{4}$

2. *What if* you subtract $53 - 48\frac{3}{8}$? Is it necessary to do any renaming? Why or why not?

TRY OUT
Subtract. Write the difference in simplest form.

3. $6\frac{3}{5} - 4\frac{4}{5}$
4. $9\frac{1}{3} - 2\frac{5}{12}$
5. $5\frac{3}{4} - 1\frac{7}{8}$
6. $5 - 2\frac{3}{10}$
7. $8 - 5\frac{2}{3}$

PRACTICE
Subtract. Write the difference in simplest form.

8. $4\frac{1}{4} - 2\frac{3}{4}$
9. $9\frac{1}{3} - \frac{2}{3}$
10. $6\frac{1}{8} - 4\frac{5}{8}$
11. $8\frac{2}{5} - 3\frac{4}{5}$
12. $10\frac{3}{7} - 5\frac{5}{7}$

13. $3\frac{1}{6} - 2\frac{2}{3}$
14. $5\frac{1}{2} - 3\frac{3}{4}$
15. $12\frac{3}{10} - 8\frac{2}{5}$
16. $3\frac{4}{9} - 2\frac{2}{3}$
17. $13\frac{1}{8} - 2\frac{3}{4}$

18. $7 - 3\frac{1}{2}$
19. $5 - 4\frac{1}{3}$
20. $8 - 2\frac{3}{5}$
21. $9 - 4\frac{5}{6}$
22. $6 - 1\frac{4}{9}$

23. $4\frac{1}{5} - 3\frac{2}{5}$
24. $7\frac{4}{9} - 6\frac{2}{3}$
25. $12\frac{1}{6} - 5\frac{5}{12}$
26. $6\frac{5}{6} - 6\frac{1}{2}$
27. $10 - 4\frac{8}{9}$

28. $6\frac{1}{6} - 1\frac{1}{3}$
29. $4 - 1\frac{3}{7}$
30. $7\frac{2}{9} - 5\frac{4}{9}$
31. $8\frac{3}{8} - 7\frac{3}{4}$

Mixed Applications

32. Between January and June, Pablo grew $3\frac{3}{4}$ in. His brother, Hector, grew $4\frac{5}{8}$ in., and their little sister, Rose, grew $5\frac{1}{6}$ in. Who grew the most? Who grew the least?

33. **Write a problem** involving subtraction of mixed numbers. Use the information found on page 521 of the Databank. Ask others to solve your problem.

Mixed Review
Measure to the nearest 1 in., $\frac{1}{2}$ in., $\frac{1}{4}$ in., and $\frac{1}{8}$ in.

34. ─────────────────
35. ─────────────────
36. ──────────────────────────

EXTRA Practice, page 431; Practice **PLUS**, page 433

DECISION MAKING

Problem Solving: Redecorating a Room

SITUATION

Alice Andrea wants her room to have a new look. She has thought of redecorating or just repainting the room. Her grandfather is willing to help her make new curtains and a bedspread if she decides to redecorate.

PROBLEM

What should Alice do? Should she repaint or redecorate her room?

DATA

- **PAINT FOR WALLS** $8.59 per gallon
- **PAINTBRUSHES** $2.69 small / $3.59 medium
- **THROW PILLOWS** $4.95 each
- **CLOCK** $10.29
- **PLASTIC DROP CLOTH** $1.79
- **PAINT TRAY** $1.52
- **AREA RUG** $16.50
- **PAINT ROLLER** $2.95

USING THE DATA

Alice is thinking of two different plans—
Plan A: paint the room; Plan B: redecorate the room.

Find the cost.

Plan A:

1. 2 gallons of paint and 2 drop cloths
2. paint roller, tray, a small brush, and a medium brush

Plan B:

3. fabric at $4.50 per yd for a bedspread and curtains
4. 2 throw pillows
5. an area rug
6. **What if** she makes 2 throw pillows? Each will take a $\frac{1}{2}$ yd of fabric at $4.50 per yd. How much will Plan B cost now?

MAKING DECISIONS

7. **What if** Alice's grandfather has the roller, tray, and brushes she needs? How might this affect her plans?
8. **What if** she wants a clock in her room? How might she adjust Plan B to avoid an increase in cost?
9. Tell how Alice could give her room a new look without repainting or redecorating it.
10. **What if** she has $40 to spend? What could she choose to do? List reasons for your answer.
11. **Write a list** of other things Alice should think about before deciding to paint or to redecorate her room.
12. Would you redecorate or repaint the room? Why?

Paint needed
2 gallons

Materials needed:
bedspread 5½ yd
curtains 1½ yd
$4.50 per yd
$3.00 per yd
$3.00 per yd

Adding and Subtracting Fractions

CURRICULUM CONNECTION

Math and Music

Notes show length. "The Farmer in the Dell" has 6 beats in a measure. An eighth note (♪) is 1 beat. A quarter note (♩) is 2 beats. A dotted note is the value of the note plus half of its value. A tie (⌣) adds values together. In measure 4, the dotted quarter note tied to a quarter note (♩．♩) equals ♩ + ♪ + ♩ or 2 + 1 + 2 = 5 beats.

$\frac{6}{8}$ | ♪ | ♩ ♪ ♪ ♪ | ♩． ♩ | ♩ | ♩ ♪ ♪ ♪ | ♩． ♩ 𝄽 |
     The | farm-er in the | dell, — The | farm-er in the | dell (rest) |

♩． ♩ ♪ | ♩ ♪ ♪ ♪ | ♩ ♪ ♪ ♪ | ♩． ♩ ‖
Heigh-ho, the | der-ry-o, the | farm-er in the | dell.

The song below has two beats in a measure. A quarter note now is 1 beat. An eighth note is $\frac{1}{2}$ beat. A sixteenth note (♬) is $\frac{1}{4}$ beat.

$\frac{2}{4}$ ♬ | ♪ ♪ ♪． ♬ | ♪ ♪ ♪． ♬ | ♪ ♪ ♪ ♪ | ♩． |
I — | come from Al-a- | ba-ma with my | ban-jo on my | knee. |

What if you want to know the length of the third note in measure 2?

Think: The dotted eighth note (♪．) equals an eighth note plus a sixteenth note, or $\frac{1}{2}$ plus $\frac{1}{4}$.

ACTIVITIES

1. Work with a partner. Write combinations of quarter and eighth notes in $\frac{2}{4}$.

2. Work with a partner. One should tap the beat of "The Farmer in the Dell" as the other says the words in rhythm. Write the number of beats of each note.

Calculator: Find the Path

You can use a calculator to add and subtract fractions.

Add: $\frac{1}{2}$ to $\frac{3}{4}$ *Think:* $\frac{3}{4}$ means $3 \div 4$ $\frac{1}{2}$ means $1 \div 2$

Follow these steps.

Enter	Calculator Display
3 ÷ 4 = M+	0.75
1 ÷ 2 M+ MRC	1.25

$\frac{3}{4} + \frac{1}{2} = 1.25$

USING THE CALCULATOR

Use a calculator to find the path to each final number in the maze.

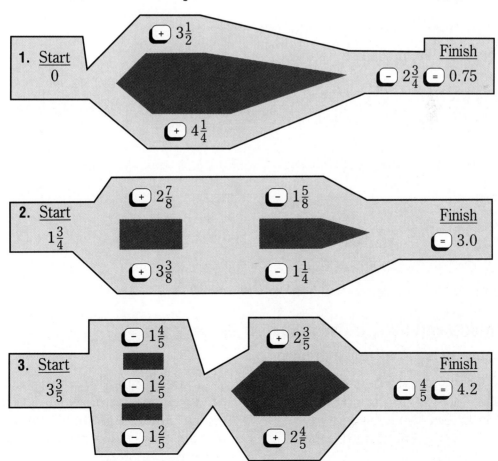

Adding and Subtracting Fractions 427

EXTRA PRACTICE

Adding Fractions: Like Denominators, page 399
Add. Write the sum in simplest form.

1. $\frac{1}{3} + \frac{1}{3}$
2. $\frac{2}{7} + \frac{5}{7}$
3. $\frac{1}{8} + \frac{5}{8}$
4. $\frac{1}{12} + \frac{3}{12}$

5. $\frac{1}{3} + \frac{2}{3} + \frac{1}{3}$
6. $\frac{1}{5} + \frac{2}{5} + \frac{4}{5}$
7. $\frac{3}{4} + \frac{1}{4} + \frac{2}{4}$
8. $\frac{8}{10} + \frac{3}{10} + \frac{4}{10}$

9. $\frac{5}{6} + \frac{5}{6}$
10. $\frac{3}{8} + \frac{5}{8}$
11. $\frac{5}{9} + \frac{4}{9}$
12. $\frac{2}{5} + \frac{1}{5}$

Adding Fractions: Unlike Denominators, page 401
Add. Write the sum in simplest form.

1. $\frac{1}{6} + \frac{5}{12}$
2. $\frac{4}{5} + \frac{3}{10}$
3. $\frac{2}{3} + \frac{1}{6}$
4. $\frac{1}{8} + \frac{3}{4}$

5. $\frac{2}{3} + \frac{1}{6} + \frac{1}{3}$
6. $\frac{1}{3} + \frac{7}{9} + \frac{2}{3}$
7. $\frac{5}{7} + \frac{3}{14} + \frac{1}{7}$
8. $\frac{3}{10} + \frac{4}{5} + \frac{7}{10}$

9. the sum of three-eighths and nine-sixteenths
10. three-fourths plus eleven-twelfths

Problem-Solving Strategy: Using Number Sense, page 403
Use number sense to solve the problem.

1. Mrs. Ross bought $\frac{1}{3}$ pound of cashew nuts and $\frac{3}{8}$ pound of almonds. Did she buy more or less than 1 pound of nuts? How do you know?

2. Mia used $\frac{5}{8}$ quart of orange juice and $\frac{1}{2}$ quart of grape juice to make some punch. Did she make more or less than 1 quart of punch? How do you know?

Extra Practice

Subtracting Fractions, page 405
Subtract. Write the difference in simplest form.

1. $\dfrac{7}{9} - \dfrac{4}{9}$
2. $\dfrac{9}{10} - \dfrac{7}{10}$
3. $\dfrac{5}{8} - \dfrac{1}{8}$
4. $\dfrac{1}{2} - \dfrac{1}{10}$

5. $\dfrac{4}{5} - \dfrac{1}{5}$
6. $\dfrac{5}{6} - \dfrac{1}{3}$
7. $\dfrac{2}{3} - \dfrac{2}{9}$
8. $\dfrac{4}{5} - \dfrac{7}{10}$

9. $\dfrac{5}{12} - \dfrac{1}{6}$
10. $\dfrac{7}{8} - \dfrac{3}{8}$
11. $\dfrac{3}{4} - \dfrac{7}{12}$
12. $\dfrac{7}{10} - \dfrac{1}{5}$

13. $\dfrac{5}{10} - \dfrac{2}{5}$
14. $\dfrac{6}{8} - \dfrac{1}{8}$
15. $\dfrac{8}{12} - \dfrac{1}{3}$
16. $\dfrac{5}{9} - \dfrac{1}{3}$

Estimating Sums and Differences, page 409
Estimate the sum or the difference.

1. $1\dfrac{1}{3} + 2\dfrac{1}{5}$
2. $3\dfrac{1}{4} + 2\dfrac{1}{2}$
3. $1\dfrac{3}{7} + 1\dfrac{1}{5}$
4. $2\dfrac{3}{4} + 2\dfrac{3}{5}$

5. $4\dfrac{3}{4} - 1\dfrac{1}{3}$
6. $2\dfrac{7}{10} - 1\dfrac{1}{5}$
7. $6\dfrac{1}{3} - 4\dfrac{1}{2}$
8. $3\dfrac{4}{7} - 1\dfrac{1}{4}$

9. $4\dfrac{1}{3} - 3\dfrac{1}{8}$
10. $4\dfrac{3}{5} - 1\dfrac{1}{10}$
11. $3\dfrac{4}{9} - 1\dfrac{1}{3}$
12. $3\dfrac{1}{2} + 6\dfrac{1}{3}$

Adding Mixed Numbers, page 413
Add. Write the sum in simplest form.

1. $2\dfrac{1}{5} + 1\dfrac{3}{5}$
2. $6\dfrac{3}{7} + 2\dfrac{1}{7}$
3. $1\dfrac{1}{6} + 1\dfrac{1}{6}$
4. $5\dfrac{1}{10} + 3\dfrac{3}{10}$

5. $2\dfrac{1}{2} + 1\dfrac{1}{4}$
6. $3\dfrac{1}{6} + 2\dfrac{1}{3}$
7. $4\dfrac{2}{5} + 2\dfrac{2}{15}$
8. $8\dfrac{7}{10} + 4\dfrac{3}{100}$

9. $3\dfrac{7}{8} + 3\dfrac{1}{16}$
10. $4 + 3\dfrac{1}{8}$
11. $3\dfrac{2}{5} + 4\dfrac{1}{10}$
12. $6\dfrac{1}{10} + 3\dfrac{5}{100}$

Adding and Subtracting Fractions

Extra Practice

Adding Mixed Numbers with Renaming, page 415

Add. Write the sum in simplest form.

1. $3\frac{3}{4} + 2\frac{1}{4}$
2. $5\frac{1}{4} + 1\frac{7}{8}$
3. $5\frac{1}{2} + 5\frac{1}{2}$
4. $3\frac{8}{11} + 4\frac{3}{11}$

5. $2\frac{1}{5} + 2\frac{2}{5} + 3\frac{7}{10}$
6. $1\frac{3}{8} + 5\frac{5}{8} + 2\frac{3}{4}$
7. $5\frac{5}{6} + 1\frac{2}{3} + 3\frac{1}{12}$
8. $3\frac{3}{4} + 1\frac{1}{8} + 4\frac{1}{2}$

9. $3\frac{1}{3} + 4\frac{7}{9}$
10. $5\frac{3}{4} + 2\frac{1}{2}$
11. $4\frac{3}{5} + 1\frac{1}{5} + 2\frac{7}{10}$
12. $6\frac{2}{3} + 5\frac{4}{12}$
13. $8\frac{3}{4} + 6\frac{2}{8}$
14. $7\frac{1}{2} + 4\frac{3}{8}$

Problem-Solving Strategy: Solving a Simpler Problem, page 417

Think about solving a simpler problem using lesser numbers. Then solve the original problem.

1. Shawn bought $31\frac{1}{3}$ yd of blue silk, $25\frac{2}{3}$ yd of red silk, and $14\frac{5}{6}$ yd of white silk. How many yd of silk did Shawn buy in all?

2. Joanne had $\frac{7}{8}$ yd of cotton. She used $\frac{3}{4}$ yd to make a pillow. How much cotton fabric did she have left?

Subtracting Mixed Numbers, page 419

Subtract. Write the difference in simplest form.

1. $4\frac{5}{6} - 1\frac{1}{6}$
2. $4\frac{9}{10} - 1\frac{7}{10}$
3. $4\frac{5}{7} - 2\frac{3}{21}$
4. $8\frac{3}{7} - 4$

5. $8\frac{7}{10} - 3\frac{1}{5}$
6. $13\frac{1}{7} - 6$
7. $3\frac{4}{5} - 1\frac{1}{10}$
8. $5\frac{11}{12} - 3\frac{1}{12}$

9. $6\frac{4}{5} - 5\frac{1}{10}$
10. $6\frac{7}{8} - 4\frac{1}{8}$
11. $3\frac{5}{12} - 1\frac{1}{4}$

EXTRA PRACTICE

Renaming to Subtract Mixed Numbers, page 421

Find the missing number.

1. $3\frac{1}{3} = 2\frac{\blacksquare}{3}$
2. $5\frac{1}{5} = 4\frac{\blacksquare}{5}$
3. $11 = 10\frac{\blacksquare}{9}$
4. $3\frac{1}{2} = 2\frac{\blacksquare}{2}$

5. $3\frac{3}{5} = 2\frac{\blacksquare}{5}$
6. $5 = \blacksquare\frac{3}{3}$
7. $12 = 11\frac{\blacksquare}{3}$
8. $16\frac{3}{4} = \blacksquare\frac{11}{4}$

9. $6 = 5\frac{\blacksquare}{10}$
10. $3\frac{5}{12} = 2\frac{\blacksquare}{12}$
11. $6 = 5\frac{3}{\blacksquare}$
12. $6\frac{1}{2} = 5\frac{\blacksquare}{2}$

13. $19 = \blacksquare\frac{12}{6}$
14. $6\frac{5}{12} = 5\frac{\blacksquare}{12}$
15. $6\frac{3}{5} = 4\frac{\blacksquare}{5}$
16. $16\frac{2}{9} = 15\frac{\blacksquare}{9}$

17. $21 = \blacksquare\frac{28}{7}$
18. $28\frac{3}{4} = \blacksquare\frac{43}{4}$
19. $16\frac{3}{8} = 14\frac{\blacksquare}{8}$
20. $12\frac{1}{3} = 10\frac{\blacksquare}{3}$

21. $8\frac{3}{16} = \blacksquare\frac{35}{16}$
22. $11\frac{4}{9} = \blacksquare\frac{13}{9}$
23. $5\frac{9}{12} = \blacksquare\frac{21}{12}$
24. $6\frac{4}{9} = 5\frac{\blacksquare}{9}$

Subtracting Mixed Numbers with Renaming, page 423

Subtract. Write the difference in simplest form.

1. $5\frac{2}{5} - 3\frac{4}{5}$
2. $10\frac{4}{9} - 2\frac{2}{3}$
3. $5\frac{1}{6} - 2\frac{2}{3}$
4. $8\frac{7}{10} - 2\frac{1}{5}$

5. $3 - 2\frac{1}{3}$
6. $8 - 3\frac{7}{8}$
7. $7 - 2\frac{1}{2}$
8. $5 - 2\frac{4}{5}$

9. $13\frac{1}{6} - 3\frac{5}{12}$
10. $9 - 2\frac{8}{9}$
11. $6\frac{1}{3} - 2\frac{4}{9}$
12. $11\frac{3}{10} - 8\frac{2}{5}$

13. $2\frac{1}{5} - 1\frac{2}{5}$
14. $10\frac{1}{7} - 3\frac{5}{7}$
15. $5 - 1\frac{6}{7}$
16. $10 - 4\frac{9}{10}$

Adding and Subtracting Fractions

Practice PLUS

KEY SKILL: Adding Mixed Numbers with Renaming (Use after page 415.)

Level A

Add. Write the sum in simplest form.

1. $3\frac{1}{2}$
   $+ 5\frac{1}{2}$

2. $4\frac{1}{4}$
   $+ 3\frac{3}{8}$

3. $3\frac{1}{3}$
   $+ 5\frac{1}{6}$

4. $6\frac{1}{4}$
   $+ 2\frac{5}{8}$

5. $5\frac{3}{8} + 4\frac{3}{4}$

6. $3\frac{7}{9} + 2\frac{1}{9}$

7. $6\frac{1}{2} + 4\frac{1}{6}$

8. Bonnie buys $1\frac{1}{4}$ pounds of Swiss cheese and $2\frac{3}{4}$ pounds of cheddar cheese. How much cheese does she buy?

Level B

Add. Write the sum in simplest form.

9. $5\frac{3}{4}$
   $+ 2\frac{5}{12}$

10. $6\frac{5}{9}$
    $+ 2\frac{2}{3}$

11. $7\frac{3}{7}$
    $+ 3\frac{5}{7}$

12. $4\frac{5}{12}$
    $+ 1\frac{1}{3}$

13. $6\frac{2}{5} + 3\frac{7}{10}$

14. $4\frac{1}{3} + 5\frac{2}{9}$

15. $2\frac{3}{7} + 6\frac{1}{14}$

16. Stan assembles his scuba gear in $2\frac{1}{2}$ minutes. It takes him $3\frac{3}{8}$ minutes to put on the equipment. How long does it take Stan before he is ready to dive?

Level C

Add. Write the sum in simplest form.

17. $6\frac{3}{8}$
    $2\frac{5}{8}$
    $+ 1\frac{3}{4}$

18. $4\frac{9}{10}$
    $2\frac{2}{5}$
    $+ 3\frac{1}{5}$

19. $3\frac{1}{12}$
    $2\frac{5}{6}$
    $+ 1\frac{1}{3}$

20. $2\frac{5}{8}$
    $1\frac{1}{4}$
    $+ 3\frac{5}{8}$

21. $5\frac{1}{4} + 3\frac{1}{8}$

22. $3\frac{3}{4} + 5\frac{1}{2}$

23. $8\frac{1}{6} + 2\frac{2}{3}$

24. $7\frac{1}{2} + 4\frac{5}{12}$

25. $6\frac{2}{3} + 9\frac{7}{9}$

26. $12\frac{3}{7} + 3\frac{9}{14}$

27. Don bought $6\frac{1}{8}$ pounds of turkey, $2\frac{3}{4}$ pounds of ham, and $5\frac{1}{2}$ pounds of chicken. How much meat did he buy?

Practice PLUS

KEY SKILL: Subtracting Mixed Numbers with Renaming (Use after page 423.)

Level A

Subtract. Write the difference in simplest form.

1. $6\frac{1}{4} - 3\frac{3}{4}$
2. $8\frac{1}{5} - \frac{2}{5}$
3. $7\frac{1}{7} - 3\frac{2}{7}$
4. $8\frac{2}{9} - 2\frac{5}{9}$

5. Last month, Rhonda's sunflower plant was $2\frac{2}{7}$ in. tall. This month, the sunflower measures $5\frac{5}{7}$ in. tall. How much did it grow?

Level B

Subtract. Write the difference in simplest form.

6. $10\frac{4}{7} - 3\frac{6}{7}$
7. $8\frac{1}{9} - 2\frac{2}{3}$
8. $4\frac{1}{10} - 1\frac{3}{5}$
9. $12\frac{3}{8} - 1\frac{1}{4}$

10. $6 - 2\frac{1}{2}$
11. $4 - 1\frac{1}{3}$
12. $6 - 2\frac{2}{5}$
13. $9 - 3\frac{7}{9}$

14. Joe's tomato plant was $8\frac{1}{4}$ in. last week. It is now $10\frac{3}{8}$ in. tall. How much did it grow?

Level C

Subtract. Write the difference in simplest form.

15. $13\frac{5}{6} - 5\frac{1}{12}$
16. $11 - 3\frac{7}{9}$
17. $8\frac{3}{5} - \frac{9}{10}$
18. $10 - 4\frac{6}{7}$

19. $8 - 2\frac{1}{3}$
20. $7\frac{2}{3} - 2\frac{1}{12}$
21. $3\frac{1}{5} - 1\frac{2}{5}$
22. $13 - 8\frac{2}{3}$

23. Mike bought stocks for $13\frac{1}{8}$ a share last week. The stock is now selling for $11\frac{3}{4}$. How much did it fall?

Adding and Subtracting Fractions

Chapter Review

LANGUAGE AND MATHEMATICS
Complete the sentences. Use the words in the chart on the right.

1. $\frac{3}{12}$ and $\frac{5}{12}$ have a ■ of 12. *(page 398)*
2. The fraction $\frac{6}{10}$ written in ■ is $\frac{3}{5}$. *(page 398)*
3. **Write a definition** or give an example of the words you did not use from the chart.

VOCABULARY
simplest form
common denominator
numerator
denominator

CONCEPTS AND SKILLS
Add. Write the sum in simplest form. *(pages 398–401)*

4. $\frac{1}{8} + \frac{3}{8}$
5. $\frac{5}{12} + \frac{2}{3}$
6. $\frac{7}{10} + \frac{4}{5}$
7. $\frac{7}{9} + \frac{2}{9} + \frac{1}{18}$
8. $\frac{2}{3} + \frac{5}{18} + \frac{1}{6}$

9. $\frac{1}{4} + \frac{3}{4}$
10. $\frac{7}{8} + \frac{3}{4}$
11. $\frac{11}{14} + \frac{3}{7}$
12. $\frac{5}{12} + \frac{2}{3}$

13. $\frac{4}{9} + \frac{2}{9}$
14. $\frac{7}{16} + \frac{1}{4}$
15. $\frac{3}{8} + \frac{1}{4} + \frac{3}{16}$
16. $\frac{2}{6} + \frac{2}{3} + \frac{1}{3}$

17. $\frac{5}{10} + \frac{8}{10}$
18. $\frac{7}{9} + \frac{2}{3} + \frac{1}{18}$
19. $\frac{5}{12} + \frac{1}{6} + \frac{2}{3}$
20. $\frac{1}{4} + \frac{1}{2} + \frac{7}{8}$

Subtract. Write the difference in simplest form. *(page 404)*

21. $\frac{7}{8} - \frac{3}{8}$
22. $\frac{13}{16} - \frac{7}{16}$
23. $\frac{7}{16} - \frac{3}{8}$
24. $\frac{3}{4} - \frac{5}{16}$
25. $\frac{3}{10} - \frac{1}{5}$

26. $\frac{5}{12} - \frac{3}{12}$
27. $\frac{3}{14} - \frac{1}{7}$
28. $\frac{7}{8} - \frac{1}{2}$
29. $\frac{5}{6} - \frac{2}{3}$

30. $\frac{6}{7} - \frac{2}{7}$
31. $\frac{9}{16} - \frac{3}{8}$
32. $\frac{10}{20} - \frac{2}{5}$
33. $\frac{5}{6} - \frac{3}{4}$

34. $\frac{10}{11} - \frac{4}{11}$
35. $\frac{10}{15} - \frac{3}{5}$
36. $\frac{15}{18} - \frac{5}{9}$
37. $\frac{16}{21} - \frac{2}{3}$

Estimate the sum or difference. Which method did you use? *(page 408)*

38. $2\frac{3}{8} + 1\frac{1}{10}$
39. $1\frac{3}{16} + 3\frac{1}{4}$
40. $3\frac{7}{12} + 2\frac{7}{8}$
41. $6\frac{1}{10} + 2\frac{2}{5}$

42. $4\frac{1}{4} + 4\frac{3}{4}$
43. $2\frac{5}{8} - 1\frac{1}{4}$
44. $5\frac{6}{7} - 2\frac{1}{14}$
45. $5\frac{8}{9} - 4\frac{3}{18}$

Add. Write the sum in simplest form. *(pages 412–415)*

46. $2\frac{7}{9}$
   $+ 1\frac{2}{9}$

47. $8\frac{5}{6}$
   $+ 2\frac{5}{6}$

48. $4\frac{7}{12}$
   $+ 5\frac{2}{3}$

49. $7\frac{7}{10}$
   $+ 4\frac{2}{5}$

50. $1\frac{5}{12}$
   $+ 4\frac{5}{6}$

51. $5\frac{3}{5} + 2\frac{1}{15}$
52. $3\frac{1}{4} + 3\frac{1}{8}$
53. $8\frac{3}{8} + 5\frac{7}{8} + 2\frac{1}{8}$
54. $1\frac{1}{4} + 3\frac{3}{4} + 2\frac{7}{8}$
55. $5\frac{7}{8} + 2\frac{1}{4}$
56. $6\frac{3}{5} + 2\frac{2}{5}$
57. $4\frac{3}{8} + 7\frac{3}{4}$
58. $1\frac{1}{3} + 3\frac{5}{9} + 2\frac{2}{3}$

Subtract. Write the difference in simplest form. *(pages 418, 422)*

59. $9\frac{3}{4}$
   $- 4\frac{1}{8}$

60. $3\frac{5}{12}$
   $- 1\frac{1}{6}$

61. $5\frac{11}{16}$
   $- 2\frac{1}{4}$

62. $7\frac{5}{7}$
   $- 3\frac{5}{28}$

63. 6
   $- 1\frac{7}{10}$

64. $7\frac{3}{8} - 2\frac{3}{4}$
65. $6\frac{1}{12} - 4\frac{1}{3}$
66. $5\frac{7}{8} - 1\frac{1}{4}$
67. $4\frac{1}{9} - 2\frac{2}{27}$
68. $9\frac{1}{2} - 5\frac{4}{5}$
69. $3 - 1\frac{1}{6}$
70. $6\frac{5}{12} - 2\frac{5}{6}$
71. $7\frac{2}{5} - 2\frac{3}{5}$

Find the missing number. *(page 420)*

72. $6\frac{2}{5} = 4\frac{\blacksquare}{5}$
73. $12\frac{4}{9} = \blacksquare\frac{13}{9}$
74. $9\frac{1}{6} = 6\frac{19}{\blacksquare}$
75. $16\frac{5}{8} = \blacksquare\frac{21}{8}$

CRITICAL THINKING

76. When renaming a mixed number, what happens to the numerator of the fraction? What happens to the denominator?

MIXED APPLICATIONS

77. Jack baled hay for three days. The first day he worked $6\frac{1}{2}$ hours. On each of the next two days, he worked $4\frac{3}{4}$ hours. How many hours did he work in all? *(page 412)*

78. Sergio bought fresh seafood. He bought $7\frac{3}{4}$ pounds of oysters and $2\frac{5}{8}$ pounds of lobster. Did he buy more or less than 10 pounds of seafood? How do you know? *(page 402)*

79. To make costumes for the school play, the drama club needed $33\frac{1}{8}$ yards of brown cotton, $26\frac{1}{4}$ yards of white cotton, and $16\frac{3}{8}$ yards of yellow silk. How many yards of cloth did they need in all? *(page 416)*

Adding and Subtracting Fractions

Chapter Test

Add. Write the sum in simplest form.

1. $\dfrac{5}{12} + \dfrac{3}{12}$

2. $\dfrac{2}{7} + \dfrac{5}{7}$

3. $\dfrac{3}{8} + \dfrac{1}{4}$

4. $\dfrac{2}{3} + \dfrac{5}{6}$

Subtract. Write the difference in simplest form.

5. $\dfrac{7}{9} - \dfrac{2}{9}$

6. $\dfrac{5}{12} - \dfrac{1}{4}$

7. $\dfrac{7}{10} - \dfrac{2}{5}$

8. $\dfrac{9}{11} - \dfrac{6}{11}$

Estimate the sum or difference by rounding.

9. $1\dfrac{1}{3} + 2\dfrac{1}{6}$

10. $2\dfrac{3}{4} + 3\dfrac{1}{8}$

11. $3\dfrac{7}{20} - 1\dfrac{2}{5}$

12. $4\dfrac{7}{9} - 2\dfrac{1}{3}$

Add. Write the sum in simplest form.

13. $5\dfrac{3}{10} + 2\dfrac{1}{10}$

14. $6\dfrac{2}{5} + 2\dfrac{3}{10}$

15. $3\dfrac{5}{7} + 2\dfrac{3}{7}$

16. $4\dfrac{9}{10} + 5\dfrac{2}{5}$

Subtract. Write the difference in simplest form.

17. $5\dfrac{7}{12} - 2\dfrac{1}{12}$

18. $4\dfrac{5}{6} - 2\dfrac{1}{2}$

19. $8 - 4\dfrac{5}{6}$

20. $12\dfrac{1}{8} - 4\dfrac{3}{4}$

Solve.

21. David worked for $6\dfrac{1}{2}$ hours on Friday, $6\dfrac{7}{8}$ hours on Saturday and $7\dfrac{3}{4}$ hours on Sunday. Did he work more or less than 20 hours?

22. Rob bought a $60\dfrac{1}{2}$ lb bag of sand and a $75\dfrac{3}{4}$ lb bag of cement. How much sand and cement did he buy in all?

23. Karen spent $\dfrac{3}{4}$ hour on math homework, $\dfrac{1}{2}$ hour on social studies, and $\dfrac{1}{2}$ hour on reading homework. Did she spend more or less than 2 hours on homework?

24. Mark bagged $112\dfrac{5}{16}$ lb of groceries in the first hour. He bagged $258\dfrac{7}{8}$ lb of groceries the next hour. How many pounds of groceries did he bag in the two hours?

25. A geologist collected $129\dfrac{1}{8}$ lb of rocks. Her two companions collected $62\dfrac{3}{4}$ lb and $57\dfrac{1}{2}$ lb. The geologists' equipment can hold 250 lb of rocks. Did they collect more or less than their equipment can hold?

Enrichment For All

PREDICTING FROM A SAMPLE

Sure Stop Market asked 50 shoppers to name the kind of milk they buy. The results of this poll, or **sample**, can be used to **predict** daily sales. The results are shown in the table.

Kind of Milk	Number of Shoppers in Sample	Fraction of Sample
Whole milk	20	$\frac{20}{50}$ or $\frac{2}{5}$
Skim milk	5	■/■ or ■/■
1% low-fat milk	10	■/■ or ■/■
2% low-fat milk	15	■/■ or ■/■

1. Which kind of milk was chosen by the greatest number of shoppers? by the least number of shoppers?

2. Write the fractions that complete the table.

3. What is the sum of the fractions? Why?

4. **What if** 200 shoppers bought milk at Sure Stop Market in one day? How many times greater than the sample would be the total number of shoppers who bought milk?

5. Complete the table to predict how many containers of each kind of milk would be sold to 200 shoppers.

Kind of Milk	Number of Shoppers in Sample	Estimated Containers Sold to 200 Shoppers
Whole milk	20	80
Skim milk	5	■
1% low-fat milk	10	■
2% low-fat milk	15	■

6. **What if** 350 shoppers bought milk at Sure Stop Market in one day? Predict the number of containers of each kind of milk that would be sold that day.

Adding and Subtracting Fractions **437**

Cumulative Review

Choose the letter of the correct answer.

1. Identify the figure.

   a. $\vec{EB}$ c. $\overline{BE}$
   b. $\overleftrightarrow{BE}$ d. not given

2. Estimate by rounding.
   $$4.753 \times 8$$
   a. 28   c. 32
   b. 40   d. not given

3. $1.606 \div 20$
   a. 0.803   c. 0.008
   b. 0.0803   d. not given

4. Round $25\frac{4}{15}$ to the nearest whole number.
   a. 24   c. 26
   b. 25   d. not given

5. Which is the decimal for $\frac{43}{100}$?
   a. 0.4   c. 43.0
   b. 0.43   d. not given

6. $\frac{5}{12} + \frac{4}{12}$
   a. $\frac{10}{12}$ c. $\frac{3}{4}$
   b. $\frac{1}{12}$ d. not given

7. $\frac{1}{2} - \frac{3}{10}$
   a. $\frac{3}{10}$ c. $\frac{1}{4}$
   b. $\frac{1}{5}$ d. not given

8. Beth needs an enlargement of a family photo. Is the enlargement congruent to the original photo?
   a. yes   c. depends on size
   b. no   d. not given

9. $2\frac{1}{5} + 4\frac{7}{15}$
   a. $7\frac{1}{3}$ c. $6\frac{2}{3}$
   b. $7\frac{2}{9}$ d. not given

10. Estimate by rounding: $1\frac{1}{4} + 3\frac{2}{5}$
    a. 4   c. 7
    b. 5   d. not given

11. Subtract. What is the difference in simplest form?
    $$4\frac{7}{10} - 2\frac{3}{10}$$
    a. $2\frac{4}{10}$ c. $2\frac{2}{5}$
    b. $2\frac{1}{2}$ d. not given

12. Find the missing number.
    $$3\frac{1}{4} = 2\frac{\blacksquare}{4}$$
    a. 1   c. 5
    b. 3   d. not given

13. Carlos's sunflower plant was $3\frac{5}{8}$ in. tall on Saturday. Two weeks later it measured $5\frac{1}{4}$ in. How much did it grow?
    a. $1\frac{5}{8}$ in.   c. $2\frac{5}{8}$ in.
    b. $2\frac{1}{2}$ in.   d. not given

14. Identify the angle.
    a. acute   c. right
    b. obtuse   d. given

Multiplying and Dividing Fractions

CHAPTER 11

MATH CONNECTIONS: MEASURING CAPACITY AND WEIGHT
• AREA • PROBLEM SOLVING

1. What information do you see in this picture?
2. What does the information tell you?
3. How can you use this information?
4. Write a problem using the information.

MANIPULATIVES
EXPLORING A CONCEPT
Multiplying Fractions

Mr. Jackson plowed $\frac{1}{2}$ of his field. He then planted corn in $\frac{1}{4}$ of the part that he plowed. What part of the field did he plant with corn?

WORKING TOGETHER

Explore the problem by making a model. Then answer the questions below.

1. How does your model show the whole field?
2. How does your model show that $\frac{1}{2}$ of the field is plowed?
3. How does your model show that $\frac{1}{4}$ of the plowed field is planted with corn?
4. What part, or fraction, of the whole field is planted with corn?

Here is one way to solve the problem using a paper-folding model.

Step 1 Let a rectangular piece of paper represent the whole field. Fold the paper in half horizontally. Color one of the halves yellow.

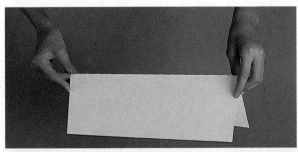

5. What part, or fraction, of the whole field does the yellow section represent?

Step 2 Now fold the paper in fourths vertically. Color red one of the fourths in the yellow part.

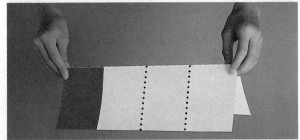

Step 3 Now unfold the paper.

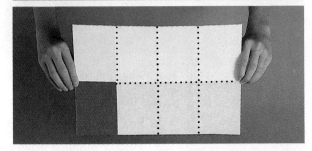

6. What part, or fraction, of the whole field does the orange section represent?

Lesson 11-1

SHARING IDEAS

7. Did you use a paper-folding model to solve the problem? If not, why did you use a different model?

8. Compare your model and results to those of others. How are they different? How are they the same?

ON YOUR OWN

Solve using a model. Explain how you solved the problem.

9. Mr. Jackson asked Jeff to paint $\frac{1}{2}$ of the barn door before lunch. By ten o'clock Jeff had finished $\frac{2}{3}$ of the job. What part of the barn door had Jeff painted?

10. Mrs. Jackson walked $\frac{7}{8}$ mile to the farm stand. The path she used was paved only $\frac{1}{2}$ of the way to the stand. For what part of a mile was the path paved?

11. It took Mrs. Jackson $\frac{3}{4}$ hour to fix the tractor. She spent $\frac{1}{2}$ of that time gathering her tools. For what part of an hour did Mrs. Jackson gather her tools?

12. Jeff picked 8 flowers, of which 4 were roses. Only $\frac{1}{4}$ of the roses had long stems. What fraction of the flowers were long-stemmed roses?

Multiplying and Dividing Fractions

DEVELOPING A CONCEPT

More Multiplying Fractions

The Magic Carpet Store lets you design your own carpet. Fred decided that $\frac{2}{3}$ of his carpet would be yellow and that $\frac{4}{5}$ of the yellow part would have green stripes. What part of Fred's carpet would be yellow with green stripes?

WORKING TOGETHER

You can use diagrams to solve the problem.

a.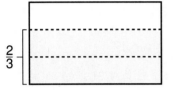

b.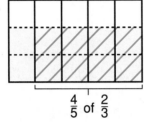

1. What fraction of the carpet in diagram a. is yellow?

2. What fraction of the yellow part in diagram b. has green stripes?

3. What fraction of the whole carpet in diagram b. is yellow with green stripes?

4. Record your answers in a table like the one below by completing the multiplication sentence.

fraction of yellow part with green stripes	of (×)	fraction of carpet that is yellow	is (=)	fraction of carpet that is yellow with green stripes
$\frac{4}{5}$	×	■	=	■

Write a multiplication sentence in your table for the pair of diagrams.

5.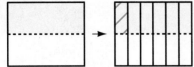

6.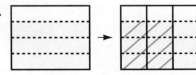

7. Draw your own pairs of diagrams. Write a multiplication sentence in your table for each.

SHARING IDEAS

Look at the multiplication sentences in your table.

8. What do you notice about the numerators in each sentence?

9. What do you notice about the denominators in each sentence?

10. Use what you have noticed to write a rule for multiplying fractions. Compare your rule to those of others.

11. Are there any products that you can simplify? Write those products in simplest form.

PRACTICE

Use your rule for multiplying fractions to find the product. Draw a diagram to check your answers in Exercises 12–16.

12. $\frac{1}{2} \times \frac{1}{5}$
13. $\frac{1}{3} \times \frac{1}{6}$
14. $\frac{3}{8} \times \frac{1}{2}$
15. $\frac{1}{6} \times \frac{2}{3}$
16. $\frac{2}{5} \times \frac{1}{4}$

17. $\frac{5}{6} \times \frac{4}{5}$
18. $\frac{5}{8} \times \frac{2}{3}$
19. $\frac{3}{4} \times \frac{5}{6}$
20. $\frac{3}{4} \times \frac{3}{4}$
21. $\frac{1}{3} \times \frac{1}{3}$

22. $\frac{4}{5} \times \frac{3}{5}$
23. $\frac{3}{5} \times \frac{2}{3}$
24. $\frac{3}{4} \times \frac{3}{5}$
25. $\frac{5}{10} \times \frac{3}{10}$
26. $\frac{9}{10} \times \frac{7}{10}$

27. $\frac{1}{4} \times \frac{1}{3}$
28. $\frac{3}{5} \times \frac{2}{8}$
29. $\frac{5}{6} \times \frac{1}{2}$
30. $\frac{4}{9} \times \frac{2}{3}$
31. $\frac{2}{3} \times \frac{5}{6}$

Mixed Applications

32. Maura is covering $\frac{3}{4}$ of her bedroom with wallpaper. She wants $\frac{1}{3}$ of the wallpaper to be blue. What part of Maura's bedroom will be covered with blue wallpaper?

33. Steven used $\frac{4}{5}$ of a can of paint to paint a storage shed and a doghouse. He used $\frac{3}{4}$ of this amount to paint the storage shed. What part of the can of paint did he use to paint the storage shed?

34. Rickie removed $\frac{3}{8}$ of the tiles from the kitchen floor. What part of the tiles does Rickie have left to remove?

35. **Write a problem** that can be solved by multiplying fractions. Ask others to solve it.

Mike has less than 50 tiles. When he arranges the tiles into rows of 2, 3, 5, or 6, there is always one tile left over. How many tiles does Mike have?

UNDERSTANDING A CONCEPT
Multiplying Fractions and Whole Numbers

A. Blue and white tiles cover $\frac{7}{10}$ of the kitchen floor. Of these tiles, $\frac{2}{3}$ are blue. What part of the floor is covered with blue tiles? You can multiply to solve the problem.

Multiply: $\frac{2}{3} \times \frac{7}{10}$

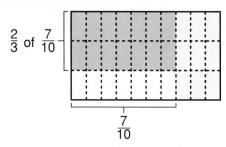

Step 1	Step 2	Step 3
Multiply the numerators.	Multiply the denominators.	Write the answer in simplest form.
$\frac{2}{3} \times \frac{7}{10} = \frac{14}{}$	$\frac{2}{3} \times \frac{7}{10} = \frac{14}{30}$	$\frac{14 \div 2}{30 \div 2} = \frac{7}{15}$

So $\frac{7}{15}$ of the floor is covered with blue tiles.

1. What is the product of $\frac{3}{4} \times \frac{1}{6}$?

B. You can multiply fractions and whole numbers similarly.

Multiply: $\frac{3}{4} \times 6$

Step 1	Step 2	Step 3
Rename the whole number as a fraction.	Multiply.	Write the answer in simplest form.
$\frac{3}{4} \times 6 = \frac{3}{4} \times \frac{6}{1}$	$\frac{3}{4} \times \frac{6}{1} = \frac{18}{4}$	$\frac{18}{4} = 4\frac{2}{4} = 4\frac{1}{2}$

Think: $6 = \frac{6}{1}$

2. What is the product of $6 \times \frac{3}{4}$? How do you know?

TRY OUT Write the letter of the correct answer. Find the answer in simplest form.

3. $\frac{2}{5} \times \frac{2}{5}$ **a.** $\frac{4}{5}$ **b.** $\frac{4}{25}$ **c.** $\frac{2}{5}$ **d.** $\frac{2}{25}$

4. $9 \times \frac{2}{3}$ **a.** $\frac{2}{3}$ **b.** $\frac{18}{27}$ **c.** 6   **d.** 18

PRACTICE

Multiply. Write the answer in simplest form.

5. $\frac{1}{9} \times \frac{1}{7}$
6. $\frac{1}{2} \times \frac{1}{2}$
7. $\frac{1}{3} \times \frac{1}{5}$
8. $\frac{4}{7} \times \frac{1}{3}$
9. $\frac{1}{9} \times \frac{3}{8}$

10. $\frac{5}{9} \times \frac{2}{3}$
11. $\frac{3}{10} \times \frac{7}{10}$
12. $\frac{2}{9} \times \frac{2}{9}$
13. $\frac{3}{7} \times \frac{2}{9}$
14. $\frac{9}{10} \times \frac{3}{100}$

15. $\frac{1}{3} \times 5$
16. $\frac{1}{8} \times 4$
17. $6 \times \frac{3}{8}$
18. $7 \times \frac{3}{9}$
19. $8 \times \frac{3}{4}$

20. $\frac{1}{4} \times 15$
21. $\frac{4}{5} \times 6$
22. $\frac{1}{32} \times 32$
23. $\frac{1}{6} \times 48$
24. $\frac{3}{10} \times 10$

25. $\frac{1}{4} \times \frac{1}{6} \times \frac{1}{2}$
26. $\frac{4}{5} \times \frac{2}{3} \times \frac{7}{8}$
27. $\frac{3}{4} \times \frac{2}{9} \times 7$
28. $\frac{7}{10} \times 10 \times \frac{5}{6}$

Critical Thinking Tell if the statement is *always true, sometimes true,* or *never true.*

29. When you multiply two proper fractions, the product is less than either fraction.

30. When you multiply a whole number and a proper fraction, the product is a whole number.

Mixed Applications

31. A mosaic design has red, yellow, and orange tiles. There are 40 tiles, of which $\frac{3}{4}$ are red. How many tiles are red?

32. Mr. Wong has $\frac{2}{3}$ of a box of floor tiles. Of the tiles in the box, $\frac{1}{4}$ are white. What part of the box contains white tiles?

33. Mrs. Jones used $\frac{1}{8}$ of a can of tile adhesive to tile the bathroom floor. She used $\frac{1}{4}$ of a can to tile the kitchen floor. How much adhesive did she use to tile the two rooms?

34. **Write a problem** involving multiplying two fractions or multiplying a fraction and a whole number. Ask others to solve it.

Mixed Review

Find the answer. Which method did you use?

MENTAL MATH
CALCULATOR
PAPER/PENCIL

35. $\frac{7}{8} + \frac{3}{4}$
36. 5.1×0.36
37. $1,004 - 498$

38. $8,069 \div 16$
39. $340 \div 10$
40. $4\frac{1}{3} - 2\frac{5}{6}$

EXPLORING A CONCEPT
Division: Whole Numbers by Fractions

The students in Dale's social studies class are making a time line. They have a strip of oaktag 6 yd long. Each section of the time line will be $\frac{1}{4}$ yd long. How many sections will there be?

WORKING TOGETHER

Explore the problem by making a model. Then answer the questions below.

1. How does your model show the 6-yd strip?
2. How does your model show the $\frac{1}{4}$-yd sections?
3. How many $\frac{1}{4}$-yd sections do you have?

Ariane solves the problem this way. She lets 4 squares of graph paper represent each yard.

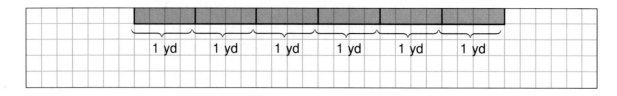

4. What fraction of a yard does each square represent?
5. How many $\frac{1}{4}$-yd sections are there?
6. **What if** Dale's class started with a strip 8 yd long with $\frac{1}{2}$-yd sections? How many sections would there be? Make a model to find out.

SHARING IDEAS

7. Did you use a graph-paper model to solve the problems? If not, why did you use a different model?

8. Compare your models and results to those of others. How are they the same? How are they different?

ON YOUR OWN

Solve using a model. Explain how you solved the problem.

9. Dale uses $\frac{1}{8}$ of a roll of tape for each section of the time line. How many sections can he tape with 4 rolls of tape?

10. There are 3 bulletin boards in the social studies classroom. Each board is divided into $\frac{1}{6}$ sections. How many $\frac{1}{6}$ sections are there in all?

11. Dale's teacher has 5 yd of string. If each student gets $\frac{1}{4}$ yd, how many students get string?

12. Dale spent 2 hours doing homework last night. If he spent $\frac{1}{2}$ hour on each subject, how many subjects did he spend time on?

PROBLEM SOLVING

Strategy: Using Different Strategies

A. Lori is on the Willowbrook School track team. She practices every day after school to get ready for the championship meet. Lori speed walks $\frac{1}{2}$ mile in 5 minutes. At this rate how far can she walk in 1 hour?

You can often use more than one strategy to solve a problem. One way to solve the problem is by making a table.

1. Copy and complete the table.

Number of miles	$\frac{1}{2}$	1	$1\frac{1}{2}$	■	■	■	■	■	■	■	■	■
Minutes	5	10	■	■	■	■	■	■	■	■	■	■

2. Describe any patterns you used to complete the table.

3. How far can Lori walk in 1 hour?

B. You can use a different strategy to check your answer. Solve the problem as a multistep problem. Look at subproblems.

4. How many 5-minute intervals are there in 1 hour?

5. How far could Lori walk in twelve 5-minute intervals?

6. Are your answers to Problems 3 and 5 the same? If not, what do you need to do next?

7. **What if** Lori had walked $2\frac{1}{4}$ miles at this rate? How long would it have taken her?

448 Lesson 11–5

PRACTICE

Solve. Then use a different strategy to check your answer. Which strategies did you use?

8. Juana rode a total of 36 miles during a 3-day bicycle race. On Sunday she covered 12 miles. On Saturday she rode half the total distance of the race. How far did she ride on the other day?

9. Each member of the track team is given 1 pair of gray and 1 pair of blue running shorts. They are also given 4 different tops: blue, red, gray, and green. How many different outfits can each team member wear?

10. Dana tried out for the swim team. The swimmers who finished first and second in a race made the team. At the end of the race, the swimmers were in these positions:
    Allison was behind Dana.
    Barbara was ahead of Carla.
    Juana was ahead of Allison and Dana.
    Dana and Allison were ahead of Barbara.
    Which two swimmers made the team?

11. Sam practiced his backstroke on Monday. On Tuesday he practiced twice as long as on Monday. On Wednesday he practiced three times as long as on Monday. His total practice time for the three days was 3 hours. How long did Sam practice on Monday?

Solve. Use mental math, estimation, a calculator, or paper and pencil.

12. The hiking club met at the school. The members then hiked $3\frac{1}{2}$ miles south. Next, they hiked 2 miles east and then $3\frac{1}{2}$ miles north. They stopped to rest. How far were they from the school?

13. The tennis club travels by bus to and from tennis meets. Their Friday meet was 25 miles away. Their Saturday meet was 32 miles away. How many miles did the team travel on Saturday?

14. The coach of the football team ordered 240 jerseys for the team. He put them on shelves that each hold 25 jerseys. How many shelves did he completely fill?

15. **Write a problem** that can be solved by using either of two different strategies. Solve the problem. Ask others to solve the problem and name some strategies that could be used.

EXTRA Practice, page 470   Multiplying and Dividing Fractions **449**

FRACTURED FRACTIONS

Using Number Concepts

A. Tear a piece of paper into four squares.

Label the squares: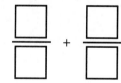

On another sheet of paper, draw four boxes arranged as shown below. The boxes should be about the same size as your number squares.

How can you place the squares into the boxes to get:

1. the greatest possible sum?
2. the least possible sum?
3. a sum of 1?
4. a sum closest to 2?
5. Show all the ways you can arrange the squares to make other whole-number sums.

B. Try a different arrangement. Suppose three boxes form a mixed number like this:

Make a copy of these boxes.
How can you place three of your squares into the boxes to get:

6. the greatest possible mixed number?
7. the least possible mixed number?
8. a mixed number equivalent to 8?

C. In the next arrangement, four boxes form a single fraction like this:

Make a copy of these boxes.
How can you place the squares into the boxes to get:

9. the greatest possible fraction?
10. the least possible fraction?
11. a fraction equivalent to $\frac{1}{4}$?
12. Form some other fraction problems with your squares. Experiment using different squares and different arrangements. Try your problems out on a partner.

DEVELOPING A CONCEPT

Interpreting Circle Graphs

Jim took a movie-preference survey among the students in his class. From that data he learned what part of the class liked a particular kind of movie. He showed the results in a circle graph.

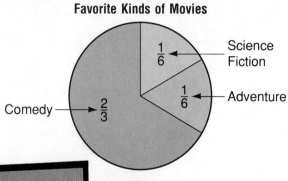

1. Which kind of movie is the most preferred?

2. Which two kinds of movies are preferred by the same number of students?

3. What part of the class does the circle graph represent? How do you know?

4. **What if** there were 24 students in Jim's class? How could you use multiplication to find how many students liked each kind of movie?

5. How many students like comedy? science fiction? adventure?

SHARING IDEAS

6. Why is a circle graph a good way to display Jim's data?

7. What makes the data in a circle graph easy to read?

PRACTICE

The circle graph at the right shows what part of Jim's class likes the kinds of books shown.

8. Which kind of book is the least preferred? the most preferred?

9. Do more students prefer mysteries or adventure books?

10. What part of the class prefers biographies? adventure books? What part of the class do these students make up? How do you know?

11. Find how many students like each kind of book if there are 24 students in Jim's class.

Favorite Kinds of Books

The circle graph at the right shows how Sara spent her weekly allowance.

12. On what item did she spend the greatest part of her allowance?

13. On what three items did Sara spend the same part of her allowance?

14. **What if** Sara's allowance were $10? Find how much money she would spend on each item.

Allowance Spent

The table at the right shows how Josh spent his monthly allowance.

15. Make a circle graph to show the data.

- Begin by tracing the bottom of a can and cutting out the circle that you get.
- Fold the circle in half, then in half again two more times.
- Mark the appropriate fold lines with a pencil and label the sections to represent the data.

16. **What if** Josh's allowance were $40? Find how much money he would spend on each item.

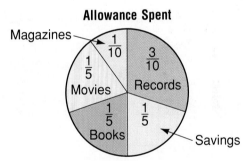

Fraction of Allowance	Items
$\frac{1}{8}$	Savings
$\frac{1}{4}$	Tapes
$\frac{3}{8}$	Magazines
$\frac{1}{4}$	Movies

DEVELOPING A CONCEPT

Area of a Right Triangle

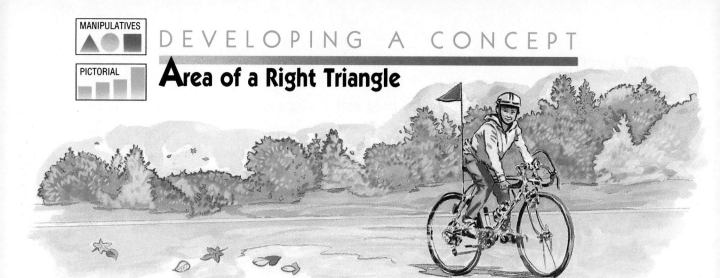

Jake made a flag in the shape of a right triangle to attach to the back of his bicycle. He used half of a rectangular piece of fabric that was 6 in. wide and 9 in. long. What was the area of his flag?

WORKING TOGETHER

Make a graph-paper model to solve the problem.

Step 1 Draw the rectangle on inch graph paper. Then draw a diagonal to form two triangles.

Step 2 Cut out the two triangles and fit them over each other to see if they are congruent.

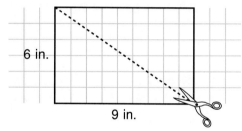

1. What fraction of the rectangle is each triangle? Why?

2. What is the area of the rectangle? How can you use this to find the area of each triangle?

3. What is the area of each triangle?

SHARING IDEAS

4. **What if** the diagonal of the rectangle had been drawn from the other two corners? Would the area of each triangle still be 27 in.²? How do you know?

5. Describe a method for finding the area of a right triangle that is labeled as shown.

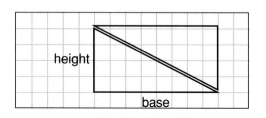

PRACTICE

Use your method to find the area of the shaded triangle.

6.
8 ft, 7 ft

7.
9 ft, 8 ft

8.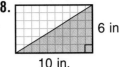
6 in., 10 in.

9.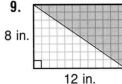
8 in., 12 in.

10.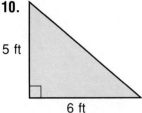
5 ft, 6 ft

11.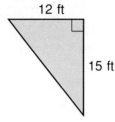
12 ft, 15 ft

12.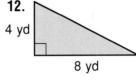
4 yd, 8 yd

13.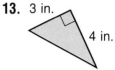
3 in., 4 in.

Mixed Applications

Solve. Which method did you use?

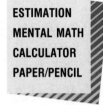

ESTIMATION
MENTAL MATH
CALCULATOR
PAPER/PENCIL

14. Marta cut a rectangular piece of plywood into two right triangles. The plywood was 24 in. long and 12 in. wide. What is the area of each triangle?

15. Dan's garden is a rectangle 18 ft long and 10 ft wide. He divided it in half diagonally and planted flowers in one half and vegetables in the other half. What is the area of Dan's vegetable garden?

16. Jean has a piece of lumber $3\frac{5}{8}$ ft long. From that piece she cuts a piece $1\frac{3}{4}$ ft long to make a shelf. How much lumber does she have left?

17. George has 3 ft² of contact paper. He needs to cut two pieces, each $1\frac{2}{3}$ ft² in area, to line two shelves. Does he have enough contact paper for this project? If not, how much more contact paper does he need?

Mixed Review

Find the answer. Which method did you use?

18. $\frac{1}{4} + \frac{3}{4}$

19. $\frac{7}{8} - \frac{3}{8}$

20. $\frac{2}{3} + \frac{5}{6}$

21. $\frac{11}{12} - \frac{1}{4}$

22. $2 + 2\frac{7}{8}$

23. $5\frac{7}{9} - 2\frac{1}{3}$

24. $6 - 2\frac{4}{5}$

25. $3\frac{1}{2} + 3\frac{1}{4}$

DEVELOPING A CONCEPT
Area of a Parallelogram

The parallelogram below has a base of 10 cm and a height of 5 cm. What is the area of the parallelogram?

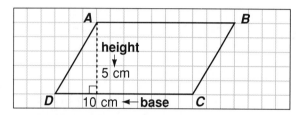

WORKING TOGETHER

You can find the area of the parallelogram by doing the following activity.

Step 1 Use centimeter graph paper to draw the parallelogram described above. Label it as shown. Then cut out parallelogram ABCD.

Step 2 Fold the figure so that point D is on side DC and the fold line goes through point A.

Step 3 Unfold the paper and cut along the fold line. Fit the triangular piece on side BC to form a four-sided figure.

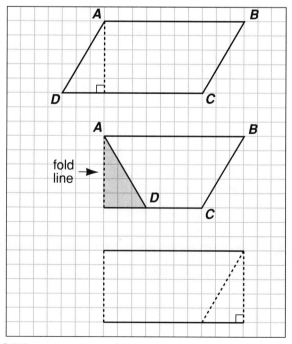

1. What four-sided figure did you form in Step 3?

2. What is the area of the figure you formed? How did you find the area?

3. What is the area of parallelogram ABCD?

456 Lesson 11-9

SHARING IDEAS

4. Describe a method for finding the area of a parallelogram if you know its base and height.

PRACTICE

Use your method to find the area of the parallelogram.

5.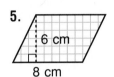
6.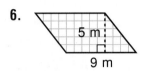
7.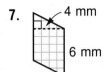
8.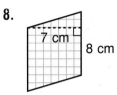

9.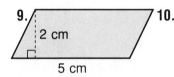
10.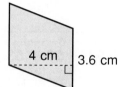
11.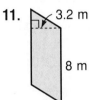
12.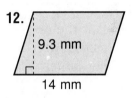

Critical Thinking

13. Use a graph-paper model of a parallelogram to show why the area of any triangle is equal to $\frac{1}{2}$ the area of a parallelogram.

Mixed Applications

Solve. Which method did you use?

14. A sheet of glass is a parallelogram. Its base is 20 cm and its height is 45 cm. What is the area of the sheet of glass?

15. Mike has a rectangular piece of canvas 30 cm long by 45 cm wide. He cut it in half diagonally to make a triangular sail for his model boat. What is the area of the sail?

16. There are 36 members in the model-boat club. Of these, $\frac{3}{4}$ made their own boats. How many members made their own boats?

17. A garden that is a parallelogram has an area that is 18 m². The base of the parallelogram is 6 m. What is its height?

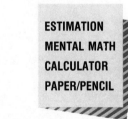

PROBLEM SOLVING

✓ UNDERSTAND
✓ PLAN
✓ TRY
✓ CHECK
✓ EXTEND

Strategy: Solving a Simpler Problem

On each of Rosa's birthdays, her father deposits money in her savings account. The amount he deposits is always equal to her age in years. He deposited $1 when she was one year old, $2 when she was two, $3 when she was three, and so on. How much will her father have deposited after Rosa's twentieth birthday?

Solve a simpler, similar problem to better understand the problem.

Find how much was deposited all together after Rosa's fourth birthday and after her sixth birthday.

- Pair the numbers to make two equal sums. After Rosa's fourth birthday, her father will have deposited $10.

    $1 + $2 + $3 + $4 = 2 × $5, or $10

- Pair the numbers to make three equal sums. After Rosa's sixth birthday, her father will have deposited $21.

    $1 + $2 + $3 + $4 + $5 + $6
    = 3 × $7, or $21

1. How much will have been deposited all together after Rosa's eighth birthday? after Rosa's tenth birthday?

2. Now solve the original problem. How much will Rosa's father have deposited after her twentieth birthday?

3. Use a calculator to check your answer. Add all the whole numbers from 1 through 20. Do you get the same result?

4. **What if** Rosa's father continues making deposits in the same way on every birthday? How much will he have deposited all together after her thirtieth birthday?

Lesson 11-10

PRACTICE

Solve a simpler, similar problem. Then solve the original problem.

5. Fifteen officials were at the Prime Bank's grand opening party. Each official shook hands with each of the others. How many handshakes did they make?

6. From January through December, the Beta Company wrote checks numbered 369–983. How many checks did the company write in this period of time?

7. The bank uses postage stamps that come in sheets of 100. There are 10 rows on each sheet, with 10 stamps in each row. The stamps are separated by perforated tearing lines the length and width of the sheet. How many tearing lines are there?

8. The wall of the bank is covered with wallpaper. The design shows 25 squares that are placed in a row. Squares that are next to one another share a full side. Each side of each square is 1 foot long. What is the perimeter of the design?

Strategies and Skills Review

Solve. Use mental math, estimation, a calculator, or paper and pencil.

9. Marie is a bank teller. She has $7,520.76 in her cash drawer. There are $1 bills, $5 bills, $10 bills, $100 bills, and coins. She has 171 one-dollar bills. She has 56 five-dollar bills and twice that number of ten-dollar bills. The number of $100 bills is $\frac{1}{3}$ the number of $1 bills. How much does Marie have in coins?

10. Each month since she started working, Martha has added $60 to what she had in her savings account. When she started she had $120 in her checking account and $218.15 in her savings account. How much does she now have in her savings account? Solve the problem if possible. If it cannot be solved, tell what else you need to know.

11. Ralph has a checking account and a savings account in the bank. The total amount in the two accounts is $3,162. He has twice as much in the savings account as in the checking account. How much does Ralph have in each account?

12. There are 10 tellers at the bank. The first takes a break at 9:20 A.M., the second takes a break at 9:40 A.M., the third takes a break at 10:00 A.M., and so on. At what times does the tenth teller take a break?

13. Rosa wants to save at least $400 to buy presents. She has $275 in her savings account. If she saves $50 this month, $60 next month, and $35 the month after that, will she reach her goal?

14. **Write a problem** that can be solved by first solving a simpler, similar problem. Solve the problem. Ask other students to solve the problem.

UNDERSTANDING A CONCEPT

Measuring Customary Capacity and Weight

A. Tracy made 1 gallon of fruit punch for a party. She used 1 quart each of orange juice, cranberry juice, apple juice, and club soda. The **cup (c)**, **pint (pt)**, **quart (qt)**, and **gallon (gal)** are customary units of capacity.

about 1 cup   about 1 pint   about 1 quart   about 1 gallon

1. Tracy wanted to be sure she had enough punch for each guest. What would be the most appropriate unit of capacity to use to find out if she has enough? Why?

B. The **ounce (oz)**, **pound (lb)**, and **ton (T)** are customary units of weight.

about 1 ounce   about 1 pound   about 1 ton

2. **What if** you had to find the weight of a box of books you wanted to mail? What customary unit of weight would you use? Why?

3. What unit would you use to measure the weight of an egg?

TRY OUT   Write the letter of the correct answer.
Choose the appropriate customary unit of measure.

4. capacity of a fish tank   **a.** cup   **b.** pint   **c.** quart   **d.** gallon
5. capacity of a glass   **a.** cup   **b.** pint   **c.** quart   **d.** gallon
6. weight of a chair   **a.** ounce   **b.** pound   **c.** ton
7. weight of an apple   **a.** ounce   **b.** pound   **c.** ton

Lesson 11-11

PRACTICE

Which unit of capacity would you use to measure? Write *c, pt, qt,* or *gal.*

8. swimming pool
9. drinking glass
10. gasoline tank
11. juice pitcher
12. bathtub
13. carton of cream

Which unit of weight would you use to measure? Write *oz, lb,* or *T.*

14. slice of cheese
15. bag of apples
16. turkey
17. elephant
18. truck
19. four postcards

Which is the most reasonable estimate of capacity?

		a.	b.	c.	d.
20.	large bowl	4 c	4 pt	4 qt	4 gal
21.	large juice carton	2 c	2 pt	2 qt	2 gal
22.	fish pond	200 c	200 pt	200 qt	200 gal
23.	can of soup	2 c	2 pt	2 qt	2 gal

Which is the most reasonable estimate of weight?

		a.	b.	c.
24.	package of rubber bands	1 oz	1 lb	1 T
25.	bag of flour	5 oz	5 lb	5 T
26.	whale	2 oz	2 lb	2 T
27.	package of hot dogs	14 oz	14 lb	14 T

Mixed Applications

28. Tracy bought a bag of ice cubes to keep the punch cold. Would the bag of ice weigh 10 oz or 10 lb?

29. Eddie made 6 lb of potato salad for the party. If each of the 24 guests got an equal amount, how much did each one get?

30. Mai used $3\frac{1}{2}$ c of orange slices and $2\frac{1}{4}$ c of blueberries to make a fruit salad. How many cups of fruit did she use?

31. **Write a problem** comparing the weights of three record-breaking fruits or vegetables. Use the Databank on page 522. Ask others to solve your problem.

UNDERSTANDING A CONCEPT
Renaming Customary Measures

A. On a camping trip Roy caught a trout that was $\frac{3}{4}$ ft long. He wants to rename the measure in inches.

To rename a larger unit with a smaller unit, you can multiply.

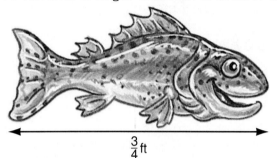

$\frac{3}{4}$ ft = ■ in.

Think: 1 ft = 12 in.

$\frac{3}{4} \times 12 = 9$

$\frac{3}{4}$ ft = 9 in.

1 ft	= 12 in.
1 yd	= 3 ft
1 mi	= 5,280 ft
1 pt	= 2 c
1 qt	= 2 pt
1 gal	= 4 qt
1 lb	= 16 oz
1 T	= 2,000 lb

1. A sailboat is 9 yd long. How many feet long is it? What did you do to rename?

2. Complete. Tell how you renamed each measure.

   a. 2 mi = ■ ft   **b.** $\frac{1}{2}$ gal = ■ qt   **c.** 13 lb = ■ oz

3. Why do you multiply when you are renaming a larger unit with a smaller unit?

B. To rename a smaller unit with a larger unit, you can divide.

51 ft = ■ yd   25 pt = ■ qt   112 oz = ■ lb
Think: 1 yd = 3 ft   **Think:** 1 qt = 2 pt   **Think:** 1 lb = 16 oz
51 ÷ 3 = 17   25 ÷ 2 = $12\frac{1}{2}$ 112 ÷ 16 = 7
51 ft = 17 yd   25 pt = $12\frac{1}{2}$ qt   112 oz = 7 lb

4. Why do you divide when you are renaming a smaller unit with a larger unit?

TRY OUT Write the letter of the correct answer.

5. $\frac{2}{3}$ yd = ■ ft   **a.** $\frac{2}{9}$ **b.** 2   **c.** 3   **d.** 9
6. 31 pt = ■ c   **a.** 124   **b.** 93   **c.** 62   **d.** $15\frac{1}{2}$
7. 80 oz = ■ lb   **a.** 5   **b.** 20   **c.** 320   **d.** 1,080
8. 78 in. = ■ ft   **a.** 936   **b.** 234   **c.** 26   **d.** $6\frac{1}{2}$

PRACTICE

Complete.

9. Feet	Inches
2	■
1	■
■	8

10. Pints	Cups
4	■
$\frac{1}{2}$	■
■	12

11. Quarts	Pints
3	■
■	10
$\frac{1}{2}$	■

12. Pounds	Tons
4,000	■
1,000	■
1,500	■

13. 4 ft = ■ in.   **14.** 9 gal = ■ qt   **15.** $\frac{1}{2}$ lb = ■ oz   **16.** 8 T = ■ lb

17. 4 pt = ■ c   **18.** 10 yd = ■ ft   **19.** 36 oz = ■ lb   **20.** 5 c = ■ pt

21. 5,000 lb = ■ T   **22.** 18 qt = ■ gal   **23.** 88 oz = ■ lb   **24.** 22 pt = ■ qt

25. 96 ■ = 8 ft   **26.** $\frac{1}{2}$ ■ = 1,000 lb   **27.** 48 qt = ■ gal   **28.** 7 ■ = $3\frac{1}{2}$ pt

29. 20 qt = ■ gal   **30.** 54 in. = ■ ft   **31.** 1 mi = ■ yd   **32.** 72 in. = ■ yd

Mixed Applications

33. At the campfire one night, Roy and Mario figured out that the length of 16 pennies placed side by side equals 1 ft. How much would a yard of pennies be worth? a mile of pennies?

34. *Write a problem* involving the length, capacity, or weight of two objects in your classroom. The measurements should be in customary units. Ask others to solve the problem.

Mixed Review

Find the volume of the rectangular prism.

35.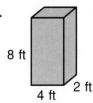

36.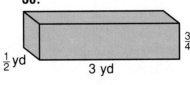

37.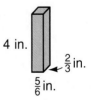

Find the area of the square and rectangles.

38.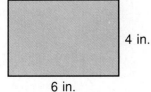

39.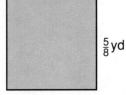

40.

UNDERSTANDING A CONCEPT
Temperature

A. This thermometer measures temperature in **degrees Fahrenheit (°F)**. The thermometer shows a normal indoor temperature of 68°F.

1. At what temperature does water freeze? boil?

If the temperature is 10° below zero, you write ⁻10°F.

If you know the temperature to start and you know how many degrees it rises or falls, you can find the final temperature.

Starting Temperature	Change	Final Temperature	
45°F	rises 12°	57°F	
65°F	falls 21°	44°F	*Think:*
17°F	falls 20°	⁻3°F	from 17° to 0° → 17° from 0° to ⁻3° → + 3° total change 20°

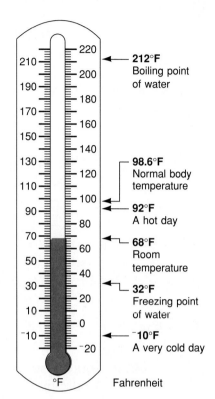

Fahrenheit

B. Temperature can also be measured in **degrees Celsius (°C)**. This thermometer shows a comfortable indoor temperature of 20°C.

2. At what temperature would rain change to snow?

If you know the temperature to start and the final temperature, you can find the amount of temperature change.

Starting Temperature	Final Temperature	Change	
15°C	28°C	rose 13°	
100°C	84°C	fell 16°	*Think:*
11°C	⁻7°C	fell 18°	from 11° to 0° → 11° from 0° to ⁻7° → + 7° total change 18°

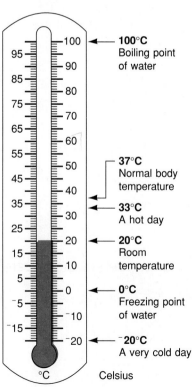

Celsius

TRY OUT   Choose the most likely temperature.

3. green salad   **a.** 18°F   **b.** 45°F   **c.** 90°F
4. your body   **a.** 10°C   **b.** 98°C   **c.** 40°C

Copy and complete the table.

	Temperature to Start	Change	Final Temperature
5.	52°F	rises 14°	■
6.	15°C	falls 21°	■

PRACTICE

Copy and complete the table.

	Temperature to Start	Change	Final Temperature
7.	35°F	rises 17°	■
8.	3°F	falls 26°	■
9.	48°F	■	16°F
10.	−12°C	■	−4°C
11.	■	rises 29°	18°C
12.	■	falls 8°	15°C

Choose the most likely temperature.

13. hot tea   **a.** 212°F   **b.** 120°F   **c.** 55°F
14. good day to ice skate   **a.** −25°F   **b.** 60°F   **c.** 25°F
15. frozen yogurt   **a.** 10°C   **b.** 2°C   **c.** −25°C
16. good day to play baseball   **a.** 5°C   **b.** 21°C   **c.** 40°C
17. good day to swim   **a.** 92°F   **b.** 40°F   **c.** 130°F
18. tuna fish sandwich   **a.** 0°C   **b.** 37°C   **c.** 15°C

Mixed Applications

19. When Shelley left for school at 7:45 A.M., the temperature was 12°F. At 12 noon the thermometer read 28°F. By how many degrees had the temperature changed? How much time had elapsed?

20. Record the outside temperature at the same time each morning and evening for five days. Find the average temperature for each day. Then find the average temperature for the five days.

DECISION MAKING

COOPERATIVE LEARNING

Problem Solving: Planning a Barbecue

SITUATION

Mr. Garcia's class is planning a barbecue for a week from Sunday. They took a survey to find out what people wanted to eat. Of the 24 people surveyed, 12 wanted hamburgers, 8 wanted hot dogs, and 4 wanted chicken. Everyone wanted fruit punch.

PROBLEM

How much food should the class buy, and where should they buy it?

DATA

The average serving of meat per person:

hamburger	1/4 lb
hot dogs	2
chicken	1/2 lb

fruit punch (6 servings):
- 3/4 cup pineapple juice
- 1/2 cup orange juice
- 1/4 cup cranberry juice
- 4 cups water

	Food Palace	Shop 'N Pak*	K&B Market
Hamburger	$2.10 per lb	$3.20 for 4 lb	$1.88 per lb
Hot dogs	$2.60 for 8	$2.89 for 20	$1.89 for 10
Chicken	$2.19 per lb	$3.60 for 4 lb	$1.89 per lb
Orange Juice	$1.35 per qt	$2.20 for ½ gal	$1.10 per qt
Cranberry Juice	$2.25 per qt	$3.60 for ½ gal	$1.25 per qt
Pineapple Juice	$1.75 per qt	$2.50 for ½ gal	$1.25 per qt
Rolls	$.89 for 6	$1.09 per doz	$1.59 per doz

*Shop 'N Pak sells food only in large quantities called "Family-Paks."

Lesson 11–14

USING THE DATA

How much of each item should the class buy?

1. hamburger
2. hot dogs
3. chicken
4. rolls

How much will it cost at each store to buy the required amount of each item?

5. hamburger
6. hot dogs
7. chicken
8. rolls

The fruit-punch recipe serves 6. Each person will have 2 drinks. How much of each ingredient is needed?

9. water
10. pineapple juice
11. orange juice
12. cranberry juice

MAKING DECISIONS

13. Where should the class shop to get the food at the lowest price?

14. **What if** the class wants to have the food delivered? Which supermarket would they choose? Why?

15. **What if** the class wants the food delivered on the day of the barbecue? Which supermarket would they choose? Why?

16. What other factors should the class consider in deciding where to shop?

17. **Write a list** of other things the committee needs to think about to plan the barbecue.

18. At which store would you shop for the barbecue? Why?

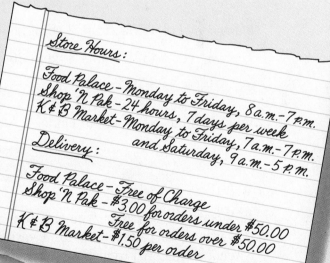

Store Hours:
Food Palace – Monday to Friday, 8 a.m. – 7 p.m.
Shop 'N Pak – 24 hours, 7 days per week
K & B Market – Monday to Friday, 7 a.m. – 7 p.m. and Saturday, 9 a.m. – 5 p.m.

Delivery:
Food Palace – Free of Charge
Shop 'N Pak – $3.00 for orders under $50.00
Free for orders over $50.00
K & B Market – $1.50 per order

Math and the Consumer

Buying the correct amount of fabric is sometimes confusing. Fabric is sold by the length in yards, but it comes in different widths. The widths are given in inches, using the symbol ″ to indicate inches. A pattern will tell you how much you need of one width, but the fabric you want may come in a different width. You need to convert, or change, the length given in one width to the length you would need in another width. Most patterns provide a table such as the one above.

FABRIC CONVERSION

Width	36″	45″	60″
Length (in yards)	$1\frac{3}{4}$	$1\frac{3}{8}$	1
	2	$1\frac{5}{8}$	$1\frac{1}{4}$
	4	$3\frac{7}{8}$	$2\frac{7}{8}$

What if you want to make two items from the same fabric. One item calls for 2 yards of 36″ fabric. The other calls for 4 yards of 36″ fabric. The fabric you want is 45″. How much fabric should you buy?

Think: $1\frac{5}{8}$ yd of 45″ fabric = 2 yd of 36″ fabric

$3\frac{7}{8}$ yd of 45″ fabric = 4 yd of 36″ fabric

$1\frac{5}{8} + 3\frac{7}{8} = 4\frac{12}{8}$, or $5\frac{1}{2}$ yd

You should buy $5\frac{1}{2}$ yd of 45″ fabric.

ACTIVITIES

1. Work with a partner. You find a piece of 60″ fabric measuring $2\frac{1}{2}$ yards. Your pattern calls for 4 yards of 36″ fabric. Do you have enough material to make the pattern? If not, how much more do you need?

2. Work in a group. On a piece of construction paper, make a pattern for a design. Use 2 one-yard pieces of butcher paper, one 36″ wide and one 45″ wide. Experiment with placement of the pattern on the paper to see how many more you can get from the 45″ width.

Computer Graphing: Circle Graph

A circle graph is useful for showing how a whole quantity is divided into parts. The computer makes it very easy for you to draw a circle graph. All you have to do is enter the data; the computer will draw and label the circle graph correctly. For this activity you will use the computer program NAME GRAPH.

To gather data for your circle graph, take a survey of the first and last names of all the members of your class. (Don't forget to include yourself.) On a separate sheet of paper, make a tally showing the number of first names that begin with each letter of the alphabet. Make another similar tally for last names.

AT THE COMPUTER

1. Use your first-name tally. Find the totals for each of these letter groups:

   A–F   G–L   M–R   S–Z

   Enter these totals in the correct columns as shown on the computer screen.

2. Have the computer draw the circle graph. Estimate about what fractional part of the circle each letter group represents. Which group is the largest? Which group is the smallest?

3. Try the activity again using the same letter groups and the last-name tally. Have the computer draw the circle graph.

4. Estimate the size of each fractional part this time. Does the last-name graph look the same as or very different from the first-name graph? Explain.

5. How could you change the letter groupings so that the last-name graph could be divided into four parts about the same size?

6. Summarize your results about first and last names. Compare with a friend in another class. What statements can you make?

EXTRA PRACTICE

More Multiplying Fractions, page 443

Use your rule for multiplying fractions to find the product. Draw a diagram to check your answers in Questions 1–8.

1. $\frac{1}{4} \times \frac{1}{2}$
2. $\frac{1}{3} \times \frac{1}{3}$
3. $\frac{2}{5} \times \frac{1}{3}$
4. $\frac{1}{2} \times \frac{1}{3}$
5. $\frac{2}{7} \times \frac{1}{3}$
6. $\frac{5}{7} \times \frac{1}{4}$
7. $\frac{2}{3} \times \frac{2}{3}$
8. $\frac{4}{5} \times \frac{2}{3}$
9. $\frac{1}{10} \times \frac{3}{10}$
10. $\frac{2}{3} \times \frac{5}{6}$
11. $\frac{1}{2} \times \frac{1}{2}$
12. $\frac{3}{5} \times \frac{3}{5}$
13. $\frac{3}{10} \times \frac{7}{10}$
14. $\frac{1}{5} \times \frac{1}{5}$
15. $\frac{1}{7} \times \frac{2}{7}$
16. $\frac{1}{6} \times \frac{5}{6}$
17. $\frac{4}{10} \times \frac{3}{10}$
18. $\frac{3}{8} \times \frac{1}{4}$
19. $\frac{2}{5} \times \frac{3}{4}$
20. $\frac{3}{5} \times \frac{1}{3}$
21. $\frac{5}{6} \times \frac{4}{5}$
22. $\frac{2}{3} \times \frac{7}{8}$
23. $\frac{5}{8} \times \frac{2}{5}$
24. $\frac{5}{7} \times \frac{2}{3}$

Multiplying Fractions and Whole Numbers, page 445

Multiply. Write the answer in simplest form.

1. $\frac{1}{3} \times \frac{1}{7}$
2. $\frac{1}{4} \times \frac{1}{4}$
3. $\frac{3}{7} \times \frac{2}{7}$
4. $\frac{1}{8} \times \frac{5}{8}$
5. $\frac{7}{9} \times \frac{1}{3}$
6. $5 \times \frac{1}{4}$
7. $\frac{3}{8} \times 6$
8. $\frac{2}{3} \times 9$
9. $\frac{1}{49} \times 49$
10. $\frac{1}{5} \times 16$
11. $\frac{1}{8} \times 32$
12. $\frac{3}{4} \times 16$
13. $\frac{3}{5} \times 10$
14. $\frac{5}{8} \times 40$
15. $28 \times \frac{2}{7}$
16. $10 \times \frac{2}{7}$
17. $\frac{1}{3} \times \frac{1}{2} \times \frac{1}{4}$
18. $\frac{2}{5} \times \frac{1}{5} \times \frac{2}{3}$
19. $\frac{2}{7} \times \frac{1}{2} \times \frac{1}{3}$
20. $\frac{1}{4} \times 10 \times \frac{2}{3}$
21. $3 \times \frac{4}{9} \times \frac{3}{8}$
22. $\frac{1}{4} \times \frac{3}{8} \times \frac{2}{3}$
23. $\frac{1}{5} \times \frac{3}{9} \times 20$
24. $\frac{1}{2} \times 3 \times \frac{2}{5}$

Problem-Solving Strategy: Using Different Strategies, page 449

Solve. Then use another strategy to check your answer. Which strategies did you use?

1. Kevin earned $75 in three days washing cars. On Friday, he earned $15. On Saturday, he earned one-third of his total. How much did he earn on Sunday?

2. Laura works at a local supermarket. She can stack 7 shelves in $\frac{1}{3}$ of an hour. At this rate, how many shelves can she stack in 2 hours?

EXTRA PRACTICE

Interpreting Circle Graphs, page 453

The circle graph at the right shows what kinds of TV shows are preferred by the students in Ellen's class.

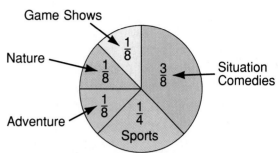

1. Which kind of show is the most preferred?

2. What part of the class prefers game shows?

3. What part of the class prefers shows other than situation comedies?

4. Find how many students like each kind of show if there are 32 students in Ellen's class.

Area of a Right Triangle, page 455

Use your method to find the area of the right triangle.

1. 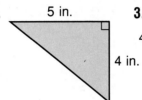 (4 ft, 6 ft)

2. (5 in., 4 in.)

3. 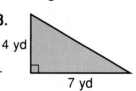 (4 yd, 7 yd)

4. 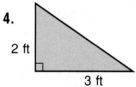 (2 ft, 3 ft)

Area of a Parallelogram, page 457

Use your method to find the area of the parallelogram.

1. 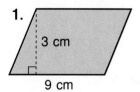 (3 cm, 9 cm)

2. 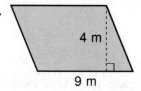 (4 m, 9 m)

3. 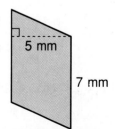 (5 mm, 7 mm)

4. 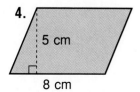 (5 cm, 8 cm)

5. 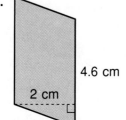 (4.6 cm, 2 cm)

6. 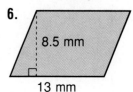 (8.5 mm, 13 mm)

Multiplying and Dividing Fractions

EXTRA PRACTICE

Problem-Solving Strategy: Solving a Simpler Problem, page 459
Solve a simpler, similar problem. Then solve the original problem.

1. Eighteen ambassadors were at the party in honor of the opening of the new embassy. Each person shook hands with each of the others. How many handshakes were made at the party?

2. From July through December, the Alpha Appliance Company wrote checks numbered 256–414. How many checks did the company write in this period of time?

Measuring Customary Capacity and Weight, page 461
Which unit of capacity would you use to measure?
Write *c, pt, qt,* or *gal.*

1. bucket of water
2. pitcher of lemonade
3. can of paint
4. bowl of soup
5. carton of milk
6. bottle of cooking oil

Which unit of weight would you use to measure?
Write *oz, lb,* or *T.*

7. slice of ham
8. bag of potatoes
9. hippopotamus
10. banana
11. newborn baby
12. car

Which is the most reasonable estimate of capacity?

13. large milk carton   **a.** 2 c   **b.** 2 pt   **c.** 2 qt   **d.** 2 gal
14. fishbowl   **a.** 7 c   **b.** 7 pt   **c.** 7 qt   **d.** 7 gal
15. mixing bowl   **a.** 5 c   **b.** 5 pt   **c.** 5 qt   **d.** 5 gal
16. blender   **a.** 2 c   **b.** 2 pt   **c.** 2 qt   **d.** 2 gal

Which is the most reasonable estimate of weight?

17. container of yogurt   **a.** 8 oz   **b.** 8 lb   **c.** 8 T
18. bag of onions   **a.** 3 oz   **b.** 3 lb   **c.** 3 T
19. elephant   **a.** 1 oz   **b.** 1 lb   **c.** 2 T
20. package of meat   **a.** 3 oz   **b.** 3 lb   **c.** 3 T

Extra Practice

Renaming Customary Measures, page 463

Complete.

1.
Gallons	Quarts
2	■
■	1
8	■
$\frac{1}{2}$	■

2.
Feet	Yards
6	■
■	3
2	■
■	$\frac{1}{3}$

3.
Pounds	Ounces
3	■
■	16
6	■
$\frac{1}{4}$	■

4. 2 ft = ■ in.
5. 6 gal = ■ qt
6. $\frac{1}{2}$ c = ■ pt
7. 3 T = ■ lb
8. 2 pt = ■ c
9. 6 yd = ■ ft
10. 72 oz = ■ lb
11. 3 c = ■ pt
12. 3,500 lb = ■ T
13. 16 qt = ■ gal
14. 56 oz = ■ lb
15. 64 in. = ■ ft
16. 6 ■ = $\frac{1}{2}$ ft
17. 16 qt = 4 ■
18. 24 oz = ■ lb

Temperature, page 465

Choose the most likely temperature.

1. hot soup   a. 315°F   b. 110°F   c. 35°F
2. good day to ski   a. ⁻15°F   b. 70°F   c. 32°F
3. iced tea   a. 20°C   b. 4°C   c. ⁻4°C
4. good day for a picnic   a. 23°C   b. ⁻23°C   c. 70°C

Copy and complete the table.

	Temperature to Start	Change	Final Temperature
5.	25°F	rises 15°F	■
6.	2°F	falls 19°F	■
7.	74°F	■	52°F
8.	⁻10°C	■	⁻3°C
9.	■	rises 38°C	28°C
10.	■	falls 6°C	46°C
11.	82°F	falls 14°F	■

Multiplying and Dividing Fractions

Practice PLUS

KEY SKILL: Multiplying Fractions and Whole Numbers (Use after page 445.)

Level A

Multiply. Write the answer in simplest form.

1. $\frac{1}{3} \times \frac{1}{5}$
2. $\frac{1}{3} \times \frac{1}{3}$
3. $\frac{1}{4} \times \frac{3}{5}$
4. $\frac{1}{8} \times \frac{2}{5}$
5. $\frac{3}{8} \times \frac{1}{4}$
6. $\frac{1}{10} \times \frac{9}{10}$
7. $\frac{4}{5} \times \frac{2}{7}$
8. $\frac{1}{10} \times \frac{7}{100}$
9. $\frac{1}{2} \times 5$
10. $\frac{1}{3} \times 3$
11. $6 \times \frac{1}{8}$
12. $8 \times \frac{1}{2}$

13. José spent $\frac{3}{4}$ of an hour studying for a test. Luís studied twice as long. How long did Luís study?

Level B

Multiply. Write the answer in simplest form.

14. $\frac{2}{7} \times \frac{1}{5}$
15. $\frac{3}{8} \times \frac{2}{3}$
16. $\frac{5}{9} \times \frac{1}{4}$
17. $\frac{7}{10} \times \frac{1}{10}$
18. $\frac{2}{5} \times 7$
19. $\frac{2}{9} \times 8$
20. $7 \times \frac{2}{3}$
21. $5 \times \frac{3}{4}$
22. $\frac{1}{3} \times 12$
23. $\frac{7}{9} \times 5$
24. $\frac{1}{40} \times 6$
25. $\frac{1}{7} \times 49$
26. $\frac{1}{10} \times 10$
27. $\frac{3}{7} \times 9$
28. $\frac{1}{15} \times 3$
29. $\frac{8}{9} \times 10$

30. Joanne has a collection of 24 glass frogs. Of these, $\frac{1}{6}$ are brown. How many frogs are brown?

Level C

Multiply. Write the answer in simplest form.

31. $\frac{1}{8} \times \frac{3}{4}$
32. $5 \times \frac{3}{8}$
33. $9 \times \frac{1}{8}$
34. $\frac{3}{4} \times \frac{2}{3}$
35. $\frac{1}{5} \times 15$
36. $\frac{2}{7} \times 42$
37. $\frac{1}{6} \times 36$
38. $\frac{7}{10} \times 10$
39. $\frac{1}{3} \times \frac{1}{6} \times \frac{1}{2}$
40. $\frac{3}{5} \times \frac{2}{5} \times \frac{3}{8}$
41. $\frac{1}{8} \times \frac{1}{2} \times \frac{1}{4}$
42. $\frac{7}{10} \times 10 \times \frac{1}{6}$
43. $\frac{4}{5} \times \frac{1}{10} \times 2$
44. $\frac{5}{6} \times \frac{2}{3} \times 3$
45. $\frac{1}{6} \times \frac{1}{2} \times \frac{5}{6}$
46. $10 \times \frac{3}{7} \times \frac{1}{5}$

47. Kevin has 60 glass bricks. Of these, $\frac{4}{5}$ are frosted. How many bricks are frosted?

Practice Plus

KEY SKILL: Renaming Customary Measures (Use after page 463.)

Level A
Complete.

1. 3 ft = ■ in.
2. 8 gal = ■ qt
3. $\frac{1}{3}$ lb = ■ oz
4. 6 T = ■ lb
5. 3 pt = ■ c
6. 8 yd = ■ ft
7. 3 c = ■ pt
8. 2,000 lb = ■ T
9. 16 qt = ■ gal
10. 22 oz = ■ lb
11. 13 pt = ■ c
12. 12 qt = ■ gal
13. 4 pt = ■ c
14. 4 ft = ■ in.
15. 2 c = ■ pt

16. Eric has 22 oz of pebbles for his fish tank. How many pounds is this?

Level B
Complete.

17. 8 yd = ■ ft
18. 6 c = ■ pt
19. 6 pt = ■ c
20. 18 qt = ■ gal
21. 38 qt = ■ gal
22. 36 in. = ■ yd
23. 16 in. = ■ ft
24. 2 mi = ■ ft
25. 3,500 lb = ■ T
26. 1 c = ■ pt
27. 1 pt = ■ qt
28. 6,000 lb = ■ T

29. There are 31 students in Mr. Reynold's fifth-grade class. Each student is given a 1-pt container of juice on the bus trip to the aquarium. How many quarts is this?

Level C
Complete.

30. $\frac{1}{2}$ qt = ■ pt
31. 9 yd = ■ ft
32. 30 in. = ■ ft
33. ■ oz = $4\frac{1}{2}$ lb
34. 7,000 lb = ■ T
35. 20 pt = ■ c
36. 96 oz = ■ lb
37. $\frac{1}{4}$ ft = ■ in.
38. 10 gal = ■ qt
39. $3\frac{1}{2}$ T = ■ lb
40. $\frac{1}{2}$ ■ = 1,000 lb
41. 128 oz = ■ lb
42. 9 T = ■ lb
43. 5 pt = ■ c
44. $\frac{1}{2}$ qt = ■ c

45. Brandon's mother drives a 4,250-lb sports car. How many tons is this?

Multiplying and Dividing Fractions

Chapter Review

LANGUAGE AND MATHEMATICS

Write the letter from Column B that best describes the items in Column A.

Column A

1. ton (page 460)
2. °C
3. °F (page 464)
4. quart (page 460)
5. area of a parallelogram (page 456)

Column B

A. A = base × height
B. customary unit of capacity
C. customary unit of length
D. degrees Celsius
E. customary unit of weight
F. degrees Fahrenheit

CONCEPTS AND SKILLS

Multiply. Write the answer in simplest form. (pages 442–445)

6. $\frac{1}{3} \times \frac{5}{8}$
7. $\frac{3}{4} \times 7$
8. $\frac{11}{16} \times \frac{1}{2}$
9. $\frac{3}{8} \times \frac{5}{6}$
10. $\frac{1}{8} \times \frac{3}{16}$
11. $\frac{3}{8} \times \frac{1}{2} \times 4$
12. $\frac{8}{9} \times 6 \times \frac{1}{4}$
13. $\frac{1}{3} \times \frac{2}{7} \times \frac{7}{8}$
14. $\frac{1}{3} \times \frac{1}{6} \times \frac{1}{4}$

The circle graph at the right shows which vegetables were preferred by 32 students. (page 452)

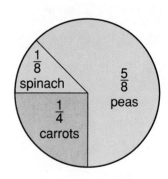

15. Which vegetable is most preferred? least preferred?

16. How many students like each vegetable?

17. What part of the class prefers peas and carrots?

Find the area. (pages 454–457)

18.

19.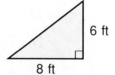

20.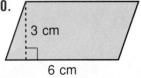

21.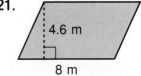

Which is the most reasonable estimate of capacity? *(page 460)*

22. glass of milk   **a.** 2 c   **b.** 2 pt   **c.** 2 qt   **d.** 2 gal
23. can of motor oil   **a.** 1 c   **b.** 1 pt   **c.** 1 qt   **d.** 1 gal
24. a bathtub   **a.** 50 c   **b.** 50 pt   **c.** 50 qt   **d.** 50 gal

Which is the most reasonable estimate of weight? *(page 460)*

25. a Sunday newspaper   **a.** 2 oz   **b.** 2 lb   **c.** 2 T
26. a bus   **a.** 4 oz   **b.** 4 lb   **c.** 4 T
27. a bag of potatoes   **a.** 10 oz   **b.** 10 lb   **c.** 10 T

Complete. *(page 462)*

28. 6 ft = ■ in.
29. $2\frac{1}{2}$ lb = ■ oz
30. 24 qt = ■ gal
31. 7 yd = ■ ft
32. 6 pt = ■ gal
33. 4,500 lb = ■ T
34. ■ in. = 3 yd
35. 8 qt = 2 ■

Choose the most likely temperature. *(page 464)*

36. water boils   **a.** 33°C   **b.** 100°C   **c.** 212°C
37. a nice fall day   **a.** 97°F   **b.** 7°F   **c.** 70°F
38. the sidewalk in summer   **a.** 110°F   **b.** 10°F   **c.** 35°F

CRITICAL THINKING

39. Why is a circle graph a good way to display fractional data?

MIXED APPLICATIONS

40. Dana is covering $\frac{7}{8}$ of her kitchen with wallpaper. She wants $\frac{2}{3}$ of the wallpaper to be yellow. What part of the kitchen will be covered with yellow wallpaper? *(page 442)*

41. Ralph spent 4 hours practicing the piano. If he spent $\frac{1}{2}$ hour on each song that he will play in the recital, how many songs did he practice? *(page 446)*

42. There are 225 dinner rolls for the banquet. An equal number of rolls are placed in 9 baskets. How many dinner rolls are in each basket? *(page 458)*

43. From January through June, Marcy wrote checks numbered 237–396. How many checks did she write in this period of time? *(page 458)*

Multiplying and Dividing Fractions

Chapter Test

Multiply. Write the answer in simplest form.

1. $\frac{1}{3} \times \frac{1}{5}$
2. $\frac{9}{10} \times \frac{7}{10}$
3. $\frac{2}{3} \times \frac{1}{6}$
4. $\frac{1}{4} \times \frac{3}{5}$
5. $\frac{1}{4} \times 6$
6. $8 \times \frac{2}{3}$
7. $\frac{9}{10} \times 10$
8. $6 \times \frac{2}{9}$

Find the area.

9. 15 ft, 18 ft
10. 2 in., 3 in.
11. 6 m, 10 m
12. 6.5 mm, 12 mm

Which is the most reasonable estimate?

13. weight of a bus   **a.** 3 oz   **b.** 3 lb   **c.** 3 T
14. weight of a hamster   **a.** 1 oz   **b.** 1 lb   **c.** 1 T
15. capacity of a juice glass   **a.** 2 c   **b.** 2 qt   **c.** 2 gal

Complete.

16. 60 in. = ■ ft
17. 15 qt = ■ gal
18. 72 oz = ■ lb
19. 3 pt = ■ c

Choose the most likely temperature.

20. hot soup   **a.** 312°F   **b.** 100°F   **c.** 45°F
21. frozen juice pops   **a.** 20°C   **b.** 0°C   **c.** −23°C

Copy and complete the table.

	Temperature to Start	Change	Final Temperature
22.	45°C	falls 17°	■
23.	■	rises 32°	62°F

Solve.

24. Monica read from page 171 through page 219 in her book. How many pages did she read?

25. Mike has a strip of wood that is 36 inches long. He wants to cut the strip into 12 pieces of equal length. How many cuts does he have to make to do this?

ENRICHMENT FOR ALL

AREA OF IRREGULAR-SHAPED REGIONS

You can estimate the area of an irregular-shaped region. Look at the shaded region at the right.

1. How many squares are completely shaded?
2. How many squares are partially shaded?

   Think: Finding $\frac{1}{2}$ of this number is a reasonable way to estimate the area of the partially shaded regions.

3. What is the estimated area of the shaded region?

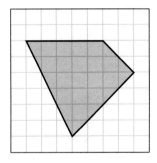

Here is one way to find the area of the shaded region. Divide the shaded region into smaller regions whose areas you can calculate.

4. What is the area of each smaller region?
5. What is the total area of the shaded region? How does this answer compare with your estimate?
6. Trace the shaded region. Show another way to divide it into smaller regions.
7. What is the total area?
8. Why should your answers to Problems 5 and 7 be the same?

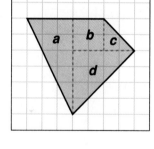

Trace the grid and the shaded region. Then find the area of the shaded region.

9.

10.

11.

12.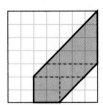

13.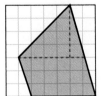

14.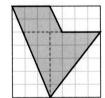

Multiplying and Dividing Fractions

Cumulative Review

Choose the letter of the correct answer.

1. Compare: $2\frac{1}{3}$ ● $3\frac{1}{6}$
   a. <
   b. >
   c. =
   d. not given

2. How many lines of symmetry?

   a. 2
   b. 4
   c. 6
   d. not given

3. What is $\frac{8}{48}$ in simplest form?
   a. $\frac{4}{24}$
   b. $\frac{2}{12}$
   c. $\frac{1}{6}$
   d. not given

4. How many sides and how many angles does a decagon have?
   a. 6
   b. 8
   c. 10
   d. not given

5. $2\frac{3}{8} + 3\frac{3}{4}$
   a. $6\frac{1}{8}$
   b. $7\frac{7}{10}$
   c. $8\frac{1}{5}$
   d. not given

6. $\frac{3}{5} \times 5$
   a. $\frac{3}{25}$
   b. 3
   c. 5
   d. not given

7. Complete the pattern.
   8, 4, 2, 1, 0.5, ■
   a. 0.25
   b. 0.025
   c. 0.75
   d. not given

8. $13\frac{3}{10} - 5\frac{3}{5}$
   a. 8
   b. $7\frac{7}{10}$
   c. $8\frac{1}{5}$
   d. not given

9. $\frac{3}{4} \times \frac{2}{5}$
   a. $\frac{5}{9}$
   b. $\frac{3}{10}$
   c. $\frac{6}{20}$
   d. not given

10. Find the area of the rectangle.

    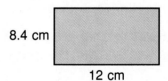

    a. 10.08 cm²
    b. 100.8 cm²
    c. 10.8 cm²
    d. not given

11. Which unit of weight would you use to measure a bag of onions?
    a. oz
    b. lb
    c. T
    d. not given

12. Bob ordered $\frac{1}{2}$ qt of hot-and-spicy soup at the Chinese restaurant. How many 1-c servings did he order?
    a. 2 servings
    b. 3 servings
    c. 1 serving
    d. not given

13. $(5 \times 2) + 3$
    a. 10
    b. 11
    c. 13
    d. not given

Ratio and Probability

CHAPTER 12

MATH CONNECTION: PROBLEM SOLVING

1. What do you see in this picture?
2. What does the picture tell you?
3. Are there more counselors than campers?
4. Write a problem using the information.

UNDERSTANDING A CONCEPT

Ratios

A pet store owner wants to compare the number of cats he sold to the number of dogs he sold.

A **ratio** is used to compare two quantities.

The ratio of cats to dogs sold is 5 to 7. This can be written in three ways.

Write: 5 to 7 $\frac{5}{7}$ 5:7

Read: five to seven or five out of seven

5 and 7 are called the **terms** of the ratio.

Kind of Pet	Number Sold
Cats	5
Dogs	7
Birds	3

1. What is the ratio of birds to dogs sold? of dogs to birds sold?

2. Are the ratios the same? Why or why not?

3. What is the ratio of dogs to all the pets sold as shown in the chart?

TRY OUT   Write the letter of the correct answer. What is the ratio?

4. red fish to yellow fish
   a. 19 to 20   c. 20 to 17
   b. 20 to 19   d. 17 to 20

   20 red fish
   17 blue fish
   19 yellow fish

5. blue fish to yellow fish
   a. $\frac{17}{20}$ c. $\frac{20}{19}$
   b. $\frac{17}{19}$ d. $\frac{19}{17}$

6. yellow fish to all the fish
   a. 19:39   c. 19:56
   b. 19:37   d. 56:17

PRACTICE

Write the ratio in three ways.

7. blue bowls to red bowls
8. red bowls to blue bowls
9. yellow bowls to blue bowls
10. red bowls to all bowls
11. small bowls to large bowls
12. large bowls to all bowls

Write the ratio as a fraction.

13. 8 to 10
14. 13 to 25
15. 20 to 100
16. 1 to 100
17. 7:9
18. 18:36
19. 40:70
20. 100:1

Mixed Applications Solve. You may need to use the Databank on page 522.

21. Which rabbit home has the greater volume, the Deluxe Rabbit Cage or the Peter Rabbit Home? How much greater is the volume of the roomier home?

22. The price of birdseed is $2 a box. The price of fish food is $1 a box. Write a ratio to compare the price of birdseed to the price of fish food.

23. The tropical birds section occupies $\frac{1}{6}$ of the pet store. The tropical fish section occupies another $\frac{1}{3}$ of the store. What part of the store remains for other pets and supplies?

24. Mr. González set aside 1 shelf of fish food for every 5 shelves of fish tanks. What is the ratio of fish food to fish tanks?

Mixed Review

Find the answer. Which method did you use?

MENTAL MATH
CALCULATOR
PAPER/PENCIL

25. $\frac{1}{2} \times \frac{5}{6}$
26. $\frac{2}{3} + \frac{1}{6}$
27. $1\frac{7}{8} - \frac{5}{8}$
28. $7.3 - 4$
29. 10×16.8
30. $32.8 \div 8$
31. $\frac{2}{3} \times 6$
32. $4.92 + 15.785$

EXTRA Practice, page 506

Ratio and Probability 483

UNDERSTANDING A CONCEPT

Equal Ratios

A. Samantha uses 1 egg to make 3 waffles. She uses 2 eggs to make 6 waffles. How many waffles does she make with 3 eggs?

Use the table to find out.

Eggs	1	2	3
Waffles	3	6	■

■ = 9

Samantha uses 3 eggs to make 9 waffles.

The ratio of eggs to waffles is 1:3.
It is also 2:6 and 3:9.
The ratios 1:3, 2:6, and 3:9 are **equal ratios.**

1. Name the equal ratios above using fractions.

2. Are the fractions equivalent? Why or why not?

3. Which of the ratios you named is in simplest form?

B. Equivalent fractions name equal ratios. You can find equal ratios by multiplying or dividing each term by the same nonzero number.

How many waffles does Samantha make if she uses 5 eggs?

eggs → $\frac{1}{3} = \frac{5}{\blacksquare}$ *Think:* $5 = 1 \times 5$ $\frac{1}{3} = \frac{1 \times 5}{3 \times 5} = \frac{5}{15}$
waffles →

Samantha makes 15 waffles.

4. **What if** Samantha makes 21 waffles? How many eggs does she use? How did you find this number?

TRY OUT Write the letter of the correct answer.
Complete to name equal ratios.

5. 5 to 12 = 10 to ■
   a. 7   c. 17
   b. 15   d. 24

6. 28:8 = ■:2
   a. 7   c. 14
   b. 20   d. 28

7. $\frac{3}{12} = \frac{12}{\blacksquare}$
   a. 3   c. 15
   b. 4   d. 48

PRACTICE

Write three equal ratios.

8. $\frac{1}{2} = \frac{\blacksquare}{4} = \frac{\blacksquare}{6} = \frac{\blacksquare}{8}$
9. $\frac{2}{3} = \frac{\blacksquare}{6} = \frac{\blacksquare}{9} = \frac{\blacksquare}{12}$
10. $\frac{3}{5} = \frac{\blacksquare}{10} = \frac{\blacksquare}{15} = \frac{\blacksquare}{20}$
11. $\frac{4}{9} = \frac{\blacksquare}{18} = \frac{\blacksquare}{27} = \frac{\blacksquare}{36}$
12. $\frac{5}{6} = \frac{10}{\blacksquare} = \frac{15}{\blacksquare} = \frac{20}{\blacksquare}$
13. $\frac{3}{4} = \frac{6}{\blacksquare} = \frac{12}{\blacksquare} = \frac{18}{\blacksquare}$

Find the missing number.

14. $\frac{6}{10} = \frac{\blacksquare}{5}$
15. $\frac{3}{8} = \frac{\blacksquare}{24}$
16. $\frac{10}{20} = \frac{1}{\blacksquare}$
17. $\frac{7}{9} = \frac{21}{\blacksquare}$
18. $\frac{4}{16} = \frac{\blacksquare}{4}$
19. $\frac{18}{27} = \frac{2}{\blacksquare}$
20. $\frac{6}{5} = \frac{\blacksquare}{15}$
21. $\frac{24}{8} = \frac{3}{\blacksquare}$
22. $\frac{12}{36} = \frac{\blacksquare}{9}$
23. $\frac{20}{5} = \frac{\blacksquare}{1}$
24. $\frac{15}{12} = \frac{5}{\blacksquare}$
25. $\frac{10}{25} = \frac{20}{\blacksquare}$

Copy and complete the table.

26.

Cups of flour	2	■	6	■	10
Ounces of raisins	3	6	■	12	■

27.

Cans of juice	1	3	■	7	■
Cans of water	3	■	15	■	27

Mixed Applications

28. Ryan's recipe for macaroni and cheese calls for $1\frac{1}{2}$ cups of shredded cheese. If he reserves $\frac{1}{4}$ cup for the topping, how much of the cheese does he actually mix into the ingredients?

29. Megan's recipe for vegetable soup calls for 5 cups of water for every 2 cups of chopped vegetables. What is the ratio of vegetables to water?

CALCULATOR

You can use a calculator to determine whether the ratios $\frac{3}{5}$ and $\frac{18}{32}$ are equal.

3 ÷ 5 = 0.6 $0.6 \neq 0.5625$, so $\frac{3}{5}$ and $\frac{18}{32}$
18 ÷ 32 = 0.5625   are not equal ratios.

Use a calculator to determine if the ratios are equal.

1. $\frac{6}{7}$ and $\frac{72}{84}$
2. $\frac{5}{11}$ and $\frac{57}{121}$
3. $\frac{2}{9}$ and $\frac{35}{157}$
4. $\frac{60}{8}$ and $\frac{90}{12}$

DEVELOPING A CONCEPT

Scale Drawings

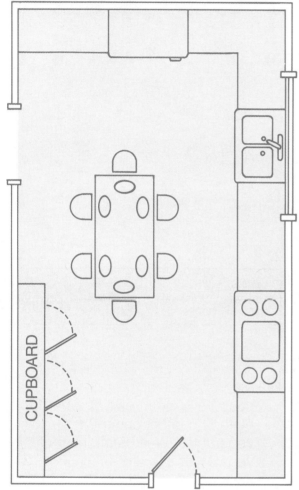

This blueprint of Mr. Jones's kitchen is a **scale drawing.** The kitchen in the scale drawing is the exact same shape as the actual kitchen. Only the size is different.

WORKING TOGETHER

Use measurement and what you know about ratios to learn how to use a scale drawing.

Step 1 Measure the width of the kitchen in the drawing to the nearest inch.

1. Write a ratio to compare the measurement in the scale drawing to the actual width if the actual width is 9 ft.

Step 2 Measure the length of the kitchen in the drawing to the nearest inch.

2. Write a ratio to compare the measurement in the scale drawing to the actual length of 15 ft.

3. Are the ratios you wrote in steps 1 and 2 equal? How do you know?

The **scale** of the drawing is the ratio of the length in the drawing to the actual length.

4. Complete the scale for the blueprint above. 1 in.: ■ ft

5. The actual length of the cupboard is 6 ft. Without measuring, find the length of the cupboard in the drawing. How did you determine the answer? Measure the length of the cupboard to check.

SHARING IDEAS

6. If you know the scale and the length of an object in a scale drawing, how can you find the actual length?

7. If you know the actual length and the scale of a scale drawing, how can you find the length of an object in the drawing?

PRACTICE

Use the scale drawing at the right to find the length and the width of the room.

8. living room
   $\ell =$ ■ $w =$ ■

9. dining room
   $\ell =$ ■ $w =$ ■

10. kitchen
    $\ell =$ ■ $w =$ ■

11. pantry
    $\ell =$ ■ $w =$ ■

12. den
    $\ell =$ ■ $w =$ ■

13. hallway
    $\ell =$ ■ $w =$ ■

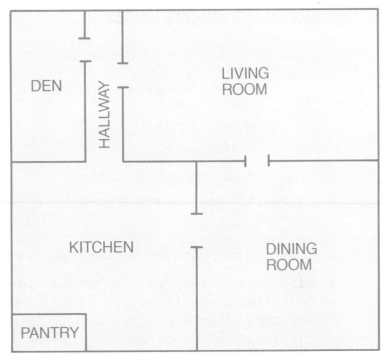

Scale: 1 cm : 2 m

14. Make a scale drawing of a floor of a house that is 18 m long and 15 m wide. Use the scale 1 cm : 3 m. Include the rectangular rooms shown in the chart. Then copy and complete the chart to tell the lengths and widths of the room in your scale drawing.

Room	Actual		Scale	
	Length	Width	Length	Width
Living room	13.5 m	6 m	15. ■	16. ■
Dining room	13.5 m	9 m	17. ■	18. ■
Kitchen	4.5 m	15 m	19. ■	20. ■

Mixed Applications

21. Mrs. Jones paid $5.75 for each place mat and $69 for each chair. How much did she spend on the place mats and chairs?

22. **Write a problem** using the information in the scale drawing at the top of this page. Ask another student to solve the problem.

Problem Solving

✓ UNDERSTAND
✓ PLAN
✓ TRY
✓ CHECK
✓ EXTEND

Strategies Review

Here are some of the strategies you have used to solve problems.

- Guess, Test, and Revise
- Making a Table
- Making an Organized List
- Using a Simpler Problem
- Looking for a Pattern
- Drawing a Diagram
- Working Backward
- Solving a Simpler Problem

Solve the problem. Tell which strategy you used.

1. Keith's hobby is bird-watching. He saw 80 birds in three days. The number he saw on the second day was 5 fewer than the number he saw on the first day. The number he saw on the third day was 15 more than the number he saw on the second day. How many birds did Keith see each day?

2. Jake, Marie, Pedro, and Amy enjoy bird-watching while cross-country skiing. Jake skied 1.5 kilometers. Pedro skied $\frac{1}{3}$ the distance that Marie did. Amy skied 1 less kilometer than Pedro. Jake skied twice the distance that Amy did. What distance did Marie ski?

3. Mr. Katz has $5 in $1 bills and $25 in $5 bills. He is going to give his son an amount of money to spend at the zoo using 3 bills. How many amounts are possible?

4. Gabriella is making an eagle needlepoint wall hanging that is $17\frac{1}{2}$ inches by $17\frac{1}{2}$ inches. She is making a blue border that is 2 inches wide at the top, bottom, and sides. What is the area of the border?

5. Tony plans to go bird-watching every 5 days and sky-watching every 3 days. If he starts by bird-watching on a Saturday and sky-watching on the next day, on what day of the week will he do both things on the same day?

488 Lesson 12-4

6. Jane displayed her ceramic animal collection on 4 shelves. There were 9 animals on the top shelf. The second shelf had $\frac{1}{4}$ the number that were on the bottom shelf. The third shelf had 1 fewer than the number on the second shelf. The top shelf had 3 times the number that were on the third shelf. How many animals were on the bottom shelf?

7. Marianna left her house to go for a walk. She walked 7 blocks east, 2 blocks north, and 5 more blocks east. She bought some cat food. Next she walked 2 blocks south and 3 blocks west. There she bought some magazines. How far was the magazine store from her house?

8. Ferdinand is connecting a set of circus electric trains. He has 4 animal box cars marked Lions, Tigers, Monkeys, and Elephants. In how many different ways can he line up the four cars after the engine?

9. Boris and Bobby were playing a safari game. There were 22 pieces left on the playing board. Boris had 1 more than twice the pieces that Bobby had. How many pieces did each player have on the board?

10. Betty is making a chain leash for her dog. She plans to use 5 blue links for every 3 white links. If she uses 48 links all together, how many are blue and how many are white?

11. Ella has a long piece of knitting yarn she wants to cut into 30 equal lengths for her cats to play with. How many cuts must she make?

12. Kate can buy small clay animals at the zoo in sets of 5 of the same animal. A set of giraffes sells for $2.25 and a set of tigers sells for $1.75. She wants twice as many tigers as giraffes. How many can she get of each if her spending limit is $25?

13. Ken's hobby is making ceramic animals in unusual colors. He is going to make a gorilla, a lion, or an elephant. He can paint his animals any one of 4 different colors. In how many ways can he decide the kind and color of the piece?

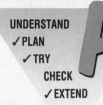

PROBLEM SOLVING

Strategy: Conducting an Experiment

Some problems can be solved by using information given in the problem to draw conclusions and find answers. But not all problems can be solved this way. Sometimes you need to conduct an experiment to get the answers.

Suppose that you have a golf ball, a tennis ball, and a basketball. If you drop these balls from the same height, which ball will bounce the greatest number of times before coming to rest?

You cannot solve a problem such as this by just thinking about it. You have to conduct an experiment.

First plan your strategy. Make a guess about which ball will bounce most often. Then drop each ball 5 times, each time from the same height. Record the number of bounces for each try in a table like this.

Ball	Number of Bounces				
	1st Try	2nd Try	3rd Try	4th Try	5th Try
Golf ball					
Tennis ball					
Basketball					

Find the average number of bounces for each kind of ball. Plan how to show your results. You might add another column to the table. Another way might be to make a bar graph. You decide. Try the experiment and show your results.

1. How does your guess compare with the averages you obtained?

2. To compare your results with the results of the group, what do you need to know?

3. If you had used 6 or 7 different kinds of balls, which would be the most helpful way to show your results, in a table or in a bar graph? Why?

4. Which kind of surface do you think would give more bounces, a hard floor or a carpeted floor?

490 Lesson 12-5

PRACTICE

Conduct an experiment to solve the problem.

5. For this experiment you will need a red cube, a yellow cube, and a paper bag. Put the cubes into the bag. In 20 picks, how many times will you get the red cube? (Both cubes must be in the bag each time you pick.) Make a guess before you start, then conduct the experiment. Record your results in a table.

6. Which word length occurs most frequently? Try this experiment. From the front page of a newspaper, cut out a column that is 10 centimeters long. (Do not include headlines.) Count how many letters long each word is. Which word length occurs most frequently? Make a bar graph to show your results.

7. How long does it take you to write "I like math" 10 times? Try the following experiment. Have another student time you, in seconds, as you write "I like math" 10 times. Do the experiment four times. Then find the average time it takes to write "I like math" 10 times.

8. Which baseball team do most students prefer? Try the following experiment. Make a list of the baseball teams that you think the other students like. Ask 20 students to choose their favorite team from the list. Make a bar graph to show the results of your experiment.

Strategies and Skills Review

Solve. Use mental math, estimation, a calculator, or paper and pencil.

9. Eight students ran in a school track meet. Before the meet started, each student shook hands with each of the others. How many handshakes did they make?

10. For sandwiches for a class trip, Mrs. Clark bought $\frac{2}{3}$ pound of cheese and $\frac{5}{8}$ pound of turkey. Did she buy more than 1 pound of food? How do you know?

11. What color eyes do most students in your class have—blue, brown, black, or another color? Survey 20 other students. Keep track of your results in a table. Make a bar graph to show your results.

12. **Write a problem** that needs to be solved by conducting an experiment. Conduct the experiment to solve the problem. Ask other students to solve the problem. How can you compare the results?

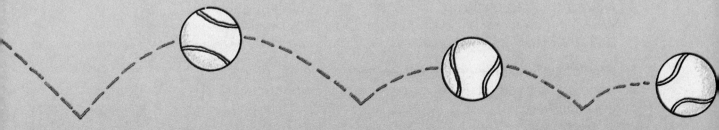

EXPLORING A CONCEPT
Experiments in Probability

Suppose that you pick a cube from this bag without looking. What is the likelihood of you picking a red cube? a white cube? a blue cube?

WORKING TOGETHER

Do this experiment to explore the problem. Use centimeter cubes.

Step 1 Put one blue cube, two white cubes, and three red cubes in a bag.

Step 2 Pick a cube from the bag without looking. Record the result in a frequency table like the one shown. Then return the cube to the bag.

Color	Tally	Frequency
Red		
White		
Blue		

Step 3 Repeat the second step 24 times.

Look at your frequency table.

1. Complete the statement to tell how many times you picked each color cube.
   a. red: ■ times out of 24
   b. white: ■ times out of 24
   c. blue: ■ times out of 24

2. What color cube did you pick most often? least often? Would you expect the same results if you repeated the experiment? Why?

SHARING IDEAS

3. Were the chances of picking each color cube **equally likely** (the same)? Why or why not?

4. ***What if*** the bag contained the cubes shown at the right? Would the chances of picking each color cube be equally likely? Why or why not?

5. How could you change the experiment at the top of the page so that your chances of picking any color cube would be equally likely?

492 Lesson 12-6

ON YOUR OWN

Do one of the probability experiments described below and answer the questions.

Use a spinner like the one shown to do this experiment. Spin the pointer. Record the result in a frequency table. Do this 20 times.

6. Are the chances of spinning each color equally likely? Why or why not?

7. Complete the statement to tell how many times you spin the color.
  a. red: ■ times out of 20
  b. green: ■ times out of 20
  c. blue: ■ times out of 20
  d. yellow: ■ times out of 20

8. Did you spin one color more often than others? If so, which color was it? If you repeated the experiment, would you expect the same results? Why or why not?

Use a number cube numbered from 1 through 6 to do this experiment. Roll the number cube. Record the result in a frequency table. Do this 60 times.

9. Are the chances of rolling each number equally likely? Why or why not?

10. Complete the statement to tell how many times you rolled the number.
  a. 1: ■ times out of 60
  b. 4: ■ times out of 60
  c. 2: ■ times out of 60
  d. 5: ■ times out of 60
  e. 3: ■ times out of 60
  f. 6: ■ times out of 60

11. Answer the questions in Problem 8 about this number cube experiment.

Use five centimeter cubes in any colors you choose to do the experiment on page 492. Do the second step 25 times.

12. Are the chances of picking each color in your bag equally likely? Why or why not?

13. Complete the statement for each color you used: ■ times out of 25.

14. Did you pick one color more often than others? If so, which color was it? If you repeated the experiment, would you expect the same results? Why or why not?

DEVELOPING A CONCEPT

Probability

What is the probability of spinning a number less than 3 on this spinner?

Probability is the likelihood of something happening.
Spinning the spinner is the **experiment.**
The possible results are the **outcomes.**
In this experiment spinning a number less than 3 is the **favorable** outcome.
The probability of a favorable outcome is this ratio:

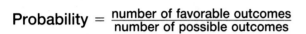

Probability = number of favorable outcomes / number of possible outcomes

1. How many possible outcomes are there for this experiment? How many favorable outcomes are there? What is the probability of spinning a number less than 3?

2. What is the probability of spinning:
   a. a 1?   **b.** a 7?   **c.** a 3?   **d.** a 4?   **e.** a 9?
   f. a number greater than 9?   **g.** a number 9 or less?

WORKING TOGETHER

Use a spinner like the one shown above.

Step 1 Spin the pointer. Record the result of the spin in a frequency table.

Step 2 Repeat the first step 50 times.

3. Complete this statement to tell how many times you spun the number in Problem 2: ■ times out of 50.

4. How do these ratios compare to the probabilities in Problem 2? Are they the same? If not, are they close?

SHARING IDEAS

5. What if you repeated the experiment 100 times? Would you expect the same results? What can you conclude?

6. ***What if*** you used the spinner at the right? Answer Problem 2 for this spinner.

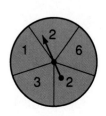

Lesson 12-7

PRACTICE

Suppose that you roll a number cube that is numbered from 1 through 6 on each side. What is the probability of rolling:

7. a 2?   **8.** a number greater than 4?

9. a 5?   **10.** a number less than 5?

11. a 7?   **12.** an even number?

Suppose that you pick a card without looking.

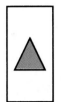    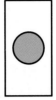   

What is the probability of picking:

13. a square?   **14.** a triangle?   **15.** a circle?

16. not a square?   **17.** not a circle?   **18.** not a triangle?

19. a polygon?   **20.** a space figure?   **21.** a plane figure?

Mixed Applications

22. Angelo bought 5 raffle tickets for the drawing. If there are 100 tickets in the drawing, what is the probability that Angelo will win?

23. Pat lives $5\frac{1}{2}$ miles from the carnival site. Angelo lives $7\frac{1}{8}$ miles from the site. How much closer to the site does Pat live than Angelo?

24. The ratio of adults to children at the carnival was 100 to 250. Write this ratio as a fraction in simplest form.

25. Angelo had $15.73. How much did he have left if he spent $2 on each of 5 raffle tickets?

Mixed Review

Find the area of the triangle or the parallelogram.

26.

3 m, 5 m

27.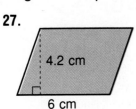

28.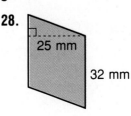

29.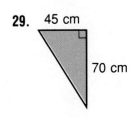

DEVELOPING A CONCEPT

Independent Events

One side of each counter is red, and the other side is yellow. Suppose that you toss both counters at the same time. What is the probability that both counters will land on yellow?

How the second counter lands does not depend on how the first counter lands. The two events are **independent** of each other.

WORKING TOGETHER

Do the experiment above to find the probability of both counters landing yellow. Use a pair of two-color counters.

Step 1 Toss both counters at one time.

Step 2 List each outcome using letters. Write the outcome of the first counter followed by the outcome of the second counter.

Step 3 Do both steps 40 times.

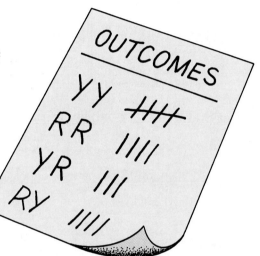

1. How many outcomes are yellow-yellow?
2. What is the probability of getting yellow-yellow in the 40 tosses?
3. What other different outcomes did you get?
4. How many times did you get each outcome?
5. What is the probability of getting each outcome?

Here is another way to find the probability of both counters landing yellow. You can use a **tree diagram** to show all the possible outcomes. The top branches show that if the first counter lands yellow, the second counter can land red or yellow.

First Counter	Second Counter	Outcomes
yellow	red	yellow-red
	yellow	yellow-yellow
red	red	red-red
	yellow	red-yellow

Lesson 12-8

6. What do the bottom branches show?
7. What are all the possible outcomes?
8. What is the probability of getting each outcome?
9. How does this probability compare to your answer to Problem 5?

SHARING IDEAS

10. What is the probability of tossing yellow with the first counter? the second counter?
11. Find the product of the probabilities you got in Problem 10. How does the product compare to the probability of getting yellow-yellow as an outcome? What can you conclude?

PRACTICE

Use the spinners.

12. Make a tree diagram showing all the possible outcomes for two independent spins.
13. How many possible outcomes did you get?

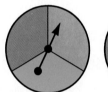

What is the probability:

14. of spinning green-green?
15. of spinning orange-blue?
16. of spinning the same color?
17. that the first spin is blue?
18. that the second spin is green?
19. that the two spins have no orange?

Mixed Applications Solve. You may need to use the Databank on page 522.

20. What is the average price of a rabbit home at Pets Galore?
21. Michelle tossed two coins at the same time. What is the probability that they will land tails-heads?
22. **Write a problem** involving the probability of flipping heads or tails when tossing two coins. Ask others to solve it.

Mixed Review
Find the answer. Which method did you use?

23. $\frac{2}{3} + \frac{2}{3}$
24. 1.3×2.97
25. $0.27 \div 3$
26. $4{,}279 + 810$
27. $2\frac{1}{8} - \frac{3}{4}$
28. $73.8 - 10.92$
29. $\frac{3}{5} \times 5$
30. $27\overline{)3{,}985}$

MENTAL MATH
CALCULATOR
PAPER/PENCIL

SPINNER'S LUCK

Experimenting and Predicting

A. Imagine that you have two weird-looking spinners. One looks like this ①②/③, and one looks like this ②③/①. For each turn, spin both spinners and add the two numbers you get.

1. What are the possible sums you can get?
2. What are the different ways you can make each sum?
3. Which of the sums do you think you will get most often?
4. Which of the sums do you think you will get least often? Why?

Use cardboard to make a pair of spinners like the ones above. (Divide each spinner into two equal parts, then divide one part in half again.)

5. Draw a frequency table like the one at the right. List all the possible sums. Then play 20 times. Spin both spinners and record the sum each time. Use tally marks.

Sum	Tally	Frequency
2		

6. Which sum was most frequent? least frequent?
7. Did the results agree with your guesses in Problems 3 and 4? Why do you think this is so?
8. Suppose you played 1,000 times. What do you think the results would be? Why?

B. Here is another pair of spinners that you can make.

9. If you spin both spinners, what are all the possible color pairs you can get?

EXPERIMENT 1

10. How many times do you think you will have to spin to get two reds? Try it. Keep playing until you spin two reds. How many times did you have to spin?

11. Draw a table like the one below. Put the number of times it took you to spin two reds in the box for Trial #1.

Trial #	1	2	3	4	5	6	7	8	9	10
Number of times it took to spin two reds										

12. Now do nine more trials. Record your results each time. What number appears most often in your table? Compare your results with those of other students.

EXPERIMENT 2

13. How many times do you think it would take you to spin two of the same color? Why?

14. Do ten trials. Record your results in a table like the one below.

Trial #	1	2	3	4	5	6	7	8	9	10
Number of times it took to spin two of the same color										

15. Are your results close to your guess? Compare your results with those of other students.

DEVELOPING A CONCEPT
Predicting Outcomes

Julia has this jar of marbles. Without looking, she picks a marble and then replaces it. How many times will she pick a green marble in 20 picks? How many times will she pick a blue marble? a red marble? a yellow marble?

1. What is the probability of her picking:
   a. a green marble?   b. a red marble?
   c. a blue marble?   d. a yellow marble?

2. How many times do you think you will pick each color in 20 picks?

WORKING TOGETHER

Do this experiment to test your prediction. Use centimeter cubes.

Step 1 Put these cubes in a bag: 3 green cubes, 1 blue cube, 2 red cubes, 4 yellow cubes.

Step 2 Without looking, pick a cube from the bag. Record the result in a frequency table. Then return the cube to the bag.

Step 3 Repeat the second step 20 times.

3. How many times did you pick:
   a. a green cube?
   b. a red cube?
   c. a blue cube?
   d. a yellow cube?

4. How do the results compare to your prediction? Are they the same? If not, are they close?

5. How did you make your predictions? Does your method of predicting outcomes work?

Bob found these results.

20 picks	
green: 5	red: 3
blue: 4	yellow: 8

500 Lesson 12-10

SHARING IDEAS

6. Multiply the probabilities in Problem 1 by 20 picks. How many times do you get:
   a. green marbles?
   b. red marbles?
   c. blue marbles?
   d. yellow marbles?

7. How do Bob's results in 20 picks compare to your answers in Problem 6? Are they the same? If not, are they close?

8. How can you use probability to help you predict outcomes?

9. **What if** the experiment were performed using 10 cubes in red, blue, and green? The results for 50 picks are shown in the table.

   a. How many red cubes do you think are in the bag? blue cubes? green cubes? How did you determine your answers?

   b. There were actually 5 red cubes, 3 blue cubes, and 2 green cubes in the bag. How close was your prediction to the actual numbers?

   c. What would be the probability of picking each color if you were to pick again?

RESULTS OF 50 PICKS

Color	Frequency
Red	23
Blue	16
Green	11

PRACTICE

Predict how many times you will spin the shape in 30 spins.

10. star
11. circle
12. triangle
13. square
14. not a circle
15. not a star

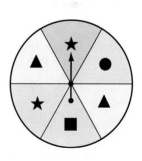

Predict how many times you will pick the color in 30 picks.

16. yellow
17. green
18. pink
19. black
20. not green
21. not yellow

The results of spinning a spinner in three colors 60 times is shown in the table.

22. Predict how many sections of the spinner are in each color if all the sections are the same size.

23. Based on your predictions, what is the probability of spinning red? yellow? blue?

RESULTS OF 60 SPINS

Color	Frequency
Red	28
Yellow	11
Blue	21

EXTRA Practice, page 509; Practice **PLUS**, page 511

DECISION MAKING

COOPERATIVE LEARNING

Problem Solving: Choosing a Summer Camp

SITUATION

Joyce is making plans for the summer. She could sign up for a sleep-away camp for a week. She could also sign up for a day camp at the Baxterville City Park for 1–4 weeks.

PROBLEM

Should Joyce choose a day camp or a sleep-away camp?

DATA

SUMMER CAMPS

	Adventure Camp	Community Day Camp
Where	20 miles west of Baxterville	Baxterville City Park
When	July 6–12, Sunday–Saturday, 1 week only	July 7 – August 1, Mondays–Fridays. May sign up for 1, 2, 3, or 4 weeks
Cost	$75	$5 per week
Activities	Swimming, Astronomy, Rowboats, Campfires, Crafts, Nature Study, Guest entertainment on Thursday, Story-telling contest on Friday night	Swimming, Crafts, Softball, Nature study, "Playground Olympics" on Fridays
Optional Activities	Horseback riding Fee $10	Tennis lessons Fee $2 per day Mon.–Wed.–Fri. equipment provided
Bring to camp	Sleeping bag, flashlight, swimsuit, towels, personal items	Lunch each day, swimsuit, towel
Bus fare	Round-trip $10	35¢ each way or ride bicycle

502 Lesson 12–11

USING DATA

1. How many days and nights is Adventure Camp?

2. How many days is two weeks of Community Day Camp?

3. How much would one week of tennis lessons cost at Community Day Camp?

4. How much would one week of Adventure Camp cost with daily horseback riding?

5. How much would four weeks of Community Day Camp cost with four weeks of tennis lessons?

MAKING DECISIONS

6. **What if** Joyce was on a softball team that practiced two evenings a week and had a game every Saturday? Which camp should she choose? Why?

7. **What if** Joyce was good at sports? Which camp might she enjoy? Why?

8. **What if** her best friend is worried about getting homesick and is going to Community Day Camp? Which camp might Joyce choose? List reasons for your answer.

9. **Write a list** of the things that Joyce should think about when deciding which camp to attend.

10. **What if** Joyce's parents have budgeted $120 for Joyce's summer camp? Can Joyce go to Adventure Camp and take 4 horseback riding sessions?

11. If you could choose, which camp would you want to attend? Why?

Ratio and Probability 503

Curriculum Connection

Math and Health

Normal vision is given as 20/20. The first 20 refers to a standard distance of 20 feet. The second 20 refers to an individual's vision. Having 20/20 vision means that you can read a line of print 20 feet away as well as a person with normal vision. If you have to be at 20 feet to read what a person with normal vision can read at 60 feet, your vision is 20/60. You are nearsighted; you need to be nearer than normal to see things clearly. Glasses or contact lenses can help correct nearsighted vision.

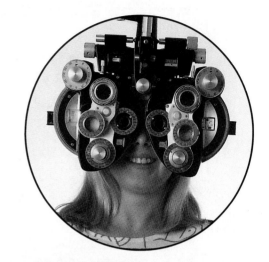

Some people are farsighted. At a distance of 20 feet they can read clearly what a person with 20/20 vision could only read by moving closer. A person with 20/15 vision reads at 20 feet what a person with 20/20 vision can read at a distance of only 15 feet. Glasses or contact lenses can help correct farsighted vision.

What if a person's vision is 20/150? What does this mean?

Think: The person is nearsighted. He or she sees at 20 feet what a person with 20/20 vision can see at 150 feet.

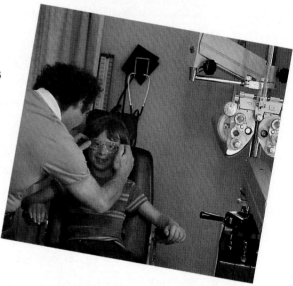

ACTIVITIES

1. Work with a partner. Measure 20 feet. Find something that you can read clearly at 20 feet with both eyes open. Be sure that you cannot read it from farther away than 20 feet. Then try covering one eye to see if you can still read it clearly. If you cannot read it clearly with one eye, have your partner move forward until you can and record your distance from the object. Then check the other eye. Compare your findings with others.

2. Survey students in your class. Count those who would use their right eyes to look through a telescope and those who would use their left eyes. Make a graphic display to show the results.

Computer Simulation: Probability Experiments

Suppose you have a number cube whose six sides are labeled 1, 2, 3, 4, 5, 6. You know that the outcomes should be equally likely, but what does this really mean? If you roll the cube six times, will you get each number exactly once? How many times will you roll each number if you roll the cube 30 times? 300 times? 3,000 times?

The computer program CUBE ROLL simulates this probability experiment. It will "roll" a number cube as many as 3,000 times. It keeps track of how the cube lands and tells you how many times each number occurs. You can use this program to test your predictions and explore probability.

AT THE COMPUTER

1. For the first round, roll the cube six times. Did you get each number once? Did the results match what you expected?

2. How many times do you think each number will appear if you roll the cube 30 times? 300 times? Run the program and compare your predictions with the actual results.

3. When you ran the program, were there any patterns in the way the numbers appeared in the six columns? Explain.

4. **What if** you ran the program several times using 300 rolls each time? Would the actual outcomes be the same in each case? Why or why not? Use the computer to check your ideas.

5. Try a few more runs, including 3,000 rolls at least once. Are the actual results closer to what you would predict for 3,000 rolls than for 30 or 300 rolls?

6. Examine the grand totals. What conclusions can you draw from using this computer experiment?

EXTRA PRACTICE

Ratios, page 483

Write the ratio in three ways.

1. yellow bowls to green bowls
2. green bowls to yellow bowls
3. green bowls to blue bowls
4. green bowls to all the bowls
5. small bowls to large bowls
6. large bowls to all the bowls

Write the ratio as a fraction.

7. 7 to 10
8. 14 to 35
9. 30 to 100
10. 1 : 100
11. 14 : 62
12. 81 : 100

Equal Ratios, page 485

Write three equal ratios.

1. $\frac{1}{3} = \frac{\blacksquare}{6} = \frac{\blacksquare}{9} = \frac{\blacksquare}{12}$
2. $\frac{2}{5} = \frac{\blacksquare}{10} = \frac{\blacksquare}{15} = \frac{\blacksquare}{20}$
3. $\frac{3}{4} = \frac{\blacksquare}{8} = \frac{\blacksquare}{16} = \frac{\blacksquare}{24}$

Find the missing number.

4. $\frac{20}{40} = \frac{1}{\blacksquare}$
5. $\frac{2}{7} = \frac{\blacksquare}{49}$
6. $\frac{4}{5} = \frac{\blacksquare}{10}$
7. $\frac{32}{8} = \frac{4}{\blacksquare}$
8. $\frac{3}{12} = \frac{\blacksquare}{4}$
9. $\frac{10}{35} = \frac{20}{\blacksquare}$
10. $\frac{18}{24} = \frac{\blacksquare}{4}$
11. $\frac{2}{3} = \frac{10}{\blacksquare}$

Scale Drawings, page 487

Use the scale drawing at the right to find the length and the width of the room.

1. bedroom
   $\ell = \blacksquare \; w = \blacksquare$
2. family room
   $\ell = \blacksquare \; w = \blacksquare$
3. living room
   $\ell = \blacksquare \; w = \blacksquare$
4. dining room
   $\ell = \blacksquare \; w = \blacksquare$
5. kitchen
   $\ell = \blacksquare \; w = \blacksquare$
6. hallway
   $\ell = \blacksquare \; w = \blacksquare$

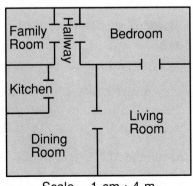

Scale – 1 cm : 4 m

EXTRA PRACTICE

Problem Solving: Strategies Review, page 489

Solve the problem. Tell which strategy you used.

1. Erik left his house to go shopping. He walked 3 blocks west, 6 blocks north, and 5 more blocks west. He bought some juice. Then he walked 4 blocks south and 8 blocks east. There he bought some yogurt at Marcus Dairy. How far north is Marcus Dairy from his house?

2. George displayed his model airplane collection on 3 shelves. There were 12 models on the bottom shelf. The middle shelf had $\frac{1}{5}$ the number that were on the top shelf. The top shelf had 2 fewer than the number on the bottom shelf. How many models were on the middle shelf?

3. Cheryl is 3 times as old as her brother David. When their ages are added together, the sum is 32. What are their ages?

4. Joanne, Pat, and Ellen want to sit together in three seats in a row on an amusement park ride. In how many different ways could they sit in the three seats?

5. Claire is making a bead necklace for her friend. She plans to use 6 green beads for every 2 yellow beads. If she uses 64 beads all together, how many green and how many yellow beads will she use?

6. Kevin and Janet were playing a treasure-hunt board game. There were 18 treasure pieces left on the game board. Kevin had 3 more than twice the number of treasure pieces that Janet had. How many treasure pieces did each player have on the game board?

7. Timothy chose a 386-page autobiography from the library to read for his book report. He has already read 197 pages. If he reads 50 pages a day for the next 3 days, will he finish the book?

Ratio and Probability

Extra Practice

Problem-Solving Strategy: Conducting an Experiment, page 491

Conduct an experiment to solve the problem.

1. How long does it take you to count to 100? Try the following experiment. Have others time you in seconds as you count to 100. Repeat the experiment five times. Then find the average time it takes you to count to 100.

2. Which type of fruit juice do students prefer? Try the following experiment. Ask 20 students to name their favorite juice. Make a list as you go and tally their answers. Make a bar graph to show the results of your experiment.

3. For this experiment you will need a red cube, a green cube, and a yellow cube. Place the cubes in a paper bag. In 30 picks, how many times do you think you will pick the yellow cube? (All three cubes must be in the bag each time you pick.) Make a guess before you start your experiment. Then conduct your experiment. Record your results in a table.

4. Suppose you roll two number cubes each numbered 1 to 6. What sums do you think will come up most often? First list all the sums possible. If you roll the cubes 30 times, how often will each sum occur? Make and record a guess. Then do the experiment. Show your results in a bar graph.

Probability, page 495

Suppose that you roll a number cube whose faces are numbered 1 through 6.

What is the probability of rolling:

1. a 4?
2. a number greater than 2?
3. a 6?
4. a number less than 3?
5. an 8?
6. an odd number?

EXTRA PRACTICE

Independent Events, page 497

Use the spinners.

1. Make a tree diagram showing all the possible outcomes for two independent spins.

2. How many possible outcomes did you get?

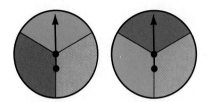

Use your tree diagram. What is the probability:

3. of spinning orange-orange?
4. of spinning green-orange?
5. of spinning the same color?
6. that the first spin will be green?
7. that the second spin will be blue?

Predicting Outcomes, page 501

Predict how many times you will spin the shape in 30 spins.

1. triangle
2. circle
3. square
4. star
5. not a triangle
6. not a square

Predict how many times you will pick the color in 40 picks.

7. green
8. black
9. pink
10. yellow

Ratio and Probability   509

Practice PLUS

KEY SKILL: Equal Ratios (Use after page 485.)

Level A

Write three equal ratios.

1. $\frac{1}{3} = \frac{\blacksquare}{6} = \frac{\blacksquare}{9} = \frac{\blacksquare}{12}$
2. $\frac{1}{2} = \frac{2}{\blacksquare} = \frac{3}{\blacksquare} = \frac{4}{\blacksquare}$
3. $\frac{2}{5} = \frac{\blacksquare}{10} = \frac{\blacksquare}{15} = \frac{\blacksquare}{20}$
4. $\frac{2}{3} = \frac{\blacksquare}{6} = \frac{\blacksquare}{9} = \frac{\blacksquare}{12}$
5. $\frac{1}{5} = \frac{2}{\blacksquare} = \frac{3}{\blacksquare} = \frac{4}{\blacksquare}$
6. $\frac{3}{4} = \frac{\blacksquare}{8} = \frac{\blacksquare}{16} = \frac{\blacksquare}{24}$
7. $\frac{3}{8} = \frac{\blacksquare}{16} = \frac{\blacksquare}{24} = \frac{\blacksquare}{32}$
8. $\frac{4}{5} = \frac{\blacksquare}{10} = \frac{\blacksquare}{15} = \frac{\blacksquare}{20}$
9. $\frac{5}{6} = \frac{10}{\blacksquare} = \frac{15}{\blacksquare} = \frac{20}{\blacksquare}$

Find the missing number.

10. $\frac{1}{2} = \frac{2}{\blacksquare}$
11. $\frac{2}{3} = \frac{\blacksquare}{6}$
12. $\frac{20}{30} = \frac{2}{\blacksquare}$
13. $\frac{5}{15} = \frac{1}{\blacksquare}$

Level B

Write three equal ratios.

14. $\frac{9}{12} = \frac{\blacksquare}{24} = \frac{\blacksquare}{36} = \frac{\blacksquare}{48}$
15. $\frac{12}{16} = \frac{\blacksquare}{48} = \frac{\blacksquare}{64} = \frac{\blacksquare}{80}$
16. $\frac{36}{48} = \frac{\blacksquare}{24} = \frac{\blacksquare}{16} = \frac{\blacksquare}{12}$
17. $\frac{12}{15} = \frac{\blacksquare}{60} = \frac{\blacksquare}{75} = \frac{\blacksquare}{120}$
18. $\frac{48}{60} = \frac{\blacksquare}{20} = \frac{\blacksquare}{30} = \frac{\blacksquare}{15}$
19. $\frac{14}{21} = \frac{\blacksquare}{42} = \frac{\blacksquare}{63} = \frac{\blacksquare}{84}$

Find the missing number.

20. $\frac{3}{16} = \frac{\blacksquare}{32}$
21. $\frac{48}{4} = \frac{\blacksquare}{16}$
22. $\frac{6}{42} = \frac{\blacksquare}{84}$
23. $\frac{24}{8} = \frac{\blacksquare}{16}$
24. $\frac{18}{24} = \frac{\blacksquare}{4}$
25. $\frac{40}{15} = \frac{\blacksquare}{3}$
26. $\frac{36}{9} = \frac{\blacksquare}{18}$
27. $\frac{20}{10} = \frac{\blacksquare}{40}$

Level C

Write three equal ratios.

28. $\frac{2}{7} = \frac{\blacksquare}{14} = \frac{\blacksquare}{28} = \frac{16}{\blacksquare}$
29. $\frac{27}{12} = \frac{\blacksquare}{4} = \frac{\blacksquare}{16} = \frac{\blacksquare}{64}$
30. $\frac{7}{12} = \frac{\blacksquare}{96} = \frac{\blacksquare}{48} = \frac{14}{\blacksquare}$
31. $\frac{14}{21} = \frac{\blacksquare}{63} = \frac{84}{\blacksquare} = \frac{168}{\blacksquare}$
32. $\frac{12}{9} = \frac{\blacksquare}{18} = \frac{36}{\blacksquare} = \frac{\blacksquare}{36}$
33. $\frac{16}{20} = \frac{\blacksquare}{60} = \frac{24}{\blacksquare} = \frac{\blacksquare}{15}$

Find the missing number.

34. $\frac{14}{24} = \frac{\blacksquare}{48}$
35. $\frac{18}{20} = \frac{36}{\blacksquare}$
36. $\frac{\blacksquare}{36} = \frac{9}{12}$
37. $\frac{54}{\blacksquare} = \frac{18}{91}$
38. $\frac{12}{18} = \frac{\blacksquare}{36}$
39. $\frac{16}{48} = \frac{48}{\blacksquare}$
40. $\frac{\blacksquare}{28} = \frac{6}{4}$
41. $\frac{45}{\blacksquare} = \frac{15}{35}$

Practice PLUS

KEY SKILL: Predicting Outcomes (Use after page 501.)

Level A

Predict how many times you will spin the letter in 24 spins.

1. A
2. B
3. C
4. not A
5. not B
6. not C

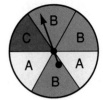

7. Spin the spinner 30 times. How do the results for the letter B compare with the results in Exercise 2?

Level B

A bag contains 4 green marbles, 2 blue marbles, 3 red marbles, 6 yellow marbles, and 1 black marble. What is the probability of picking:

8. a green marble?
9. a blue marble?
10. a red marble?
11. a yellow marble?
12. a black marble?
13. not a yellow marble?

14. How many times do you think you will pick each color in 24 picks?

Level C

Predict how many times you will pick the color in 40 picks.

15. blue
16. pink
17. green
18. black
19. not pink
20. not blue

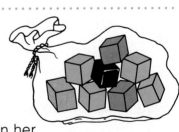

21. Rhonda has two red markers and one black marker in her pocket. Without looking, she picks up a marker and then puts it back. Predict how many times she will pick up a red marker in 30 picks.

22. Ali has a bag with 3 blue markers, 7 green markers, 4 red markers and 6 yellow markers. Without looking, Ali picks one marker at a time and then puts it back in the bag. Predict how many times he will pick a blue marker in 60 picks? in 100 picks?

Ratio and Probability **511**

Chapter Review

LANGUAGE AND MATHEMATICS

Complete the sentences. Use the words in the chart on the right.

1. The likelihood of something happening is its ■. *(page 494)*

2. A ■ is used to compare two quantities. *(page 482)*

3. A drawing that has the same shape as another, but is either smaller or larger is a ■. *(page 486)*

4. **Write a definition** or give an example of the words you did not use from the chart.

VOCABULARY
ratio
equal ratio
probability
scale drawing
outcome
independent event

CONCEPTS AND SKILLS

Write the ratio as a fraction. *(page 482)*

5. 12 to 15
6. 7 to 10
7. 27 to 100
8. 6 to 100
9. 5:6
10. 17:21
11. 42:56
12. 100:10

Write three equal ratios. *(page 484)*

13. $\frac{2}{7}$
14. $\frac{4}{5}$
15. $\frac{6}{9}$

Find the missing number. *(page 484)*

16. $\frac{7}{9} = \frac{21}{■}$
17. $\frac{9}{11} = \frac{36}{■}$

Use the scale drawing to find the length and width of the room. *(page 486)*

18. living room
    $\ell = ■$ m $w = ■$ m
19. dining room
    $\ell = ■$ m $w = ■$ m
20. kitchen
    $\ell = ■$ m $w = ■$ m
21. pantry
    $\ell = ■$ m $w = ■$ m

Scale: 1 cm = 2 m

What is the probability of picking:
(page 494)

22. a circle?
23. not a square?
24. a polygon?
25. a square?
26. not a star?
27. a triangle?
28. not a circle?
29. not a triangle?
30. a star?

512 Chapter 12

Predict how many times you will pick the shape in 30 picks. *(page 500)*

31. circle

32. square

33. triangle

34. not a triangle

35. star

36. not a star

Use the spinners. *(page 498)*

37. Make a tree diagram showing all the possible outcomes for two independent spins.

38. How many possible outcomes did you get?

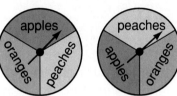

Use your tree diagram. What is the probability: *(page 496)*

39. of spinning apples–apples?

40. of spinning apples–peaches?

41. that the two spins have no peaches?

42. that the second spin is oranges?

CRITICAL THINKING

43. How can you use cross multiplication to find equal ratios? Give an example to support your answer.

MIXED APPLICATIONS

44. Mrs. Anderson is making pancakes. She uses 1 cup of mix to make 3 pancakes, 2 cups of mix to make 6 pancakes, and 3 cups of mix to make 9 pancakes. How many cups does she need to make 15 pancakes? *(page 488)*

45. The needlepoint store sells framed needlepoint pictures. There are 3 different designs in 2 different colored frames in 4 different sizes. How many different kinds of framed needlepoint pictures can be made? *(page 488)*

46. A paper bag contains a green crayon, a red crayon, and a black crayon. You reach into the bag 20 times. How many times will you pick the black crayon? (All three crayons are in the bag each time you pick.) Conduct an experiment to find out. Record your results in a table. *(page 490)*

Chapter Test

Write the ratio in three ways.

2 blue marbles
5 yellow marbles
9 pink marbles

1. blue marbles to yellow marbles
2. pink marbles to blue marbles
3. yellow marbles to all the marbles
4. pink marbles to yellow marbles

Complete to name equal ratios.

5. $\frac{5}{15} = \frac{\blacksquare}{3}$
6. $\frac{12}{16} = \frac{\blacksquare}{4}$
7. $\frac{12}{20} = \frac{3}{\blacksquare}$
8. $\frac{16}{36} = \frac{4}{\blacksquare}$

Use the scale drawing at the right to find the actual length and the width of the room.

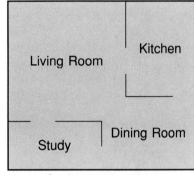

9. kitchen ℓ = ■ w = ■
10. living room ℓ = ■ w = ■
11. dining room ℓ = ■ w = ■
12. study ℓ = ■ w = ■

Scale: 1 cm = 2 m

Suppose that you roll a number cube that is numbered from 1 through 6. What is the probability of rolling:

13. a 3?
14. a number greater than 5?
15. a 4?

Predict how many times you will pick the color in 30 picks.

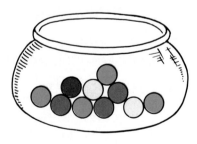

16. green
17. red
18. yellow

19. A three-colored spinner has five sections. The results of spinning it 30 times are:
red—11, yellow—12, and blue—7.
All the sections are the same size. Predict how many are in each color.

Solve.

20. Survey 20 other students to find out what their favorite school subject is. Make a bar graph to show your results.

Enrichment For All

PERCENT

A **percent** is a ratio that compares a number to 100. The symbol for percent is **%**. In this grid, 30 out of 100 squares are shaded red. So, $\frac{30}{100}$ or 30% of the grid is shaded red. What percent of the grid is shaded blue?

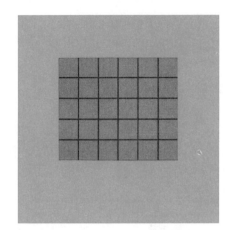

1. How many of the squares are shaded blue?
2. What is the ratio of blue squares to the total number of squares?
3. What fraction of the grid is shaded blue?
4. What percent of the grid is shaded blue?

Use graph paper to make a 10-by-10 grid. Shade the grid to make a design using the following percents of color.

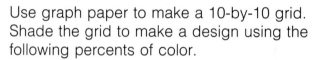

10% blue   5% green   25% red   50% yellow

5. What fraction of the grid is shaded yellow? is shaded red?
6. What percent of the grid is shaded? is unshaded?
7. **What if** you were asked to shade 0% of the grid orange? How many squares would you shade?

Make some more 10-by-10 grids. Shade them to make designs using the following percents of color.

8. 37% blue   22% green   12% red   18% yellow
9. 9% blue   45% green   45% red   22% yellow
10. 16% blue   0% green   60% red   28% yellow

11. Which design or designs were you unable to complete? Why?
12. Create your own color designs. Then ask others to tell what percents of color you used.

Ratio and Probability **515**

Cumulative Review

Choose the letter of the correct answer.

1. Which is the fraction for nineteen-twentieths?
   a. $\frac{1}{20}$
   b. $\frac{9}{20}$
   c. $\frac{19}{20}$
   d. not given

2. Which is the improper fraction for $3\frac{4}{5}$?
   a. $\frac{19}{5}$
   b. $\frac{12}{5}$
   c. $\frac{60}{5}$
   d. not given

3. $4\frac{5}{7} - 2\frac{3}{21}$
   a. $2\frac{1}{7}$
   b. $2\frac{4}{7}$
   c. $2\frac{3}{7}$
   d. not given

4. $\frac{1}{28} \times 28$
   a. $\frac{28}{28}$
   b. 1
   c. $\frac{28}{1}$
   d. not given

5. What is the likely temperature of hot soup?
   a. 110°F
   b. 50°F
   c. 215°F
   d. not given

6. What is the ratio of 30 to 100 written as a fraction?
   a. $\frac{30}{10}$
   b. $\frac{100}{30}$
   c. $\frac{30}{100}$
   d. not given

7. $49.72 \div 4$
   a. $12.18
   b. $12.43
   c. $13.13
   d. not given

8. Find the missing number.
   $\frac{20}{4} = \frac{\blacksquare}{1}$
   a. 20
   b. 5
   c. 4
   d. not given

Suppose that you roll a number cube that is numbered from 1 through 6 on each side.

9. What is the probability of rolling a 4?
   a. $\frac{1}{4}$
   b. $\frac{6}{1}$
   c. $\frac{1}{6}$
   d. not given

10. What is the probability of rolling an odd number?
    a. $\frac{1}{2}$
    b. $\frac{1}{4}$
    c. $\frac{1}{6}$
    d. not given

11. What is the probability of rolling a number greater than 4?
    a. $\frac{1}{3}$
    b. $\frac{1}{4}$
    c. $\frac{1}{5}$
    d. not given

Use the cubes to answer Questions 12 and 13.

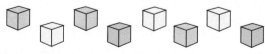

12. Predict how many times you will pick a blue cube in 40 picks.
    a. 4
    b. 5
    c. 10
    d. not given

13. Predict how many times you will pick a pink cube in 40 picks.
    a. 4
    b. 5
    c. 10
    d. not given

DATABANK

LOWEST ANNUAL RAINFALL IN THE UNITED STATES

State	Annual Rainfall (in inches)
Arizona	7
Idaho	12
Montana	11
Nevada	4
New Mexico	8

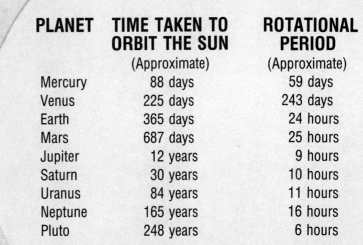

PLANET	TIME TAKEN TO ORBIT THE SUN (Approximate)	ROTATIONAL PERIOD (Approximate)
Mercury	88 days	59 days
Venus	225 days	243 days
Earth	365 days	24 hours
Mars	687 days	25 hours
Jupiter	12 years	9 hours
Saturn	30 years	10 hours
Uranus	84 years	11 hours
Neptune	165 years	16 hours
Pluto	248 years	6 hours

MONEY SPENT BY AMERICANS ON TICKETS TO SPORTS EVENTS, MOVIES, THEATER, AND OPERA IN THE UNITED STATES

Year	Amount Spent
1982	$7,800,000,000
1983	$8,600,000,000
1984	$9,500,000,000
1985	$9,500,000,000
1986	$10,200,000,000
1987	$11,100,000,000

DATABANK

SEATING CAPACITIES OF BASEBALL STADIUMS

Stadium	Seating Capacity
Astrodome	45,000
Dodger Stadium	56,000
Shea Stadium	55,300
Three Rivers Stadium	58,727
Veterans Stadium	62,382
Wrigley Field	38,040

ART

ART IN EARLY AMERICA
A classic on American folk art with many pictures.
SALE PRICE $14.79

MAKING QUILTS
A best-selling how-to book with helpful diagrams.
SALE PRICE $8.66

ASTRONOMY

BEYOND OUR GALAXY
What we know and do not know about other galaxies.
SALE PRICE $6.95

GUIDE TO THE SOLAR SYSTEM
Up-to-date facts.
SALE PRICE $5.99

U.S. POSTAL RATES
FIRST-CLASS MAIL

Letter Rates
1st ounce . . . . 29¢
Each additional ounce . . . . 23¢

Postcards
Single . . . . 19¢
Double . . . . 29¢

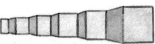

WORKERS IN MANUFACTURING IN THE UNITED STATES

Year	All Employees	Production Workers	Average Weekly Earnings
1955	16,882,000	13,288,000	$75.30
1960	16,796,000	12,586,000	89.72
1965	18,062,000	13,434,000	107.53
1970	19,367,000	14,044,000	133.33
1975	18,323,000	13,043,000	190.79
1979	21,040,000	15,068,000	269.34
1980	20,285,000	14,214,000	288.62
1985	19,260,000	13,092,000	386.37
1986	18,994,000	12,895,000	396.01
1987	19,112,000	13,021,000	406.31
1988	19,642,000	13,432,000	417.58

MILK COWS ON FARMS IN THE UNITED STATES
(in thousands)

Year	Number of Cows
1980	10,758
1982	10,896
1983	11,047
1985	10,805
1986	11,177
1987	10,502
1988	10,307

NUMBER OF PEOPLE PER SQUARE MILE IN TEXAS

Year	Population Density (per square mile)
1920	17.8
1960	36.4
1970	42.7
1980	54.3

DATABANK

WEATHER IN SOME SOUTHERN CITIES IN THE UNITED STATES IN 1987

City	Annual Rainfall (in.)	Clear Days	Cloudy Days
Houston, TX	40.60	92	150
Jackson, MS	47.31	133	144
Jacksonville, FL	46.39	98	160
Miami, FL	57.55	81	174
Mobile, AL	67.12	121	139
New Orleans, LA	60.63	103	158
Phoenix, AZ	8.40	199	61
Tampa, FL	49.08	117	111

BOYD'S HOME IMPROVEMENT SALE

Storm Doors (all 36 inches by 80 inches)	
STYLE	PRICE
Crossbuck	$199.99
Traditional	$209.99
Full View	$279.99

Rectangular Decks (all outdoor-treated wood)	
STYLE	PRICE
6 feet by 10 feet	$194.45
8 feet by 8 feet	$270.42
10 feet by 12 feet	$369.13
16 feet by 16 feet	$905.15

SIZES OF MODERN AND PREHISTORIC MAMMALS

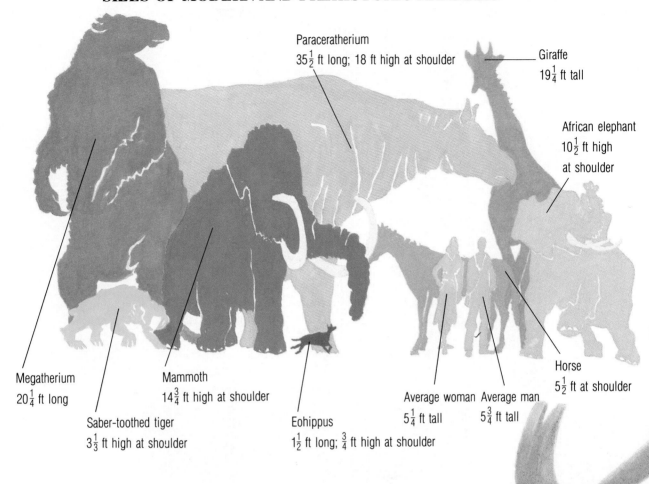

Paracheratherium — $35\frac{1}{2}$ ft long; 18 ft high at shoulder

Giraffe — $19\frac{1}{4}$ ft tall

African elephant — $10\frac{1}{2}$ ft high at shoulder

Megatherium — $20\frac{1}{4}$ ft long

Mammoth — $14\frac{3}{4}$ ft high at shoulder

Saber-toothed tiger — $3\frac{1}{3}$ ft high at shoulder

Eohippus — $1\frac{1}{2}$ ft long; $\frac{3}{4}$ ft high at shoulder

Average woman — $5\frac{1}{4}$ ft tall

Average man — $5\frac{3}{4}$ ft tall

Horse — $5\frac{1}{2}$ ft at shoulder

DO IT YOURSELF STORES
SPRING SALE

Item	Dimensions	Price
Wallboard	$\frac{1}{2}$ in. × 48 in. × 96 in.	$2.59 ea.
Wood paneling	$\frac{3}{8}$ in. × 48 in. × 96 in.	9.44 ea.
	$\frac{5}{8}$ in. × 48 in. × 96 in.	13.77 ea.
Fence pickets	$1\frac{1}{4}$ in. × $6\frac{1}{2}$ in. × 72 in.	1.15 ea.
Fiberglass insulation	$3\frac{1}{2}$ in. × 15 in. × 40 ft	6.29/roll
Wood door	$1\frac{3}{4}$ in. × 32 in. × 80 in.	87.00 ea.
Metal sink	$33\frac{1}{2}$ in. × $22\frac{1}{4}$ in. × $6\frac{1}{3}$ in.	29.99 ea.
Outdoor light	$5\frac{1}{2}$ in. × $14\frac{1}{4}$ in.	16.66 ea.

DATABANK

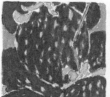  

RECORD WEIGHTS FOR FRUITS AND VEGETABLES

Fruit or Vegetable	Weight	Place
Apple	3 lb 1 oz	England, 1964
Carrot	15 lb 7 oz	New Zealand, 1978
Celery	35 lb 8 oz	England, 1986
Cucumber	59 lb	Australia, 1988
Watermelon	260 lb	Arkansas, 1985
Orange	5 lb 8 oz	South Africa, 1981
Pear	3.09 lb	Australia, 1979
Potato	18 lb 4 oz	England, 1795
Pumpkin	671 lb	New Jersey, 1986
Strawberry	8.17 oz	England, 1983
Tomato	7 lb 12 oz	Oklahoma, 1986

  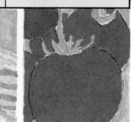

PETS GALORE SALE

Pet Homes (All boxed-shaped—All weather-proofed)

STYLE	PRICE
Deluxe Rabbit Cage (steel) 38 inches by 21 inches by 17 inches	$200
Peter Rabbit Home (pine) 42 inches by 16 inches by 18 inches	$165
Supreme Rabbit Hutch (plywood) 48 inches by 40 inches by 20 inches	$250

GLOSSARY

A

addition An operation that gives the total number, or amount in all.

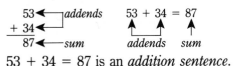

$53 + 34 = 87$ is an *addition sentence*.

angle A figure formed by two rays with the same endpoint. The rays are the *sides* of the angle and their endpoint is the *vertex* of the angle. Angles are measured in *degrees* (°). A *right angle* has a measure of 90°. An *acute angle* has a measure of less than 90°. An *obtuse angle* has a measure greater than 90° but less than 180°. A *straight angle* has a measure of 180°.

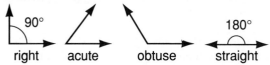

area The number of square units needed to cover a region or figure. *Square inches, square feet,* and *square yards* are units in the customary system used to measure area. *Square millimeters, square centimeters,* and *square meters* are units in the metric system used to measure area.

associative property If the grouping of addends or factors is changed, the answer remains the same.

$3 + (4 + 5) = 12 \qquad 2 \times (4 \times 3) = 24$
$(3 + 4) + 5 = 12 \qquad (2 \times 4) \times 3 = 24$

average A statistic found by adding two or more numbers and dividing by the number of numbers. (*See* mean.)

C

capacity The amount of liquid a container can hold. (*See* customary system *and* metric system.)

circle A closed plane figure having all points an equal distance from the center.

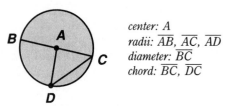

center: A
radii: $\overline{AB}, \overline{AC}, \overline{AD}$
diameter: $\overline{BC}$
chord: $\overline{BC}, \overline{DC}$

circumference The distance around a circle. The circumference is about 3 times the diameter.

common denominator The same denominator shared by two or more *fractions*. The fractions $\frac{1}{3}$ and $\frac{2}{3}$ have the common denominator 3.

common factor (*See* greatest common factor.)

common multiple (*See* least common multiple.)

commutative property If the order of addends or factors is changed, the answer remains the same.

$5 + 8 = 13 \qquad 8 \times 2 = 16$
$8 + 5 = 13 \qquad 2 \times 8 = 16$

composite number A whole number greater than 1 with more than two different factors. For example, 6 is a composite number. Its factors are 1, 2, 3, and 6.

congruent Figures that have the same shape and size are congruent. Two figures are congruent if they can be made to match exactly by *sliding, turning,* or *flipping.* Two angles or line segments are congruent if they have the same measure.

coordinates (*See* ordered pair.)

customary system A system of measurement. Units used to measure capacity are *cups, pints, quarts,* and *gallons.* Units used to measure weight are *ounces, pounds,* and *tons.* Units used to measure length are *inches, feet, yards,* and *miles.*

D

decimal A whole number, fraction, or mixed number expressed using a *decimal point,* for example, 3.0, 0.4, and 4.6.

degree (°) A unit used to measure angles.

degree Celsius (°C) A metric unit of measure for temperature.

degree Fahrenheit (°F) A customary unit of measure for temperature.

digit Any of the symbols 0, 1, 2, 3, 4, 5, 6, 7, 8, and 9.

distributive property For multiplication:

$5 \times (3 + 4) = (5 \times 3) + (5 \times 4)$
$5 \times 7 = 15 + 20$
$ 35 = 35$

division An operation on any two numbers that tells how many groups or how many in each group.

$$\text{quotient} \longrightarrow 6 \text{ R3} \longleftarrow \text{remainder}$$
$$\text{divisor} \longrightarrow 4\overline{)27} \longleftarrow \text{dividend}$$

$$27 \div 4 = 6 \text{ R3} \longleftarrow \text{remainder}$$
$$\uparrow \uparrow \uparrow$$
$$\text{dividend divisor quotient}$$

$24 \div 4 = 6$ is a *division sentence.*

E

equivalent decimals Decimals that name the same number.

$1.3 = 1.30$

equivalent fractions Two or more fractions that name the same number.

$\frac{1}{4} = \frac{2}{8} = \frac{3}{12}$

estimate To find an approximate answer for a problem. One method is to use rounded numbers.

even number A whole number that ends in 0, 2, 4, 6, or 8.

expanded form A way to write numbers that shows the value of each digit.

$456 = 400 + 50 + 6$

F

factor A number to be multiplied. In $3 \times 4 = 12$, 3 and 4 are factors of 12. 1, 2, 6, and 12 are also factors of 12.

fraction A fraction names an equal part of a whole or a part of a group.

$\frac{2}{3}$ ← *numerator:* number of equal parts (items) being counted
denominator: number of equal parts (items) in all

A *proper fraction* has a numerator that is less than the denominator. An *improper fraction* has a numerator that is equal to or greater than the denominator.

G

graph A special type of information display. Examples are *bar graph, line graph, circle graph,* and *pictograph.*

greatest common factor (GCF) The greatest number that is a *common factor* of two or more numbers. 5 is the GCF of 15 and 20.

I

identity property In addition, when 0 is added to a number, the sum is the number. In multiplication, when 1 is multiplied by a number, the product is the number.

$0 + 2 = 2 \quad 1 \times 4 = 4$

L

least common denominator The *least common multiple* of the denominators of two or more fractions. The least common denominator of $\frac{1}{3}$ and $\frac{3}{5}$ is 15 because the LCM of 3 and 5 is 15.

least common multiple (LCM) The least number that is a multiple of two or more numbers.

multiples of 3: 3, 6, 9, 12, *15,* . . .
multiples of 5: 5, 10, *15,* 20, . . .
The LCM of 3 and 5 is 15.

line A straight path that has no endpoints. It goes on forever in both directions.

A *line segment* is part of a line. It has two endpoints.

A *ray* is part of a line. It has one endpoint and goes on forever in one direction.

Parallel lines are two lines in a plane that never intersect.

Intersecting lines share a common point.

Perpendicular lines intersect and form right angles.

line of symmetry A line that divides a figure into two parts that match, or are congruent. A figure that has a line of symmetry is *symmetrical.*

lowest terms A fraction is in lowest terms when the *numerator* and the *denominator* have no *common factors* other than 1.

M

mass A measurement that tells how much of something there is. (*See* metric system.)

mean Another name for *average.*

median The number in the middle of a group of numbers arranged in order from least to greatest.

metric system A system of measurement. Units used to measure length are *millimeters, centimeters, meters,* and *kilometers.* Units used to measure mass are *milligrams, grams,* and *kilograms.* Units used to measure capacity are *milliliters* and *liters.*

mixed number A number greater than 1, such as $5\frac{1}{2}$, that has a whole number and a fraction.

mode The number or numbers in a group of numbers that occur most often.

multiples Numbers formed by multiplying a number by whole numbers. The multiples of 3 are 0, 3, 6, 9, 12, 15, . . .

multiplication An operation that gives the product of two numbers.

$$\begin{array}{r} 7 \\ \times\, 8 \\ \hline 56 \end{array} \begin{array}{l} \leftarrow factors \\ \\ \leftarrow product \end{array} \qquad 7 \times 8 = 56$$

$7 \times 8 = 56$ is a *multiplication sentence.*

N

number sentence An inequality or an equation.

$14 > 11 + 1$ or $5 + 4 = 9.$

O

odd number A whole number that ends in 1, 3, 5, 7, or 9.

ordered pair A pair of numbers that can be used to locate a point on a map, grid, or line graph. (4, 8) is an ordered pair. The 4 tells how many units to the right of 0. The 8 tells how many units up. 4 and 8 are the *coordinates* of the point.

P

perimeter The distance around a figure. To find the perimeter of a polygon, add the lengths of the sides.

place value The value of the place of a digit in a number.

plane A flat surface, like a tabletop, that goes on and on in all directions.

plane figure Any two-dimensional figure. Plane figures may have straight or curved sides. Examples are *angles, polygons,* and *circles.*

polygon A closed plane figure with sides that are line segments. Two line segments meet at a *vertex (vertices)* to form an angle. In *regular polygons* the line segments have the same length.

Polygon	Number of Sides and Vertices
triangle	3
quadrilateral	4
pentagon	5
hexagon	6
octagon	8
decagon	10

A *parallelogram* is a quadrilateral in which the opposite sides are parallel and congruent. A *rectangle* is a parallelogram with four right angles. A *square* is a rectangle with four congruent sides. A *rhombus* is a parallelogram with all sides congruent. A *trapezoid* is a quadrilateral with one pair of parallel sides.

prime number A whole number greater than 1 that has only itself and 1 as factors. 11 is a prime number because its only factors are 1 and 11.

probability The chance that something will happen.

R

range The difference between the greatest number and the least number in a group of numbers.

ratio A way to compare two numbers.

5 to 6, 5:6, or $\frac{5}{6}$

rounding Replacing a number by another number that is easier to use. You round to tell *about* how many.

S

similar Figures that have the same shape but not necessarily the same size.

space figure A three-dimensional figure. The sides of a space figure are called *faces.* The faces of a space figure meet at an *edge.* Edges can be straight or curved. The corner of a space figure is called the *vertex.*

rectangular prism   triangular prism   hexagonal prism

A *cube* is a special rectangular prism in which all 6 faces are congruent.

A *pyramid* is a space figure with 1 base that is a polygon and triangular faces that meet at a common vertex.

triangular prism   square pyramid   rectangular pyramid

A *sphere* is a space figure that has the shape of a ball.

A *cone* is a space figure with 1 curved surface, 1 flat surface, 1 curved edge, and 1 vertex.

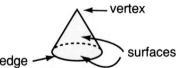

A *cylinder* is a space figure with 2 flat surfaces, 1 curved surface, 2 curved edges, and no vertices.

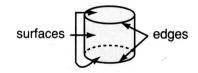

standard form The usual or common way to write a number.

subtraction An operation that gives the difference between two numbers or amounts.

$$\begin{array}{r} 11 \\ -4 \\ \hline 7 \end{array} \leftarrow \textit{difference} \qquad 11 - 4 = \underset{\uparrow}{7} \\ \textit{difference}$$

$11 - 4 = 7$ is a *subtraction sentence*.

T

time The units used to measure time are *seconds, minutes, hours, days, weeks, months,* and *years*.

V

volume The number of *cubic units* it takes to fill a figure. Units used to measure volume in the customary system are *cubic inches, cubic feet,* and *cubic yards*. Units used to measure volume in the metric system are *cubic centimeters* and *cubic meters*.

whole numbers The set of numbers 0, 1, 2, 3, 4, 5, 6, . . .

W

weight The measure of how heavy something is. (*See* customary system.)

Z

zero property In multiplication, when one factor is zero, the product is zero.

COMPUTER TERMS

cell Each individual box or square within the coordinate grid of an electronic spreadsheet.

data Information that is put into a computer.

simulate To use a computer to study the possible solutions to a problem before trying them in real life.

spreadsheet A computer program that arranges data and formulas in a grid of cells.

Table of Measures

METRIC UNITS

LENGTH
- 1 millimeter (mm) = 0.001 meter (m)
- 1 centimeter (cm) = 0.01 meter
- 1 decimeter (dm) = 0.1 meter
- 1 dekameter (dam) = 10 meters
- 1 hectometer (hm) = 100 meters
- 1 kilometer (km) = 1,000 meters

MASS/WEIGHT
- 1 milligram (mg) = 0.001 gram (g)
- 1 centigram (cg) = 0.01 gram
- 1 decigram (dg) = 0.1 gram
- 1 dekagram (dag) = 10 grams
- 1 hectogram (hg) = 100 grams
- 1 kilogram (kg) = 1,000 grams
- 1 metric ton (t) = 1,000 kilograms

CAPACITY
- 1 milliliter (mL) = 0.001 liter (L)
- 1 centiliter (cL) = 0.01 liter
- 1 deciliter (dL) = 0.1 liter
- 1 dekaliter (daL) = 10 liters
- 1 hectoliter (hL) = 100 liters
- 1 kiloliter (kL) = 1,000 liters

AREA
- 1 square centimeter (cm²) = 100 square millimeters (mm²)
- 1 square meter (m²) = 10,000 square centimeters
- 1 hectare (ha) = 10,000 square meters
- 1 square kilometer (km²) = 1,000,000 square meters

CUSTOMARY UNITS

LENGTH
- 1 foot (ft) = 12 inches (in.)
- 1 yard (yd) = 36 inches
- 1 yard = 3 feet
- 1 mile (mi) = 5,280 feet
- 1 mile = 1,760 yards

WEIGHT
- 1 pound (lb) = 16 ounces (oz)
- 1 ton (T) = 2,000 pounds

CAPACITY
- 1 cup (c) = 8 fluid ounces (fl oz)
- 1 pint (pt) = 2 cups
- 1 quart (qt) = 2 pints
- 1 quart = 4 cups
- 1 gallon (gal) = 4 quarts

AREA
- 1 square foot (ft²) = 144 square inches (in.²)
- 1 square yard (yd²) = 9 square feet
- 1 acre = 43,560 square feet
- 1 square mile (mi²) = 640 acres

TIME
- 1 minute (min) = 60 seconds (s)
- 1 hour (h) = 60 minutes
- 1 day (d) = 24 hours
- 1 week (wk) = 7 days
- 1 year (y) = 12 months (mo)
- 1 year = 52 weeks
- 1 year = 365 days
- 1 century (c) = 100 years

FORMULAS

- $P = 2(\ell + w)$ — Perimeter of a rectangle
- $P = 4s$ — Perimeter of a square
- $A = \ell \times w$ — Area of a rectangle
- $A = s^2$ — Area of a square
- $A = b \times h$ — Area of a parallelogram
- $A = \frac{1}{2}(b \times h)$ — Area of a triangle
- $C = \pi \times d$, or $2 \times \pi \times r$ — Circumference of a circle
- $A = \pi \times r^2$ — Area of a circle
- $V = \ell \times w \times h$ — Volume of a rectangular prism
- $V = B \times h$ — Volume of any prism
- $V = \pi \times r^2 \times h$ — Volume of a cylinder

SYMBOLS

$=$	is equal to
$\neq$	is not equal to
$>$	is greater than
$<$	is less than
$\geq$	is greater than or equal to
$\leq$	is less than or equal to
$\approx$	is approximately equal to
$\cong$	is congruent to
$\sim$	is similar to
...	continues without end
$1.\overline{3}$	repeating decimal 1.333...
%	percent
π	pi (approximately 3.14)
°	degree
°C	degree Celsius
°F	degree Fahrenheit
$\overleftrightarrow{AB}$	line AB
$\overline{AB}$	line segment AB
$\overrightarrow{AB}$	ray AB
$\angle ABC$	angle ABC
$\triangle ABC$	triangle ABC
$\parallel$	is parallel to
$\perp$	is perpendicular to
2:5	ratio of 2 to 5
10^2	ten to the second power
$^+4$	positive 4
$^-4$	negative 4
$|^-4|$	absolute value of $^-4$
$(^+3, ^-4)$	ordered pair 3, $^-4$
$P(E)$	probability of event E

INDEX

A

Acute angles, 314–315
Acute triangles, 318–319
Addends, 52–53
Adding and Subtracting Decimals, 93–136
Adding and Subtracting Fractions, 395–438
Addition, *see also* Sums
  compensation in, 58–59
  of decimals, 93–136
  equal additions in subtraction, 62–63
  fact families in, 52–53
  of fractions and mixed numbers, 395–438
  of fractions with like denominators, 398–399
  of fractions with unlike denominators, 400–401
  of hundredths, 104–105
  meaning of, 52–53
  of mixed numbers, 410–413
  of mixed numbers with renaming, 414–415
  properties of, 52–53
  of tenths, 104–105
  of thousandths, 114–115
  using, 51–92
  of whole numbers, 58–61
Algebra, preparation for, 22–25, 28–29, 39, 52–53, 68–69, 73, 98–99, 109, 138–139, 140–141, 142–143, 161, 164–165, 169, 179, 184–185, 187, 209, 238, 246–247, 251–252, 265, 307, 349, 372–373, 454–455, 456–457
A.M., 74–75
Angles
  acute, 314–315
  congruent, 328–329
  corresponding, 328–329
  exterior, 312–313
  in degrees, 312–313
  drawing with Logo, 339
  interior, 312–313
  measuring, 312–313
  obtuse, 314–315
  right, 314–315
  straight, 314–315
  sums of angles of quadrilaterals, 318–319
  sums of angles of triangles, 318–319
Area, 162–163
  of irregular-shaped regions, 479
  of parallelograms, 456–457
  of right triangles, 454–455
  of squares and rectangles, 162–163
Associative property
  of addition, 52–53
  of multiplication, 138–139, 142–143
Average of numbers, 202–203
Axis
  horizontal, *see* Horizontal axis
  vertical, *see* Vertical axis

B

Bar graphs
  interpreting, 118–119
  making, 9, 120–121
Base-five number system, 49
Base-ten number system, 49
Billions, 18–19

C

c (cup), 460–461
Calculation methods, choosing, 60–61, 64–65, 67, 77, 97, 103, 109, 111, 114–115, 117, 143, 151, 156–157, 159, 187, 191, 199, 203, 205, 223, 225, 234–235, 237, 243, 272–273, 275, 286–287, 289, 293, 317, 321, 329, 331, 355, 359, 367, 375, 403, 409, 417, 419, 445, 449, 455, 457, 459, 483, 491, 497
Calculators, 7, 39, 60–61, 64–65, 67, 69, 77, 81, 97, 103, 109, 111, 114–115, 117, 142–143, 151, 152–153, 156–157, 159, 164, 168, 169, 172, 186–187, 189, 191, 199, 203, 205, 209, 223, 225, 234–235, 237, 243, 264, 272–273, 275, 276–277, 278–279, 286–287, 289, 293, 297, 317, 321, 329, 331, 332–333, 355, 359, 367, 375, 403, 409, 417, 419, 427, 445, 449, 455, 457, 459, 469, 483, 485, 488–489, 491, 497
Capacity
  in customary units, 460–461
  in metric units, 290–291
Celsius thermometer, 464–465
Center of circle, 164–165
Centimeter (cm), 70–73
  cubic, 244–245
  square, 162–163
Centimeter ruler, 70–73
Challenge, 319, 357
Changing estimates, 230–231
Changing measures of time, 240–241
Chapter opener, 11, 51, 93, 137, 181, 221, 263, 309, 351, 395, 439, 481
Chapter review, 46–47, 88–89, 132–133, 176–177, 216–217, 258–259, 304–305, 346–347, 390–391, 434–435, 476–477, 512–513
Chapter test, 48, 90, 134, 178, 218, 260, 306, 348, 392, 436, 478, 514
Choosing the operation, 66–67, 110–111, 236–237
Chord of circle, 164–165
Circle graphs
  interpreting, 452–453
  drawing with Logo, 469
Circles, 164–165
  central angles, 339
Circumference, 164–165
cm (centimeter), 70–73
Common factors, 356–357
  greatest, 356–357
Common multiples, least, 370–371
Commutative property
  of addition, 52–53
  of multiplication, 138–139, 142–143
Comparing decimals, 98–99
Comparing fractions and mixed numbers, 372–373
Comparing whole numbers, 24–25
Compatible numbers, 188–189
Compensation in addition, 58–59, 104–105

Composite numbers, 357, 393
Computers
  circle graph, 469
  double line graph, 125
  Logo
    angles, 339
    quadrilaterals, 383
  powers and exponents, 39
  probability, 505
  spreadsheets, 81, 251
Conducting an experiment, 490–491, 505
Cones, 334–335
Congruent angles, 328–329
Congruent figures, 328–329
Congruent line segments, 328–329
Congruent polygons, 328–329
Cooperative learning
  chapter opener, 11, 51, 93, 137, 181, 221, 263, 309, 351, 395, 439, 481
  decision making, 36–37, 78–79, 122–123, 166–167, 206–207, 248–249, 294–295, 336–337, 380–381, 424–425, 466–467, 502–503
  sharing ideas, 12, 15, 33, 71, 121, 165, 183, 201, 203, 227, 245, 269, 277, 281, 283, 311, 319, 324, 335, 354, 363, 376, 397, 398, 405, 411, 413, 419, 421, 441, 443, 447, 452, 454, 457, 486, 492, 494, 497, 501
  thinking mathematically, 1–10, 22–23, 68–69, 106–107, 152–153, 194–195, 238–239, 278–279, 322–323, 368–369, 406–407, 450–451, 498–499; *see also* Thinking mathematically
  working together, 12, 14, 32, 70, 120, 164, 182–183, 200, 202, 244, 276, 280, 324, 362, 376, 396, 410, 440, 442, 446, 454, 456, 486, 492, 494, 496–497, 500
Coordinate graphs, 28–29
  enlarging figures on, 179
Coordinates, 28–29
Corresponding angles, 328–329
Corresponding sides, 328–329
Critical thinking, 13, 15, 25, 27, 47, 53, 55, 57, 89, 99, 105, 119, 133, 149, 163, 177, 193, 217, 229, 259, 271, 305, 319, 325, 327, 329, 331, 347, 357, 371, 379, 391, 409, 415, 435, 445, 457, 477, 513
Cubes, 334–335
Cubic centimeter, 244–245
Cubic meter, 244–245
Cumulative review, 50, 92, 136, 180, 220, 262, 308, 350, 394, 438, 480, 516
Cup (c), 460–461
Curriculum connections
  math and art, 80, 338
  math and the consumer, 468
  math and health, 504
  math and literature, 208
  math and music, 250, 426
  math and science, 124, 168, 296
  math and social studies, 38, 382
Customary units of measurement
  capacity, 460–461
  length, 376–379
  renaming, 462–463
  temperature, 464–465
  weight, 460–461

529

Cylinders, 334–335

D

d (day), 240–241
Data
    displaying, 81, 200–201, 251
    recording, 12–13
Data collection and analysis, 8–9, 12–15, 20–21, 25, 30–33, 70–71, 118–121, 125, 155, 163, 164–165, 197, 200–203, 233, 235, 265, 285, 313, 339, 361, 376–377, 415, 423, 452–453, 461, 469, 483, 492–501, 505
Databank, 517–522
Day (d), 240–241
Decagons, 316–317
Decimal points
    lining up, 98–99
    zeros before, 94–95
Decimals
    addition of, 93–136
    comparing, 98–99
    division of, 263–308
    division of decimals by whole numbers, 264–265, 280–287
    equivalent, 94–95
    estimating products of, 266–267
    estimating sums and differences of, 102–103, 112–113
    hundredths, *see* Hundredths
    mixed numbers and, 94–95
    multiplication of, 263–308
    multiplication of whole numbers and, 264–265, 268–273, 297
    ordering, 98–99
    place-value chart with, 96
    products of two decimals, 276–277
    rounding, 100–101, 125
    subtraction of, 93–136
    tenths, *see* Tenths
    thousandths, *see* Thousandths
Decision making, 36–37, 78–79, 122–123, 166–167, 206–207, 248–249, 294–295, 336–337, 380–381, 424–425, 466–467, 502–503
Degrees Celsius, 464–465
Degrees Fahrenheit, 464–465
Degrees of angles, 312–313
Denominators, 352–353
    addition of fractions with like denominators, 398–399
    addition of fractions with unlike denominators, 400–401
    in equivalent fractions, 372–373
Diagonals of rectangles, 454–455
Diagrams, tree, 496–497
Diameter, 164–165
Differences, 52–53; *see also* Subtraction
    estimating, 102–103, 112–113
    estimating differences by rounding, 54–55
    estimating differences of mixed numbers, 408–409
    of fractions, 396–397
    front-end estimation of, 56–57
Displaying data, 200–201
Distributive property of multiplication, 138–141
Dividends, 182–183

Dividing Whole Numbers—One-Digit Divisors, 181–220
Dividing Whole Numbers—Two-Digit Divisors, 221–262
Divisibility, 219
Division, *see also* Quotients
    of decimals, 263–308
    of decimals by whole numbers, 264–265, 280–287
    fact families in, 184–185
    of five-digit numbers, 192–193
    of four-digit numbers, 192–193
    of fractions, 439–480
    of large numbers, 186–187, 234–235
    meaning of, 182–183
    mental math in, 233
    multiplication and, 184–185
    short division, 197
    of three-digit numbers, 190–191
    by two-digit numbers, 226–229
    of two-digit numbers, 190–191
    using zeros in, 284–285
    of whole numbers by fractions, 446–447
    of whole numbers by one-digit divisors, 181–220
    of whole numbers by two-digit divisors, 221–262
Division facts, 184–185
Division patterns, using, 186–187, 222–223
Division properties, 184–185
Divisors, 182–183
Double-bar graphs, 118–119
Double-line graphs on a computer, 125
Doubles, 140–141
Drawing a diagram, 320–321
Drawings, scale, 486–487

E

Edges of prisms, 334–335
Egyptian number system, 91
Elapsed time, 74–75
Enlarging figures on coordinate graphs, 179
Enrichment for all
    area of irregular-shaped regions, 479
    base five, 49
    divisibility, 219
    enlarging figures on coordinate graphs, 179
    magic squares, 261
    number sequences, 135
    order of operations, 307
    other number systems, 91
    percent, 515
    predicting from a sample, 437
    prime factorization, 393
    Venn diagrams, 349
Equal additions in subtraction, 62–63, 108–109
Equal ratios, 484–485
Equilateral triangles, 318–319
Equivalent decimals, 94–95
Equivalent fractions, 354–355, 372–373, 484–485
Estimates, changing, 230–231
Estimating/estimation
    compatible numbers, 188–189
    computation
        differences, 54–57, 102–103, 112–113, 408–409

        products, 144–145, 198–199, 266–267
        quotients, 188–189, 224–225, 230–231
        sums, 54–57, 102–103, 112–113, 116–117, 408–409
    front-end, *see* Front-end estimation
    measurements, *see also* Customary units of measurement; Metric units of measurement
        capacity, 290–291, 460–461
        length, 70–73, 376–377
        mass, 290–291
        temperature, 464–465
        weight, 460–461
    rounding, *see* Rounding
Events, independent, 496–497
Expanded form, numbers in, 16–17
Experiments, 494–495
    in probability, 492–493, 505
Exponents, 39, 307
Expressions, 52–53, 68–69, 138–141, 169, 184–185, 187, 209, 251, 265, 307
Extra practice, 40–43, 82–85, 126–129, 170–173, 210–213, 252–255, 298–301, 340–343, 384–387, 428–431, 470–473, 506–509

F

Faces of prisms, 334–335
Fact families
    in addition and subtraction, 52–53
    in multiplication and division, 184–185
Factor trees, 393
Factorization, prime, 393
Factors, 138–139, 169, 356–357
    common, *see* Common factors
Fahrenheit thermometer, 464–465
Favorable outcomes, 494–495
Figures
    congruent, 328–329
    enlarging figures on coordinate graphs, 179
    flips of, 326–327
    plane, 316–317
    sides of, 162–163
    similar, 330–331
    slides of, 326–327
    space, 334–335
    turns of, 326–327
Finding needed information, 150–151
Finding a pattern, 242–243
Five-digit numbers, division of, 192–193
Flips of figures, 326–327
Foot (ft), 376–377
Four-digit numbers, division of, 192–193
Fractions, 352–353
    addition of fractions with like denominators, 398–399
    addition of fractions with unlike denominators, 400–401
    comparing mixed numbers and, 372–373
    differences of, 396–397
    division of whole numbers by, 446–447
    equivalent, 354–355, 372–373
    improper, 360–361
    mental math with, 365
    multiplication of, 440–443
    multiplication of whole numbers and, 444–445

simplifying, 358-359
subtraction of, 404-405
sums of, 396-397
understanding, 351-394
Frequency tables, 14-15, 492-493, 498
Front-end estimation
to estimate products, 144-145, 266-267
to estimate sums and differences, 56-57, 102-103, 112-113
to estimate sums and differences of mixed numbers, 408-409
ft (foot), 376-377
Functions, 22-23, 28-29, 140-141, 179, 184-185, 187, 194-195, 242-243, 251, 288-289, 463, 484-485

G

g (gram), 290-291
Gallon (gal), 460-461
GCF (greatest common factor), 356-357
Geometry, 309-350
angles, *see* Angles
basic, 310-311
center of circle, 164-165
chord of circle, 164-165
circumference, 164-165
cones, 334-335
congruence, 328-329
coordinates, 28-29
cubes, 334-335
cylinders, 334-335
diameter, 164-165
figures, *see* Figures
lines, *see* Lines
ordered pairs of numbers, 28-29
parallelograms, *see* Parallelograms
planes, 310-311
points, 310-311
polygons, *see* Polygons
prisms, *see* Prisms
pyramids, *see* Pyramids
quadrilaterals, *see* Quadrilaterals
radius, 164-165
rays, 310-311
rectangles, *see* Rectangles
rhombuses, 318-319
shapes, 316-317
sides of figures, 162-163
similarity, 330-331
spheres, 334-335
squares, *see* Squares
symmetry, 324-325
trapezoids, 318-319
triangles, *see* Triangles
Glossary, 523-527
Gram (g), 290-291
Graph-paper models, 280-281
Graphs
bar, *see* Bar graphs
circle, 452-453, 469
coordinate, *see* Coordinate graphs
double-bar, 118-119
line, *see* Line graphs
pictographs, 200-201
Greatest common factor (GCF), 356-357
Guess, test, and revise, 204-205

H

h (hour), 74-75, 240-241
Hexagonal prisms, 334-335

Hexagons, 316-317
Horizontal axis
of bar graphs, 118-119
of line graphs, 30-31
Hour (h), 74-75, 240-241
Hundred thousands, 16-17
Hundredths, 94-95
addition of, 104-105
subtraction of, 108-109

I

Identifying extra information, 76-77
Identity property
of addition, 52-53
of multiplication, 138-139
Improper fractions, 360-361
Inch (in.), 376-377
Inch ruler, 376-377
Independent events, 496-497
Interpreting bar graphs, 118-119
Interpreting circle graphs, 452-453, 469
Interpreting line graphs, 30-31
Interpreting the quotient and remainder, 374-375
Intersecting lines, 310-311
Irregular-shaped regions, area of, 479
Isosceles triangles, 318-319

K

Kilogram (kg), 290-291
Kilometer (km), 70-71

L

L (liter), 290-291
Large numbers
division of, 186-187, 234-235
multiplication of, 142-143, 156-157
lb (pound), 460-461
LCM (least common multiple), 370-371
Leap year, 240-241
Least common multiple (LCM), 370-371
Length
in customary units, 376-379
in metric units, 70-73
Line graphs, 125
interpreting, 30-31
making, 32-33
Line segments, 310-311
congruent, 328-329
Lines, 310-311
intersecting, 310-311
parallel, 310-311
perpendicular, 314-315
of symmetry, 324-325
Liter (L), 290-291
Logical reasoning, 10, 368-369, 443
Logo
drawing angles, 339
drawing quadrilaterals, 383

M

m (meter), 70-73
Magic squares, 261
Magic sums, 261
Making bar graphs, 120-121
Making line graphs, 32-33
Making an organized list, 366-367

Making a table, 288-289
Manipulative activity, 12-13, 14-15, 32-33, 70-71, 164-165, 182-183, 202-203, 226-227, 244-245, 268-269, 276-277, 280-281, 282-283, 318-319, 324-325, 326-327, 334-335, 362-363, 376-377, 396-397, 406-407, 410-411, 418-419, 440-441, 442-443, 446-447, 454-455, 456-457, 486-487, 492-493, 494-495, 496-497, 500-501, 505
Mass in metric units, 290-291
Mean of numbers, 202-203
Measurement
customary, *see* Customary units of measurement
estimating, *see* Estimating/estimation, measurements
metric, *see* Metric units of measurement
Measures of time, changing, 240-241
Measuring angles, 312-313
Measuring tape, 376-377
Median of numbers, 202-203
Mental math
adding left to right, 58-59, 104-105
adding whole numbers, 58-59
compensation, 58-59, 104-105
division, 233
equal additions, 62-63, 108-109
fractions, 365
multiplication, 149
multiplying and dividing decimals, 264-265
subtracting left to right, 62-63, 108-109
subtracting whole numbers, 62-63
using division patterns, 186-187, 222-223
using multiplication patterns, 142-143
Meter (m), 70-71
cubic, 244-245
square, 162-163
Metric units of measurement
capacity, 290-291
length, 70-73
mass, 290-291
renaming, 292-293
temperature, 464-465
mg (milligram), 290-291
Mile (mi), 376-377
Milligram (mg), 290-291
Milliliter (mL), 290-291
Millimeter (mm), 70-73
square, 162-163
Millions, 18-19
Minute (min), 74-75, 240-241
Mixed applications, 27, 47, 53, 55, 57, 59, 61, 63, 65, 73, 89, 99, 101, 103, 105, 109, 113, 115, 133, 139, 141, 143, 145, 147, 149, 155, 157, 161, 163, 165, 177, 185, 187, 189, 191, 193, 197, 203, 217, 223, 225, 229, 231, 233, 235, 241, 247, 259, 265, 267, 271, 273, 283, 285, 287, 291, 293, 305, 313, 315, 317, 329, 331, 335, 347, 355, 357, 359, 361, 365, 371, 373, 379, 391, 399, 401, 405, 409, 413, 415, 419, 421, 423, 435, 443, 445, 455, 457, 461, 463, 465, 477, 483, 485, 487, 495, 497, 513
Mixed numbers, 360-361
addition of, 410-413
addition of mixed numbers with renaming, 414-415

531

comparing fractions and, 372–373
decimals and, 94–95
estimating sums and differences of, 408–409
renaming, 362–363
rounding, 364–365
subtraction of, 410–411, 418–419
subtraction of mixed numbers with renaming, 420–423
understanding, 351–394

Mixed review, 31, 33, 59, 63, 73, 97, 109, 143, 157, 187, 191, 203, 225, 227, 235, 273, 285, 293, 325, 329, 331, 355, 359, 371, 399, 419, 423, 445, 455, 463, 483, 495, 497

mL (milliliter), 290–291
mm (millimeter), 70–73
mo (month), 240–241
Mode of numbers, 202–203
Models, graph-paper, 280–283
Money
  addition, 58–61, 104–105
  consumer math, 2–3, 34–35, 66–67, 122–123, 150–151, 158–159, 206–207, 236–237, 242–243, 248–249, 274–275, 288–289, 294–295, 320–321, 332–333, 336–337, 366–367, 374–375, 380–381, 402–403, 416–417, 424–425, 458–459, 466–467, 488–489, 502–503
  division, 192–193, 196–197, 232–235, 282–287
  estimating, 54–57, 116–117, 144–145, 198–199
  multiplying, 146–149, 154–157, 270–273
  rounding, 26–27
  subtracting, 62–65, 108–109
Month (mo), 240–241
Multiples, least common, 370–371
Multiplication, *see also* Products
  of decimals, 263–308
  division and, 184–185
  fact families in, 184–185
  of fractions, 440–443
  of fractions and whole numbers, 444–445
  of large numbers, 142–143, 156–157
  meaning of, 138–139
  mental math in, 149
  by one-digit numbers, 146–147
  by two-digit numbers, 148–149, 154–155
  using multiplication patterns, 142–143
  of whole numbers, 137–180
  of whole numbers and decimals, 264–265, 268–273, 297
  zeros in, 154–155
Multiplication facts, 140–141, 184–185
Multiplication properties, 138–139, 142–143
Multiplying and Dividing Decimals, 263–308
Multiplying and Dividing Fractions, 439–480
Multiplying Whole Numbers, 137–180

N

Number lines, 26–27, 100–101
Number sense, *see* Estimating/estimation; Mental math

Number sentences, parentheses in, 209, 307
Number sequences, 135
Number systems, 49
  Egyptian, 91
  other, 91
  Roman, 91
Numbers
  average of, 202–203
  billions, 18–19
  compatible, 188–189
  composite, 357, 393
  divisibility, 219
  in expanded form, 16–17
  five-digit, division of, 192–193
  four-digit, division of, 192–193
  hundred thousands, 16–17
  large, *see* Large numbers
  mean of, 202–203
  median of, 202–203
  millions, 18–19
  mixed, *see* Mixed numbers
  mode of, 202–203
  one-digit, *see* One-digit numbers
  ordered pairs of, 28–29
  powers of 10, 39
  prime, 357, 393
  range of, 202–203
  rounding, 26–27
  in standard form, 16–17
  three-digit, divison of, 190–191
  two-digit, *see* Two-digit numbers
  understanding, 11–50
  whole, *see* Whole numbers
  word name of, 16–19
Numerals, Roman, 91
Numerators, 352–353
  in equivalent fractions, 372–373

O

Obtuse angles, 314–315
Obtuse triangles, 318–319
Octagons, 316–317
One, identity property, *see* Identity property
One-digit numbers
  division by, 181–220
  multiplication by, 146–147
Order of operations, 209, 307
Ordered pairs of numbers, 28–29
Ordering decimals, 98–99
Ordering whole numbers, 24–25
Ounce (oz), 460–461
Outcomes, 494–495
  predicting, 500–501
oz (ounce), 460–461

P

Pairs, ordered, of numbers, 28–29
Parallel lines, 310–311
Parallelograms, 318–319
  area of, 456–457
Parentheses in number sentences, 209, 307
Patterns
  division, using, 186–187, 222–223
  multiplication, 142–143
  for number sequences, 135
  powers of 10, 39

Patterns, relations, and functions, 22–23, 28–29, 140–141, 142–143, 152–153, 184–185, 186–187, 222–223, 242–243, 251, 264–265, 354–355, 484–485
Pentagons, 316–317
Percent, 515
Perimeter, 73, 160–161
Perpendicular lines, 314–315
Pictographs, 200–201
Pint (pt), 460–461
Place-value chart with decimals, 96
Plane figures, 316–317
Planes, 310–311
P.M., 74–75
Points, 310–311
Polygons, 316–317
  congruent, 328–329
Possible outcomes, 494–495
Pound (lb), 460–461
Powers, 39, 307
Practice plus, 44–45, 86–87, 130–131, 174–175, 214–215, 256–257, 302–303, 344–345, 388–389, 432–433, 474–475, 510–511
Predicting from a sample, 437
Predicting outcomes, 500–501, 505
Prime factorization, 393
Prime numbers, 357, 393
Prisms, 334–335
  cubes, 334–335
  hexagonal, 334–335
  rectangular, *see* Rectangular prisms
  triangular, 334–335
Probability, 494–495, 505
  experiments in, 492–493, 505
  ratio and, 481–516
  computer simulation, 505
Problem formulation, 11, 17, 25, 35, 36–37, 51, 55, 67, 71, 73, 77, 78–79, 93, 107, 109, 111, 117, 122–123, 137, 141, 147, 151, 159, 166–167, 181, 185, 197, 199, 205, 206–207, 221, 229, 231, 237, 243, 248–249, 263, 275, 285, 291, 293, 294–295, 309, 321, 336–337, 351, 367, 373, 375, 377, 380–381, 395, 399, 403, 415, 417, 423, 424–425, 439, 443, 445, 449, 459, 461, 463, 466–467, 481, 487, 491, 497, 502–503
Problem solving
  consumer mathematics, 2–3, 35, 36–37, 45, 47, 48, 51, 53, 55, 56, 58–59, 65, 67, 78–79, 81, 87, 89, 90, 93, 109, 116–117, 122–123, 128, 130, 131, 133, 139, 141, 143, 145, 149, 151, 155, 157, 159, 161, 163, 177, 178, 181, 189, 193, 198–199, 203, 204–205, 206–207, 212, 213, 217, 218, 221, 225, 229, 236–237, 241, 248–249, 254, 260, 263, 264–265, 267, 272–273, 275, 278–279, 281, 283, 285, 286–287, 289, 294–295, 299, 305, 306, 315, 321, 331, 333, 336–337, 367, 370, 375, 381, 387, 395, 399, 409, 415, 417, 424–425, 439, 453, 466–467, 468, 483, 487, 489, 495, 502–503
  decision making, 36–37, 78–79, 122–123, 166–167, 206–207, 248–249, 294–295, 336–337, 380–381, 424–425, 466–467, 502–503

532

problem formulation, 11, 17, 25, 35,
36–37, 51, 55, 67, 71, 73, 77,
78–79, 93, 107, 109, 111, 117,
122–123, 137, 141, 147, 151, 159,
166–167, 181, 185, 197, 199, 205,
206–207, 221, 229, 231, 237, 243,
248–249, 263, 275, 285, 291, 293,
294–295, 309, 321, 336–337, 351,
w67, 373, 375, 377, 380–381, 395,
399, 403, 415, 417, 423, 424–425,
439, 443, 445, 449, 459, 461, 463,
466–467, 481, 487, 491, 497,
502–503
strategies and skills review, 67, 77, 111,
117, 151, 159, 199, 205, 237, 243,
275, 289, 321, 367, 375, 403, 417,
459, 491
thinking mathematically, 1–10, 22–23,
68–69, 106–107, 152–153,
194–195, 238–239, 278–279,
322–323, 368–369, 406–407,
450–451, 498–499
applying mathematics, 2–3, 278–279
collecting and interpreting data, 8–9
experimenting and predicting,
106–107, 498–499
investigating patterns, 22–23,
152–153, 194–195
logical reasoning, 10, 368–369, 443
measuring, 7
using number concepts, 4–5,
68–69, 238–239, 450–451
visual reasoning, 6, 19, 322–323,
327, 406–407, 421

Problem-solving strategies and skills
choosing the operation, 66–67,
110–111, 236–237
conducting an experiment, 490–491
drawing a diagram, 320–321
finding needed information, 150–151
finding a pattern, 242–243
guess, test, and revise, 204–205
identifying extra information, 76–77
interpreting the quotient and remainder,
374–375
making an organized list, 366–367
making a table, 288–289
solving a multistep problem, 158–159
solving a simpler problem, 416–417,
458–459
strategies review, 332–333, 488–489
using different strategies, 448–449
using estimation, 116–117, 198–199
using the five-step process, 20–21
using number sense, 34–35, 402–403
working backward, 274–275

Products, 138–139; *see also* Multiplication
estimating, 144–145, 198–199, 266–267
of two decimals, 276–277
zeros in, 270–271

Property of zero in multiplication, 138–139
Protractors, 312–313
pt (pint), 460–461
Pyramids, 334–335
rectangular, 334–335
square, 334–335
triangular, 334–335

Q

qt (quart), 460–461

Quadrilaterals, 316–319
drawing with Logo, 383
sums of angles of, 318–319
Quart (qt), 460–461
Quotients, 182–183; *see also* Division
estimating, 188–189, 224–225, 230–231
zeros in, 196–197, 232–233

R

Radius, 164–165
Range of numbers, 202–203
Ratio and Probability, 481–516
Ratios, 482–483
equal, 484–485
Rays, 310–311
Reasonableness of answers, 8–9, 34, 60,
64, 67, 114, 121, 146, 148, 196, 228,
270, 272, 291, 412, 414, 416, 461
Recording data, 12–13
Rectangles, 318–319
areas of, 162–163
diagonals of, 454–455
perimeters of, 160–161
Rectangular prisms, 334–335
volumes of, 246–247
Rectangular pyramids, 334–335
Remainders, 182–183
Renaming
addition of mixed numbers with, 414–415
subtraction of mixed numbers with
420–423
Renaming customary measures, 462–463
Renaming metric measures, 292–293
Renaming mixed numbers, 362–363
Rhombuses, 318–319
Right angles, 314–315
Right triangles, 318–319
area of, 454–455
Roman numerals, 91
Rounding
to estimate products, 144–145, 266–267
to estimate quotients, 188–189, 224–225
to estimate sums and differences,
54–55, 102–103, 112–113, 116–117
to estimate sums and differences of
mixed numbers, 408–409
Rounding decimals, 100–101
Rounding mixed numbers, 364–365
Rounding numbers, 26–27
Ruler
centimeter, 70–73
inch, 376–377

S

s (second), 240–241
Sample, predicting from a, 437
Scale, vertical, of graphs, 120–121
Scale drawings, 486–487
Scalene triangles, 318–319
Second (s), 240–241
Sequences, 135
Short division, 197
Sides of angles, 312–313
corresponding, 328–329
Sides of figures, 162–163
Similar figures, 330–331
Simplest form, 358–361, 398–401, 404–405,
412–415, 418–419, 422–423, 444–445

Simplifying fractions, 358–359
Slides of figures, 326–327
Solving a multistep problem, 158–159
Solving a simpler problem, 416–417,
458–459
Space figures, 334–335
Space perception, 349
Spheres, 334–335
Spreadsheets, 81, 251
Square centimeter, 162–163
Square meter, 162–163
Square millimeter, 162–163
Square pyramids, 334–335
Squares, 318–319
areas of, 162–163
magic, 261
perimeters of, 160–161
Standard form, numbers in, 16–17
Statistics: mean, median, mode, and range,
202–203
Straight angles, 314–315
Strategies review, 332–333, 488–489
Subtraction, *see also* Differences
of decimals, 93–136
equal additions in, 62–63
fact families in, 52–53
of fractions, 404–405
of fractions and mixed numbers, 395–438
of hundredths, 108–109
meaning of, 52–53
of mixed numbers, 410–411, 418–419
of mixed numbers with renaming,
420–423
of tenths, 108–109
of thousandths, 114–115
using, 51–92
of whole numbers, 62–65
Sums, 52–53; *see also* Addition
of angles of triangles, 319
estimating, 102–103, 112–113
estimating sums by rounding, 54–55,
116–117
estimating sums of mixed numbers,
408–409
of fractions, 396–397
front-end estimation of, 56–57
magic, 261
Symmetry, 324–325, 338
lines of, 324–325, 338

T

T (ton), 460–461
Table of measures, 528
Tables
frequency, 14–15, 492–493, 498
using, 12–13
Tallies, 492–493, 498
Tape, measuring, 376–377
Technology
calculators, 7, 39, 60–61, 64–65, 67,
69, 77, 81, 97, 103, 109, 111,
114–115, 117, 142–143, 151,
152–153, 156–157, 159, 164, 168,
169, 172, 186–187, 189, 191, 199,
203, 205, 209, 223, 225, 234–235,
237, 243, 264, 272–273, 275,
276–277, 278–279, 286–287, 289,
293, 297, 317, 321, 329, 331,
332–333, 355, 359, 367, 375, 403,

409, 417, 419, 427, 445, 449, 455, 457, 459, 469, 483, 485, 488–489, 491, 497
computers, 39, 81, 125, 251, 339, 383, 469, 505; *see also* Computers

Temperature
in customary units, 464–465
in metric units, 464–465

Tenths, 94–95
addition of, 104–105
subtraction of, 108–109

Terms of ratios, 482–483

Thermometer
Celsius, 464–465
Fahrenheit, 464–465

Thinking mathematically, 1–10, 22–23, 68–69, 106–107, 152–153, 194–195, 238–239, 278–279, 322–323, 368–369, 406–407, 450–451, 498–499
applying mathematics, 2–3, 278–279
collecting and interpreting data, 8–9
experimenting and predicting, 106–107, 498–499
investigating patterns, 22–23, 152–153, 194–195
logical reasoning, 10, 368–369, 443
measuring, 7
using number concepts, 4–5, 68–69, 238–239, 450–451
visual reasoning, 6, 19, 322–323, 327, 406–407, 421

Thousandths, 96–97
addition and subtraction of, 114–115

Three-digit numbers, division of, 190–191

Time, 74–75
A.M. and P.M., 74–75
changing measures of, 240–241
elapsed, 74–75

Ton (T), 460–461

Trapezoids, 318–319

Tree diagrams, 496–497

Trees, factor, 393

Triangles, 316–319
acute, 318–319
equilateral, 318–319
isosceles, 318–319
obtuse, 318–319
right, *see* Right triangles
scalene, 318–319
sums of angles of, 318–319

Triangular prisms, 334–335

Triangular pyramids, 334–335

Turns of figures, 326–327

Two-digit numbers
division by, 226–229
division of, 190–191
multiplication by, 148–149, 154–155

U

Understanding Fractions and Mixed Numbers, 351–394
Understanding Numbers, 11–50
Using Addition and Subtraction, 51–92
Using different strategies, 448–449
Using division patterns, 186–187, 222–223
Using estimation, 116–117, 198–199
Using the five-step process, 20–21
Using multiplication patterns, 142–143
Using number sense, 34–35, 402–403
Using pictographs, 200–201
Using tables, 12–13
Using zeros in division, 284–285

V

Vertex/vertices
of angles, 312–313
of prisms, 334–335

Vertical axis
of bar graphs, 118–119
of line graphs, 30–31

Vertical scale of graphs, 120–121

Visual reasoning, 6, 19, 322–323, 327, 406–407, 421

Volume, 244–245
of rectangular prisms, 246–247

W

Week (wk), 240–241
Weight in customary units, 460–461
Whole numbers
addition of, 58–61
comparing, 24–25
division by one-digit divisors, 181–220
division by two-digit divisors, 221–262
division of decimals by, 264–265, 280–287
division of whole numbers by fractions, 446–447
multiplication of, 137–180
multiplication of decimals and, 264–265, 268–73, 297
multiplication of fractions and, 444–445
ordering, 24–25
subtraction of, 62–65
wk (week), 240–241
Word name of numbers, 16–19
Working backward, 274–275

Y

y (year), 240–241
Yard (yd), 376–377
Year (y), 240–241

Z

Zeros
before decimal points, 94–95
in multiplication, 154–155
in products, 270–271
property of zero in multiplication, 138–139
in quotients, 196–197, 232–233
using zeros in division, 284–285